Sustainable Agriculture in Egypt

C000053366

Hassan Auda Awaad

Sustainable Agriculture in Egypt

Climate Change Mitigation

 Springer

Hassan Auda Awaad
Crop Science Department
Faculty of Agriculture
Zagazig University
Zagazig, Egypt

ISBN 978-3-030-81875-3   ISBN 978-3-030-81873-9   (eBook)
https://doi.org/10.1007/978-3-030-81873-9

This Springer imprint is published by the registered company Springer Nature Switzerland AG
The registered company address is: Gewerbestrasse 11, 6330 Cham, Switzerland

Preface

The subject of sustainable agriculture and its relation to climate change is important for Egypt and also for all countries having analogous conditions. Understanding field crop responses to climate change is necessary to deal with expected changes in temperature, drought and salinity conditions, and increased probability of extreme events. The results of long-term expectations using simulation models and different scenarios of climate change indicate that climate change and the consequent rise in global surface temperature and hence increased water consumption will negatively affect the productivity and quality of most major crops.

For that reason and in light of climate change and its negative effects on the agricultural sector, it is important to develop and release new cultivars of field crops and determine their adaptability and stability to cope with environmental changes. Also, breeding efforts, nanotechnology as well as biotechnology are vital components for adaptation and mitigation the environmental stress. Consequently, in this book, we tried to evaluate the impact of climate change on economic field crops, also focused on different approaches represented by genetic systems of morpho-physiological and biochemical traits related to environmental stress tolerance, role of biotechnology, and nanotechnology in designing new adaptive genotypes. Also, the relationship between crop production and the impact of environmental conditions on wheat, rice, faba bean, sesame, sunflower, and cotton crops are discussed. So, this work presents important information depending upon various environmental and agronomic studies related to the adaptation and stability of genotypes in the context of climatic changes and how to adapt and mitigate their effects on the crop production for the benefit of decision-makers in different sectors.

This book consists of 10 Chapters in 4 Parts. The Part I is an introduction and contains Chap. 1 to introduce the book chapters to the audiences. The Part II consists of two chapters under the name, Impact of Climate Change on Crop Production and the Physiological and Biochemical Basis for Crops Tolerance. Chapter 2 focuses on critical periods of crop plants to stress conditions and the expected impacts of climate changes mainly on the productivity and quality of field crops. While Chap. 3 highlights the foundations of crop tolerance to environmental stress and plant traits relevant to stress tolerance. It is devoted to explain and discuss mechanisms of adaptation to environmental stress conditions and addresses various plant characters related to stress environmental tolerance, i.e., phenological, morpho-physiological, and biochemical traits which could be used as selection criteria for crops improvement. The Part III consists of 5 Chapters under the theme "Improve Crop Adaptability and Stability to Climate Change and Modern Technology". Chapter 4 highlights the most important strategic food grain crop (wheat), Chap. 5 focuses on the important staple food crop (rice), while Chap. 6 deals with one of the most important food legume crops (faba bean). Chapters 7 and 8 discuss two important oil crops (sesame and sunflower). However, Chap. 9 focuses on cotton as one of the most important fiber crops. The author addresses these crops under the following headings: genotype × environment interaction and its relation to climatic change on yield production, performance of genotypes in response to environmental changes, adaptability and yield stability to environmental conditions, additive main effects and multiplicative interaction model, gene action, genetic behavior and heritability for traits related to environmental stress tolerance, role of recent approaches, biotechnology, and nanotechnology. This is besides how can measure sensitivity of genotypes to environmental stress, and finally the appropriate agricultural practices to mitigate environmental stress on crops under attention. The book ends with Chap. 10 where the editor presents an update of the book topics, present the most important conclusions and recommendations from all chapters.

This book has been prepared and supported by recent references and statistics with colored tables and illustrations for audiences interested in crop science, environment, plant breeding, genetics, and biotechnology, as well as postgraduate students and researchers in universities and research centers.

Zagazig, Egypt   Hassan Auda Awaad

Acknowledgments

The editor wishes to express their thanks to all who contributed to make this high-quality book a real source of knowledge and latest findings in the field of adaptability, stability, and modern technology to mitigate climate change and sustain crop productivity to food security. Acknowledgments must be extended to include all members of the Springer team who have worked long and hard to produce this book and make it a reality for the researchers, graduate students, and scientists in the field of crop science, environment, plant breeding, genetics, and biotechnology around the world. The editor appreciates Dr. Abdelazim M. Negm Prof. of Water and Water structures Engineering Department, Faculty of Engineering, Zagazig University, Egypt for his advice and encouragement. The editor highly grateful to appreciate the efforts made by his wife, Dr. Azza Hussein Emam in accomplishing this book. Also, the editor appreciates the help of Dr. El-Sayed Mansour El-Sayed, Assistant Professor, Department of Crop Science, Faculty of Agriculture, Zagazig University, Egypt in accomplishing some figures in some chapters of the book. The editor gives great thanks to Prof. Dr. Das et al., RRTTS Ranital Odisha University of Agriculture & Technology, India to agree to use Table 1 Combined analysis of variance for grain yield response in Chap. 5. Also, thanks to Dr. Omar, Agronomy Department, Faculty of Agriculture, Zagazig University, Egypt for giving the approval to use modified Table 3 Flag leaf area, chlorophyll content, grain yield (ton/fad), crude protein, and amylose contents as combined in Chap. 5. Also, thanks to Dr. Hamza and Dr. Safina, Agronomy Department, Faculty of Agriculture, Cairo University, Egypt for giving the approval to use Fig. 2 growing degree days for each planting date of sunflower cultivars in Chap. 8. Much thanks to Dr. Maleia et al., Mozambique Agrarian Research Institute, Av. das FPLM, 2698. C.P. 2698, Maputo, Mozambique to use Fig. 6 Graphics biplot of PC1 against PC2 for seed cotton yield of genotypes in Chap. 9. Also, great thanks to Mudada, Plant Quarantine Services Institute, Mazowe, Zimbabwe and Chitamba, Macheke, Manjeru, Agronomy Department, Midlands State University, Gweru, Zimbabwe to use Table 6 AMMI-1 Analysis of variance and decomposition of degrees of freedom for total seed cotton yields of cotton genotypes at two sites in 2011–2012 in Chap. 9.

The volume editor would be happy to receive any comments to improve future editions. Comments, feedback, suggestions for improvement are welcomed and should be sent directly to the author. The email of the editor can be found inside the book at the end of the preface.

Zagazig, Egypt   Hassan Auda Awaad
May 2021   Head of Crop Science Department

Contents

Part I
Introduction

Chapter 1
Introduction to "Sustainable Agriculture in Egypt: Climate Change Mitigation"

Abstract This chapter provides a brief overview of background of the book, purpose of the book, scope of the book, exposed to and discussed four categories about (1) Part I: Introduction, (2) Part II: Impact of Climate Change on Sustainable Crop Production and the Physiological and Biochemical Basis for Crops Tolerance, (3) Part III: Improve Crop Adaptability and Stability to Climate Change and Modern Technology, include approaches in wheat, rice, faba bean, sesame, sunflower and cotton to mitigate impact of climate change and (4) Part IV: Conclusions and recommendations.

Keywords Purpose of the book · Scope of the book · Impact of climate change on sustainable crop production · Mechanisms of adaptation to environmental stress conditions · Plant characters · Approaches to mitigate impact of climate change · Rice · Wheat · Faba bean · Sesame · Sunflower · Cotton

1.1 Background

Worldwide, recent climate change especially in Mediterranean region obstructs crop production and these changes are expected to increase (Mansour et al. 2018). Crop plants suffer from stress due to higher temperatures, water stress, salinity, pollution and others particularly during critical periods of plant life cycle. Global warming is one of the most important environmental problems of mankind in its history that negatively affect the environment and human life. The global average temperature has increased by about 0.13 °C per decade since 1950 (IPCC 2007). The global mean surface temperature change for the period 2016–2035 is will likely be in the range 0.3 °C to 0.7 °C by mid-twenty-first century, the magnitude of the expected climate change is significantly affected by the emissions scenario (IPCC 2014). Liu et al. (2016) showed that, with a 1°C global temperature increase, global wheat production is projected to decrease between 4.1% and 6.4%. Prediction studies have indicated that the expected relative temperature effects were similar for the major wheat producing countries, China, India, USA and France, but less so for Russia. This underscores the effects of climate on global food security.

Several studies have estimated the likely impacts of future climate on crop production in the developing countries including Egypt where subsistence farming is dominant, the impact of climate change is often locally specific and hard to predict (El Massah and Omran 2015; El Afandi 2017; El-Marsafawy et al. 2018 and Mahmoud 2019). The World Bank has identified Egypt as one of the countries at risk of environmental stress in the Mediterranean region (Hereher 2016). Multi-location assessment of genotypes aids the plant breeders to ascertain the adaptability of a genotype to an individual environment and also the stability of that genotype over different environments (Sreedhar et al. 2011). The variability in grain yield is characterized principally by the effects of environment and genotype under the similar conditions of management (Dingkuhn et al. 2006). The degree to which the character performs is the produce of the genotype (G) by environment (E) interaction (G x E). The assessment of genotypes for the stability under various environments for yield has been an integral aspect of any breeding program.

In the light of these considerations, the study of the interaction between the genotype x environment has a significant impact on the expression of the phenotype and the response to environmental variables. Adaptability, genetic stability and diversity are three of the key factors for the improvement of many crop plants to environmental changes. A major challenge for plant breeders is selection of high yielding genotype with wide adaptation. Stability is a key word for plant breeders who analyze G x E interaction data because it enhances the progression of selection in any single environment. This determines the decision of plant breeders in breeding programs to develop more adaptable and stable varieties. Crop cultivars are differed in their response to heat or drought stress under different environmental situations. Obviously, yield is the most important trait that might solely determine the success of a plant breeder. Studies have shown that stability analyses according to various measures can result in better identification of stable genotypes (Akter and Islam 2017; Iqbal et al. 2017 and Farag, Fatma et al. 2019; Melandri et al. 2020 and Awaad 2021). So, the negative effects of climate change on agricultural crops can be mitigated by implementing integrated adaptation strategies and appreciative the nature of the relationships between different crop traits, and cultivating more stable genotypes to environmental changes.

Hence, understanding crop responses to climate change is essential to cope with anticipated changes in environmental conditions. Response of crop yields to environmental changes varies depending on the location, the season and the growth stage and could be discussed based on the four different concepts.

1.1.1 Concept of Sustainability

Modern day world, and especially developing societies, faces numerous challenges; among which the progressive growth in the population and hence the increase of food consumption and the extension of industrial production. These factors require a shift in agriculture towards sustainability. Agriculture is considered the major profession on earth, as it occupies 40% of the land on which we are living and consumes 70% of water resources and 30% of green reserves around the world.

Sustainable agriculture is agriculture in ways that meet society's current needs for foods and textile, without compromising the ability for current or future generations to meet their requirements (Anonymous 2018). Sustainable agriculture is based on an understanding of the components of the ecosystem. Enhancing the sustainability of agriculture depends on the development of flexible commercial procedures and agricultural practices within sustainable food systems.Agriculture has a huge environmental footprint (Brown et al. 2012) and is concurrently causing environmental changes and is also affected by these changes. The best ways to alleviate climate change is to build sustainable food systems based on sustainable agriculture to meet the human population needs and offers a possible solution to support agricultural systems within the fluctuating environmental conditions (Johan et al. 2016).

Sustainable agriculture can be achieved through crop diversity, environmental and land quality conservation, smart management and consumption of water resources, and increased agricultural production in terms of quantity and quality, taking into account climate changes.

1.1.2 Concept of Vulnerability

The term vulnerability initiates from the Latin word vulnus, used in the 1970s by geographers and social scientists in risk management matters to define the fragility of certain populations facing severe environmental or socio-economic risks (Blaikie et al. 1994). In the decade after 2000, the use of the vulnerability concept increased with its adoption by the Intergovernmental Panel on Climate Change (IPCC) to evaluate the prospective impacts of global warming at regional and global levels (Mc Carthy et al. 2001). Vulnerability recently became a very central focus of the global change science research community for defining adaptation and mitigation plans (Downing et al. 2005).

According to IPCC, the vulnerability conception benefits from a highly operational framework to pronounce the relationship between the considered system and its environment (Adger 2006 and Paavola 2008). This framework discriminates between three distinct elements (Fig. 1.1): (i) the level of exposure i.e. the frequency, intensity and duration of perturbations affecting the studied system, (ii) the level of sensitivity i.e. the degree to which the system under study is affected by environmental disturbances and (iii) the adaptive capability i.e. the system's ability to overcome disturbances and increase the extent of variance to deal with those risks.

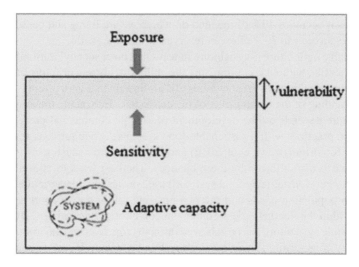

Fig. 1.1 Illustration of vulnerability and resilience concepts. (Adapted from de Goede et al. 2013 and Mumby et al. 2014)

1.1.3 Concept of Adaptation

Adaptation is defined as a modification in structure, function, or performance by which a crop species improves its chance of persistence in a specific environment. Adaptations develop as the result of natural or artificial selection operating on random genetic variations that are capable of being passed from one generation to the next (https://www.dictionary.com/browse/adaptation).

Adaptability is also defined as the capability of a crop or genotype to respond positively to changes in agricultural conditions. Adaptation is genetically controlled and provides an ability to exploit environmental attributes, both natural and agronomic. The ability of crop varieties to adaptation can be measured by the regression coefficient of the tested genotypes over the average yield under different environments (Chloupek and Hrstkova 2005).

1.1.4 Concept of Stability

The word stability meaning to the ability to maintain ecological functions despite disturbances (Turner et al. 1993) or the ability to return to the initial equilibrium state (Ives and Carpenter 2007). The stability of genotypes has been extensively used in plant breeding programs to identify genotypes that maintain specific features (Fig. 1.2) i.e. yield or grain protein content over a wide range of environments (Brancourt-Hulmel 1999 and Sabaghnia et al. 2012). So, the constancy of agricultural outputs over long periods of time or across different environments. According

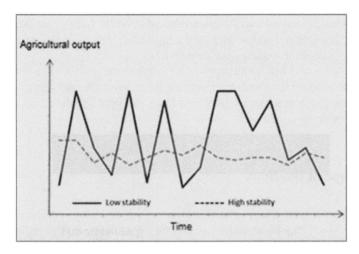

Fig. 1.2 Illustration of stability concept (Adapted from de Goede et al. 2013 and Mumby et al. 2014)

to genotype × environment interactions, two types of stability are sometimes distinguished: (i) static stability which refers to a genotype with small variance between different environments and (ii) dynamic stability which refers to a genotype for which the response to different environments is associated with the mean response of all genotypes in the experiment (Annicchiarico 2002). Yield stability of promising genotypes in diverse environments are very important in breeding approach for new cultivars that are to be released regionally or nationally. Therefore, it is of important to ascertain stable genotypes of wider adaptability or high yielding genotypes for a specific environment. In plant breeding programs and because a significant interaction with G x E can seriously obstruct the efforts of the plant breeder to select superior genotypes of new crop introductions and the breeding development program, the stability assessment is important (Shafii and Price 1998).

1.2 Purpose of the Book

This book focused on how crop plants deal with climatic change and necessary to develop and release new cultivars of field crops and determine their adaptability and stability to cope with environmental changes. Also, breeding efforts, Nano technology as well as biotechnology through molecular markers and gene transfer are considered a recent trend that helps the breeder in developing genotypes for adaptation to stress environmental conditions. Consequently, relationship between climate-crop production and the impact of environmental conditions on wheat, rice, faba bean, sesame, sunflower and cotton crops will be discussed in this book. This work offerings important information depending upon various environmental, agronomic and genetic studies associated with the adaptation and stability of genotypes in the context of climatic changes and how to adapt and mitigate their effects on the production for the benefit of decision makers in diverse sectors.

The main reasons behind this approach are due to the need to cope with climate change and mitigate its effects by producing more adaptable and stable crop genotypes for the horizontal expansion of new lands in Egypt and the world that suffer from various stress conditions.

1.3 Scope of the Book

The purpose of this book is to expand the awareness of the nature of the environmental stresses to which crop plants are exposed in the light of climate change. We observe the trends of warming in the average temperature and water stress during the growing season and then its effects over the last few decades with clear evidence of negative impacts on phenology, physiology, biochemistry and yield of various strategic crops. This scientific work aims to analysis the effect of G X E stress on crop plants and the strategies to be followed to avoid the negative impacts on the productivity and quality of agricultural crops in order to narrowing the food gap and ensuring food requirements. Halford (2011) stated that water shortage is repeatedly, associated by heat stress and both pressures induce diverse reactions in crop varieties.

This global warming will lead to changes in temperature at an unprecedented rate than in the past 50 million years. The atmospheric CO_2 concentration has increased from a pre-industrial level of 270 ppm to 390 ppm and is increasing by 1–2 ppm/year, taking it to levels not seen for 20 million years. How crop plants respond to rising CO_2 levels over the coming decades will determine the world's food security situation; in particular, will elevated CO_2 levels have useful effects on crop physiology and crop production that will balance the negative effects of increased temperature and water stress? In this respect, great research has been performed over the last twenty years on the long-term acclimation of crop plants to elevated CO_2 using free-air CO_2 enrichment. Estimates of global climate models in International Panel for Climate Change (IPCC) 4th Assessment Report indicate a 5 °C increase in global average temperature by the end of the century. In addition to the increase in average temperature, it is expected that there will be an increase in the frequency and intensity of extreme weather events, adversely affecting food production. Levels of the main long-lived greenhouse gases, carbon dioxide (CO_2), methane (CH_4) and nitrous oxide (N_2O) have reached new highs. In 2018, global CO_2 concentration was 407.8 parts per million (ppm), 2.2 ppm higher than 2017. Primary results from a subset of greenhouse gas monitoring sites for 2019 indicate that CO_2 concentrations are on track to reach or even exceed 410 ppm by the end of 2019. In 2017, globally averaged atmospheric concentrations of CO_2 were 405.6 ± 0.1 ppm, CH_4 at 1859 ± 2 parts per billion (ppb) and N_2O at 329.9 ± 0.1 ppb. These values constitute, respectively, 146, 257 and 122% of pre-industrial levels (pre-1750). The increase rate of CO_2

averaged over three successive decades growth from 1.42 ppm/yr. to 1.86 ppm/yr. and to 2.06 ppm/yr. (IPCC 2019).

Currently the agricultural activities used 75% of global water consumption and irrigation consumes over 90% of water account in many developing countries. This water deficiency might threaten sustainable crop farming (Yang et al. 2010), more by 2050 the shortage of water are expected to affect 67% of the world populations. Moreover, Reynolds and Ortiz (2010) expected that by 2030, developing countries will be most severely affected by climate change because, (1) climate change will have the greatest impact on the tropics and subtropics regions, (2) most of the projected population growth to 2030 will happen in developing countries and (3) more than half of the workforce in developing countries is involved in agricultural. Accordingly, the population growth and variable climate is expected to threaten food security on a worldwide scale. Lobell et al. (2011) indicated that forecasts related to climate change mainly effect on major crops through simulations. These important crops are in need of adaptation investments to avoid catastrophic productivity losses and to meet the food demand of a fast human population growth rate. So it is necessary to increase crop yields by at least 40% in arid and semi-arid regions (Nakashima et al. 2014; Shaar-Moshe et al. 2017). In this concern, crop plants have the ability to cope with these changes in extreme climates of CO_2 concentration and crop genotypes varied in their response to stress conditions, and showed various degrees of adaptability and stability (Golestani and Pakniyat 2015; Santhosh et al. 2017; Elbasyoni 2018; Farag, Fatma et al. 2019, Qaseem et al. 2019; Fadia Chairi et al. 2020 and Awaad 2021). So, the problem of food shortages is one of the most important challenges facing the achievement of the policy and sustainable development goals in Egypt 2020–2030, for which the efforts of all concerned sectors should be combined.

1.4 Themes of the Book and Contribution of the Chapters

In addition to the introduction (this chapter) and the conclusions (the last chapters), the book consists of 3 themes. The first is titled "Impact of Climate Change on Crop Production and the Physiological and Biochemical Basis for Crops Tolerance" which is covered in two chapters. Chapter 2 is titled "Climate Change and its Impact on Crop Production, while Chap. 3 under title "Foundations of Crop Tolerance to Environmental Stress and Plant Traits Relevant to Stress Tolerance". The third theme is written in 6 chapters and is titled "Improve Crop Adaptability and Stability to Climate Change and Modern Technology". In the next subsection, the main technical elements of the chapters under each theme are presented.

Part II Impact of Climate Change on Sustainable Crop Production and the Physiological and Biochemical Basis for Crops Tolerance

Chapter 2 focuses on the negative effects of climate change on sustainable crop productivity and quality. The Intergovernmental Panel on Climate Change (IPCC 2018) showed how keeping temperature increases below 2 °C would reduce the risks to human well-being, ecosystems and sustainable development. The report said that net zero emissions of carbon dioxide must reach zero by 2050. At the current rate of emissions, the world will reach 1.5 °C warming by between 2030 and 2052. For the strategic crops temperature increases of 2 °C or more, induced by climate change, will negatively affect crop production if there is no adaptation (Porter et al. 2014). Generally, crop production is sensitive to in climate change, hence, this will lead in the future to significant changes in the distribution of crops. Knowledge of sensitive periods in plant life is useful in developing breeding programs to deal with current environmental conditions. Egypt is vulnerable to the impacts of climate change, several publications have been delivered concerning the fluctuating climate of Egypt comprising the impact of changing climate on decrease crop production (El Massah and Omran 2015; El Afandi 2017; El-Marsafawy et al. 2018 and Mahmoud 2019). Crop varieties have been shown to exhibit contrasting responses to environmental stresses (Awaad et al. 2016; González-Barrios et al. 2017; Sahito et al. 2015; Amin et al. 2016 and Dewdar 2019). Consequently, breeding genotypes with higher tolerance to stresses is considered unlimited attention from breeders (Dossa et al. 2017 and Awaad 2021).

Chapter 3 highlights on mechanisms of adaptation to environmental stress conditions in some important field crops i.e. wheat, rice, faba bean, sesame, sunflower and cotton from the following aspects, plant characters related to environmental stress tolerance viz earliness, morpho-physiological, biochemical and molecular base. Besides, interrelationships among yield traits under different environments. Aforementioned results revealed the importance of earliness in wheat under heat and drought stress (Qaseem et al. 2019), leaf cuticular wax (Ahmed et al. 2012 and Guo et al. 2016), earliness and photoperiod response in rice (Rana et al. 2019 and Anonymous 2021), water use efficacy in rice tolerance to cold, high salinity and drought (Li et al. 2011), shoot, root and root to shoot ratios in cotton (Ashraf and Ahmad 2000, Basal et al. 2006 and Dewi 2009) and sunflower (Santhosh 2014); leaf chlorophyll content, relative water content, transpiration rate, canopy temperature, stomatal conductance osmotic adjustment in crop plants (Hirasawa et al. 1995; Guidi and Soldatini 1997; Lu et al. 1998; Pankovic et al. 1999; Roohi et al. 2015 and Bhanu et al. 2018). Biochemical traits i.e. soluble sugar, soluble protein, reduced glutathione, reduced ascorbate, catalase, phenylalanine ammonialyase activity, proline, superoxide dismutase, peroxidase activity represents the important role in tolerating of drought, salinity, cold and heat stresses (Najafi et al. 2018 and Jian et al. 2019); Heat shock proteins (HSP) in response to high temperature, water stress, salinity, oxidative, heavy metals, and high intensity irradiations (Swindell et al. 2007; Li et al. 2011; Xu et al. 2011 and Muthusamy et al. 2017).

Part III: Improve Crop Adaptability and Stability to Climate Change and Modern Technology

The proposed Chap. 4 highlights on approaches in wheat to mitigate impact of climate change on crop production. Where, in Egypt, total production 8.45 million metric tons while the Egyptian consumption of wheat grains is about 18 million tons in (USDA 2019). Therefore, the task of breeder is to screen out genotypes at different environments to enable selection of those genotypes, which are suitable for wider range of environments.

Different stability models were applied to determine the most adapted and stable genotypes of them phenotypic stability for Eberhart and Russel (1966), Genotypic stablity (Tai 1971), AMMI model (Crossa et al. 1990) and others. Cruz and Carneiro (2006) emphasized that the G x E interaction can be reduced by stratification of the production site into areas with similar environmental features so that the interaction becomes unimportant. Studies have shown that stability analyses according to various measures can result in better identification of stable genotypes (Akter and Islam 2017 and Iqbal et al. 2017). Particular attention is given to role of QTLs and gene transfer to improve stress tolerance of transgenic plants (Li et al. 2018 and Elbasyoni et al. 2018). Also, how can measure sensitivity of wheat genotypes to nvironmental stresses by using stress sensitivity meaurments as carried out by Fischer and Maurer (1978). Also, this chapter focused on mode of gene action, and heritability for wheat traits. Recent research advises that nanotechnology in wheat has positive prospects for effective nutrient use, control and management of pests and diseases, pesticides, and etc. Jasrotia et al. (2018). Besides, the application of suitable agricultural practices plays an significant role in alleviating the environmental pressure on the wheat crop.

Due to the limited water and the threats of the Grand Ethiopian Renaissance Dam, Chap. 5 was investigated to find ways to economize water use without significant loss in grain yield. The authors confirmed that it is necessity of evaluating rice genotypes under multi environments. Das et al. (2018) found maximum contribution rice grain yield was due to environments (E) followed by GxE interaction, and then genotypes valued 74.2%, 18.76% and 6.0%, respectively as due to the diverse nature of the environments representing different agroecological regimes. Evaluation promising rice genotypes for their stability and adaptability are useful in identifying the most dynamic stable genotypes to make reliable yield assessments (Hasan et al. 2014 and Das et al. 2018). Gaballah et al. (2021) stated that Egyptian rice genotypes Giza179, Hybrid 1 and Hybrid 2 exhibited the greatest grain yield with maximum water use efficiency and well adapted to water sacristy. The authors exploited stress sensitivity index and cumulative stress response index in screening rice genotypes to heat stress (Karwa et al. 2020), saline stress (Krishnamurthy et al. 2016) and drought stress (El-Hashash and EL-Agoury 2019). On the other hand, several genes and/or alleles were identified that controlled photoperiod sensitivity and length of basic vegetative growth in rice (Rana et al. 2019). On the other hand, researchers emphasized the importance of utilizing transgenic approaches in improve grain yield, Stokstad (2020) showed that the transgenic rice yielded up to 20% more grain and resist to heat waves. For maximize rice grain yield, Nanotechnology, recommendation procedures and cultivation of short-duration cultivars, which can save about 20–30% of irrigation water are essential for sustaining rice productivity.

Chapter 6 comes from the point of view that in recent years, Egypt has witnessed a severe crisis in the quantity of production of the faba bean production, which has declined significantly, due to the decrease in its cultivated area and environmental stresses. So, Chap. 6 is proposed to highlight on an important environmental stresses affecting faba bean productivity. This Chapter brings to the reader's attention of some concepts related the interaction between genotype x environment and its effects to identify stable genotypes (Tekalign 2018). From point of view that water stress is a major limiting factor in faba bean (*Vicia faba* L.) production in the Mediterranean region, which is well-known for its irregular water distribution and modest moisture levels (~500 mm rainfall). Hereby, Maalouf et al. (2015) showed that the spectral indices structure-insensitive pigment index and normalized pheophytinisation index were found to be associated positively with seed yield and might be used for selection under drought stress environments. Diverse stability parameters explained genotypic performance differently, irrespective of yield performance (Temesgen et al. 2015 and Tolessa 2015). This chapter also concentrates on SDS-PAGE examination and DNA Markers techniques and recognizes stress resistance QTLs under different agro climatic regions. Finally, the author suggested that the effects of environmental stress can be addressed through particular agronomic practices with tolerant varieties and using nanotechnology (Abo-Sedera 2016 and Almosawy et al. 2018). The differentiation between faba bean genotypes in their tolerance could be evaluated by yield sensitivity index.

Chapter 7 was prepared to focus more light on sesame as known as orphan crop. Currently, though world demand for its seeds is interestingly increasing due to its good quality oil and protein. Sesame oil is rich in polyunsaturated fatty acids and natural antioxidants like sesamin, sesamolin and tocopherol homo-logues (Anilakumar et al. 2010). Sesame grows well in tropical to temperate environments, however, severe drought is detrimental to natural plant growth and development (Anjum et al. 2011). The researchers recommended that it is necessity of evaluating sesame genotypes under multi environments in order to identify the best genetic make-up to be grown under particular environments (Awaad and Aly 2002 and Abbasali et al. 2017). Various stability parameters S^2di, Wi, σ^2i, ASV, S1 and S^2 have been reported to measure the adaptability and stability of new sesame cultivars (Chemeda et al. 2014). A survey of a genome-wide of gene family in sesame was done and Ali Mmadi et al. (2017) identified at the genome-wide level 287 MYB genes that the MYB proteins are highly active in growth and adaptation to the major abiotic stresses namely drought and waterlogging in sesame. Co-application of iron nano-chelate and fulvic acid were found to be improved seed yield under drought stress (Ayoubizadeh et al. 2019). Some measurements were utilized to recognize tolerate or sensitive genotypes i.e. Stress Sensitivity Index (SSI), Stress Tolerance Index (STI), yield stability index (YSI) and others (Boureima et al. 2016 and Hassanzadeh et al. 2009).

On the light of climate change worldwide, and in the Mediterranean region, the objectives of Chap. 8 are to identify adaptability and stability of sunflower genotypes for earliness, quality and achene yield under different environmental combinations. Egypt's edible vegetable oil production suffers from many problems due to the decrease in domestic production of oil crops, which has led to the failure to meet domestic consumption needs (Hassan and Sahfique 2010). Where, sunflower has become an imperative oil crop in the world. Santhosh et al. (2017) showed that stress affects different physiological characters and processes, it causes up to 50% decrease in biomass and more than 50% lessening in yields. The author has recommended the need for improvement of cultivars for water shortage tolerance and production are expressed as a major objective in breeding programs for a long time (Chachar et al. 2016). Many investigators estimated performance and stability and found significant genotype x environment for seed yield and oil content with different degrees of adaptability and stability for sunflower genotypes (Ali et al. 2006; Ghafoor et al. 2005 and Tabrizi 2012 and Awaad et al. 2016). The author also focused on genetic components controlling physiological, achene yield and achene oil content characters of sunflower genotypes under diverse environments (Awaad et al. 2016 and Razaq et al. 2017). The use of biotechnological techniques for improving sunflower characters is limited principally by the difficulty of regenerating plants in a reproducible and efficient way (Moghaddasi 2011). Furthermore, applying optimum agricultural practices and using Nano fertilizer are important to reduce the impact of environmental conditions on growth and sunflower productivity (Naderi and Abedi 2012; Abd El-Gwad and Salem 2013 and Yaseen and Wasan 2015).

In Egypt, because of the area of cotton cultivation is reduced into 0.09 million hectares with an average production 718 kg/ha. gave total production 0.30 million bales (USDA 2019). So it is necessary to increase crop yields by at least 40% in arid and semi-arid regions (Nakashima et al. 2014 and Shaar-Moshe et al. 2017). For this purpose Chap. 9 came to discuss ways to select suitable varieties of cotton more adapted to climate change and to avoid major impacts of the interaction of the genetic makeup with the environment. The author showed that a combination of two or more abiotic stresses, such as drought heat and salinity results in more yield loss than normal condition. Then identify the magnitude and nature of genotype × environment interaction and determine stability of yield potentiality is of great important. Results of review showed high degree of genetic differentiation in response to climate change based on mean performance and yield stability parameters in cotton varieties for economic traits (Hassan et al. 2012 and Sahat 2015). Mudada et al. (2017) showed that cotton yield and fiber quality parameters are reliant on the environment in which the crop is grown. Also, this Chapter provides detailed information's about gene action, heritability and genetic system of economic cotton traits related to stability, breeding methods, Nano-Technology and biotechnology. Hence, previous information's will serve as a foundation for developing highly adapted and stable cultivars.

References

Abbasali M, Gholipouri A, Tobeh A, Kh Sima NA, Ghalebi S (2017) Identification of drought tolerant genotypes in the Sesame (*Sesamum indicum* L.) collection of national plant gene bank of Iran. Iran J Field Crop Sci 48(1):275–289

Abd El-Gwad AM, Salem EMM (2013) Effect of biofertilization and silicon foliar application on productivity of sunflower (*Helianthus annuus* L.) under new valley conditions. Egypt J Soil Sci 53 (4):509–536

Abo-Sedera FA, Shams AS, Mohamed MHM, Hamoda AHM (2016) Effect of organic fertilizer and foliar spray with some safety compounds on growth and productivity of snap bean. Ann Agric Sci, Moshtohor 54(1):105–118

Adger WN (2006) Vulnerability. Glob Environ Chang 16:268–281

Ahmed M, Asif M, Goyal A (2012) Silicon the non-essential beneficial plant nutrient to enhanced drought tolerance in wheat. In: Goyal A (ed) IntechOpen, London, UK

Akter NR, Islam M (2017) Heat stress effects and management in wheat: a revies. Agron Sustain Dev 37(37):1–17

Ali Mmadi M, Dossa K, Wang L, Zhou R, Wang Y, Cisse N, Oureye Sy M, Zhang X (2017) Functional characterization of the versatile MYB gene family uncovered their important roles in plant development and responses to drought and waterlogging in sesame. Genes 8(12):362

Ali SS, Manzoor Z, Awan TH, Mehdi SS (2006) Evaluation of performance and stability of sunflower genotypes against salinity stress. J Anim Pl Sci 16(1–2):47–51

Almosawy AN, Alamery AA, Al-Kinany FS, Mohammed HM, Alkinani LQ, Jawad NN (2018) Effect of optimus nanoparticles on growth and yield of some broad bean cultivars (Vicia faba L.). Int J Agricult Stat Sci 14(2):1–4

Amin A, Nasim W, Mubeen M, Sarwar S, Urich P, Ahmad A, Wajid A, Khaliq T, Rasul F, Hammad HM, Rehmani MIA, Mubarak H, Mirza N, Wahid A, Ahamd S, Fahad S, Ullah A, Khan MN, Ameen A, Amanullah BS, Saud S, Alharby H, Ata-Ul-Karim ST, Adnan M, Islam F, Ali QS (2016) Regional climate assessment of precipitation and temperature in Southern Punjab (Pakistan). Theoret Appl Climatol 131(1–2):121–131

Anilakumar KR, Pal A, Khanum F, Bawas AS (2010) Nutritional, medicinal and industrial uses of sesame (Sesamum indicum L.) seeds. Agric Conspec Sci 75:159–168

Anjum AH, Xie XY, Wang LC, Saleem MF, Man C, Le W (2011) Morphological, physiological and biochemical responses of plants to drought stress. Afr J Agric Res 6:2026–2032

Annicchiarico P (2002) Genotype x environment interactions: challenges and opportunities for plant breeding and cultivar recommendations. In: FAO plant production and protection paper (FAO), No. 174, Food & Agriculture Org, Rome (Italy)

Anonymous (2018) What is sustainable agriculture | agricultural sustainability institute. 11 Dec 2018. Retrieved 20 Jan 2019, asi.ucdavis.edu

Anonymous (2021) Technical recommendations for rice crop (2020) Ministry of agriculture and land reclamation, agricultural research center. Field Crops Research Institute, Rice Research and Training Center, Egypt

Ashraf M, Ahmad S (2000) Influence of sodium chloride on ion accumulation, yield components and fiber characteristics in salt-tolerant and salt-sensitive lines of cotton (Gossypium hirsutum L.). Field Crops Res 66:115–127

Awaad HA (2021) Performance and genetic diversity in water stress tolerance and relation to wheat productivity under rural regions. In: Awaad HA, Abu-hashim M, Negm A (eds) Handbook of mitigating environmental stresses for agricultural sustainability in Egypt. Springer Nature Switzerland AG, pp 63–103

Awaad HA, Salem AH, Ali MMA, Kamal KY (2016) Expression of heterosis, gene action and relationship among morpho-physiological and yield characters in sunflower under different levels of water supply. J Plant Prod, Mansoura Univ 7(12):1523–1534

Awaad HA, Aly AA (2002) Genotype x environment interaction and interrelationship among some stability statistics in sesame (Sesamum indicum L.). Zagazig J Agric Res 29(2):385–403

Ayoubizadeh N, Laei G, Dehaghi MA, Sinaki JM, Rezvan S (2019) Seed yield and fatty acids composition of sesame genotypes as affecet by foliar application of iron Nano-chelate and fulvic acid under drought stress. Appl Ecol Environ Res 16(6):7585–7604

Basal H, Hemphill JK, Smith CW (2006) Shoot and root characteristics of converted race stocks accessions of upland cotton (Gossypium hirsutum L.) grown under salt stress conditions. Am J Plant Physiol 1:99–106

Bhanu A, Arun B, Mishra VK (2018) Genetic variability, heritability and correlation study of physiological and yield traits in relation to heat tolerance in wheat (*Triticum aestivum* L.). Biomed J Sci & Tech Res 2(1):212–216

Blaikie P, Cannon T, Davis I, Wisner B (1994) At risk: natural hazards, people's vulnerability and disasters. Routledge, New York

Boureima S, Diouf M, Amoukou AI, Van Damme P (2016) Screening for sources of tolerance to drought in sesame induced mutants: assessment of indirect selection criteria for seed yield. Int J Pure App Biosci 4(1):45–60

Brancourt-Hulmel M (1999) Expliquer l'interaction génotype/milieu par des génotypes révélateurs chez le blé tendre d'hiver, thèse de biologie, Ensar, Rennes. http://www.theses.fr/1999NSARB113

Brown LR (2012) World on the edge. Norton, Earth policy institute

Chachar MH, Chachar NA, Chachar Q, Mujtaba SM, Chachar S, Chachar Z (2016) Physiological characterization of six wheat genotypes for drought tolerance. Int J Res-Granthaalayah 4:184–196

Chemeda D, Ayana A, Zeleke H, Wakjira A (2014) Association of stability parameters and yield stability of sesame (*Sesamum indicum* L.) Genotypes in Western Ethiopia. East Afr J Sci 8(2):125–134

Chloupek O, Hrstkova P (2005) Adaptation of crops to environment. Theor Appl Genet 111(7):1316–1321

Crossa J, Gauch HG, Zobel RW (1990) Additive main effect and multiplicative interaction analysis of two international maize cultivar trials. Corp Sci 30:493–500

Cruz CD, Carneiro PCS (2006) Modelos biométricos aplicados ao melhoramento genético, vol 2, UFV, Viçosa, 585 p

Das CK, Bastia D, Naik BS, Kabat B, Mohanty MR, Mahapatra SS (2018) GGE biplot and AMMI analysis of grain yield stability and adaptability behaviour of paddy (*Oryza sativa* L.) genotypes under different agroecological zones of Odisha: ORYZA—an. Int J Rice 55(4):528–542

de Goede DM, Gremmen B, Blom-Zandstra M (2013) Robust agriculture: balancing between vulnerability and stability. NJAS Wagening J Life Sci 64–65:1–7

Dewdar MDH (2019) Productivity; cotton; stress susceptibility index; relative productivity. Int J Plant Soil Sci 27(5):1–7

Dewi ES (2009) Root morphology of drought resistance in cotton (Gossypium hirsutum L.). M.S. Thesis. Texas A & M University, College Station, TX

Dingkuhn M, Luquet D, Kim H, Tambour L, ClementVidal A (2006) Ecomeristem, a model of morphogenesis and competition among sinks in rice. 2: simulating genotype responses to phosphorus deficiency. Funct Plant Biol 33:325–337

Dossa K, Diouf D, Wang L, Wei X, Yu J, Niang M, Fonceka D, Yu J, Mmadi MA, Yehouessi LW et al (2017) The emerging oilseed crop Sesamum indicum enters the "Omics" era. Front Plant Sci 8:1154

Downing TE, Patwardhan A, Klein RJ, Mukhala E (2005) Assessing vulnerability for climate adaptation. In: Lim B, Spanger-Siegfried E (eds) Adaptation policy frameworks for climate change: developing strategies, policies and measures. Cambridge University Press, New York, pp 67–90

El-Marsafawy S, Noura B, Tamer E, Hassan El (2018) Climate. In: The soils of Egypt, pp 69–92. © Springer Nature Switzerland AG 2019

Eberhart SA, Russell WA (1996) Stability parameters for comparing varieties. Crop Sci 6:36–40

El Afandi G (2017) Impact of climate change on crop production. In: Chen WY et al (eds) Handbook of climate change mitigation and adaptation. Springer, Science + Business Media New York, pp 723–748. https://doi.org/10.1007/978-3-319-14409-2_64

El Massah S, Omran G (2015) Would climate change affect the imports of cereals? the case of Egypt. In: Leal Filho W (ed) Handbook of climate change adaptation. Springer, Heidelberg. https://doi.org/10.1007/978-3-642-38670-1_61

Elbasyoni IS, Morsy SM, Ramamurthy RK, Nassar AM (2018) Identification of genomic regions contributing to protein accumulation in wheat under well-watered and water deficit growth conditions. Plants 7(56):1–15

Elbasyoni IS (2018) Performance and stability of commercial wheat cultivars under terminal heat stress. Agronomy 8(4):5–29

El-Hashash EF, R.Y.A. EL-Agoury, (2019) Comparison of grain yield based drought tolerance indices under normal and stress conditions of rice in Egypt. J Agric Vet Sci 6(1):41–45

Chairi F, Aparicio N, Serret MD, Araus JL (2020) Breeding effects on the genotype × environment interaction for yield of durum wheat grown after the green revolution: the case of Spain. Crop J 8(4):623–634

Farag FM (2019) Breeding parameters for grain yield and some morpho-pysiological characters related to water stress tolerance in bread wheat. M.Sc. thesis, Agron Department, Faculty of Agriculture, Zagazig University, Egypt

Farid MA, Abou Shousha AA, Negm MEA, Shehata SM (2016) Genetical and molecular studies on salinity and drought tolerance in rice (*Oryza sativa* L). J Agric Res, Kafr El-Shaikh Univ 42(2):1–23

Fischer RA, Maurer R (1978) Drought resistance in spring wheat cultivars, 1. Grain yield responses. Aust J Agric Res 26(4):897–912

Gaballah MM, Metwally AM, Skalicky M, Hassan MM, Brestic M, Sabagh AEL, Fayed AM (2021) Genetic diversity of selected rice genotypes under water stress conditions. Plants 10(27):1–19

Ghafoor A, Arshad IA, Muhammad F (2005) Stability adaptability analysis in sunflower from eight locations in Pakistan. J Appl Sci 5:118–121

Golestani M, Pakniyat H (2015) Evaluation of traits related to drought stress in sesame (*Sesamum indicum* L.) genotypes. J Asian Sci Res 5(9):465–472

González-Barrios P, Castro M, Pérez O, Vilaró D, Gutiérrez L (2017) Genotype by environment interaction in sunflower (*Helianthus annuus* L.) to optimize trial network efficiency. Span J Agric Res 15(4):1–13

Guidi L, Soldatini GF (1997) Chlorophyll fluorescence and gas exchanges in flooded soybean and sunflower plants. Plant Physiol Biochem 35:713–717

Guo J, Xu W, Yu X, Shen H, Li H, Cheng D, Liu A, Liu J, Liu C, Zhao S, Song J (2016) Cuticular wax accumulation is associated with drought tolerance in wheat near-isogenic lines. Front Plant Sci 7:1809

Halford NG (2011) The role of plant breeding and biotechnology in meeting the challenge of global warming, planet earth 2011—global warming challenges and opportunities for policy and practice, Prof Carayannis E (ed), eBook (PDF) ISBN: 978–953–51–6063–2

Hasan MJ, Kulsum MU, Hossain MM, Akond Z, Rahman MM (2014) Identification of stable and adaptable hybrid rice genotypes. SAARC J Agri 12(2):1–15

Hassan ISM, Badr SSM, Hassan ISM (2012) Study of phenotypic stability of some Egptian cotton gynotypes under different environment. Egypt, J Appl Sci 27(6):298–315

Hassan MB, Sahfiqu FA (2010) Current situation of edible vegetable oils and some propositions to curb the oil gap in Egypt. Nat Sci 8:1–12

Hassanzadeh M, Asghari A, Jamaati-e-Somarin Sh, Saeidi M, Zabihi-e-Mahmoodabad R, Hokmalipour S (2009) Effects of water deficit on drought tolerance indices of sesame (*Sesamum indicum* L.) genotypes in Moghan region. Res J Environ Sci 3:116–121

Hereher ME (2016) Time series trends of land surface temperatures in Egypt: a signal for global warming. Environ Earth Sci 75:1218

Hirasawa T, Wakabayashi K, Touya S, Ishihara K (1995) Stomatal responses to water deficits and abscisic acid in leaves of sunflower plants (*Helianthus annuus* L.) grown under different conditions. Plant Cell Physiol 36:955–964

IPCC (2007) Climate change 2007: the physical science basis: contribution of wor-king group i to the fourth assessment report of the intergovernmental panel on climate change. In: Solomon S, Qin D, Manning M, Chen Z, Marquis M, Averyt KB, Tignor M, Miller HL (eds). Cambridge University Press, Cambridge, United Kingdom and New York, NY, USA, 996 p

IPCC (2014) Climate change 2014: synthesis report: contribution of working groups I, II and III to the fifth assessment report of the intergovernmental panel on climate change. In: Core Writing Team, Pachauri RK, Meyer LA (eds). IPCC, Geneva, Switzerland, 151 p

IPCC (2018) Climate change report is a "wake-up" call on 1.5°C global warming. https://pub lic.wmo.int/en/media/press-release/climate-change-report-%E2%80%9Cwake-%E2%80%9D-call-15%C2%B0c-global-warming

IPCC (2019) Landmark united in science report informs climate action summit. © 2019 World Meteorological Organization (WMO). https://public.wmo.int/en/media/press-release/landmark-united-science-report-informs-climate-action-summit.

Iqbal M, Raja NI, Yasmeen F, Hussain M, Ejaz M, Shah MA (2017) Impacts of heat stress on wheat: a critical review. Adv Crop Sci Technol 5:251

Ives AR, Carpenter SR (2007) Stability and diversity of ecosystems. Science 317:58–62

Jasrotia P, Kashyap PL, Bhardwaj AK, Kumar S, Singh GP (2018) Scope and applications of nanotechnology for wheat production: a review of recent advances. Wheat Barley Res 10(1):1–14

Jian S, XiaoWen Y, MeiWang LE, YueLiang RAO, TingXian YAN, YanYing YE, HongYing Z (2019) Physiological response mechanism of drought stress in different drought-tolerance genotypes of sesame during flowering period. Sci Agric Sin 52(7):1215–1226

Johan R, John W, Gretchen D, Andrew N, Nathanial M, Line G, Hanna W, Fabrice D, Mihir Sh (2016) Sustainable intensification of agriculture for human prosperity and global sustainability. Ambio 46(1):4–17

Karwa S, Bahuguna RN, Chaturvedi AK, Maurya S, Arya S S, Chinnusamy V, Pal M (2020) Phenotyping and characterization of heat stress tolerance at reproductive stage in rice (*Oryza sativa* L.). Acta Physiol Plant 42(1):1–16

Krishnamurthy SL, Sharma PC, Sharma SK, Batra V, Kumar V, Rao LVS (2016) Effect of salinity and use of stress indices of morphological and physiological traits at the seedling stage in rice. Indian J Exp Biol 54:843–850

Li HW, Zang BS, Deng XW, Wang XP (2011) Overexpression of the trehalose-6-phosphate synthase gene *OsTPS1* enhances abiotic stress tolerance in rice. Planta 234(5):1007–1018

Li Q, Wang W, Wang W, Zhang G, Liu Y, Wang Y, Wang W (2018) Wheat F-box protein gene TaFBA1Is involved in plant tolerance to heat stress. Front Plant Sci 9:521

Asseng LBS, Müller C, Ewert F et al (2016) Similar estimates of temperature impacts on global wheat yield by three independent methods. Nat Clim Chang 6(12):1130–1137

Lobell DB, Schlenker W, Costa-Roberts J (2011) Climate trends and global crop production since 1980. Science 333:616–620

Lu Z, Percy RG, Qualset CO, Zeiger E (1998) Stomatal conductance predicts yields in irrigated Pima cotton and bread wheat grown at high temperatures. J Exp Bot 49:453–460

Maalouf F, Miloudi N, Ed Ghanem M, Murari S (2015) Evaluation of FABA bean breeding lines for spectral indices, yield traits and yield stability under diverse environments. Crop Pasture Sci 66(10):1012–1023

Mahmoud MA (2019) Impact of climate change on the agricultural sector in Egypt. Springer International Publishing, Publisher

Mansour E, Moustafa ED, Qabil N, Abdelsalam A, Wafa HA, El Kenawy A, Casas AM, Igartua E (2018) Assessing different barley growth habits under Egyptian conditions for enhancing resilience to climate change. Field Crop Res 224:67–75

Mc Carthy J, Canziani O, Leary N, Dokken D, White K (2001) Impacts, adaptation and vulnerability–contribution of working group ii to the third assessment report of the IPCC. Cambridge University Press

Melandri G, AbdElgawad H, Riewe D, Hageman JA, Asard H, Beemster GTS, Kadam N, Jagadish K, Altmann T, Ruyter-Spira C, Bouwmeester H (2020) Biomarkers for grain yield stability in rice under drought stress. J Exp Bot 71(2):669–683

Moghaddasi MS (2011) Sunflower tissue culture. Adv Environ Biol 5(4):746–755

Mudada N, Chitamba J, Macheke TO, Manjeru P (2017) Genotype × environmental interaction on seed cotton yield and yield components. Open Access Libr J 4(11):1–22

Mumby PJ, Chollett I, Bozec Y-M, Wolff NH (2014) Ecological resilience, robustness and vulnerability: how do these concepts benefit ecosystem management? Curr Opin Environ Sustain 7:22–27

Muthusamy SK, Dalala M, Chinnusamy V, Bansal KC (2017) Genome-wide identification and analysis of biotic and abiotic stress regulation of small heat shock protein (HSP20) family genes in bread wheat. J Plant Physiol 211:100–113

Naderi MR, Abedi A (2012) Application of nanotechnology in agriculture and refinement of environmental pollutants. J Nanotech 11(1):18–26

Najafi S, Sorkheh K, Nasernakhaei F (2018) Characterization of the APETALA2/Ethylene-responsive factor (AP2/ERF) transcription factor family in sunflower. Sci Rep 8:1–16

Nakashima K, Yamaguchi-Shifnozaki K, Shinozaki K (2014) The transcriptional regulatory network in the drought response and its crosstalk in abiotic stress responses including drought, cold and heat. Front Plant Sci 5:170

Paavola J (2008) Livelihoods, vulnerability and adaptation to climate change in Morogoro, Tanzania. Environ Sci Pol 11:642–654

Pankovic D, Sakac Z, Kevresan S, Plesnicar M (1999) Acclimation to long-term water deficit in the leaves of two sunflower hybrids: photosynthesis, electron transport and carbon metabolism. J Exp Bot 50:127–138

Porter JR, Xie L, Challinor V et al (2014) Food security and food production systems. In: Field CB et al (eds) Climate change 2014: impacts, adaptation, and vulnerability. Part A: global and sectoral aspects. Contribution of working group ii to the fifth assessment report of the intergovernmental panel on climate change. Cambridge University Press, Cambridge, UK/New York, USA, pp 485–533.

Qaseem MF, Qureshi R, Shaheen H (2019) Effects of pre-anthesis drought, heat and their combination on the growth, yield and physiology of diverse wheat (*Triticum aestivum* L.) genotypes varying in sensitivity to heat and drought stress. Sci Rep 9:1–12

Rana BB, Kamimukai M, Bhattarai M, Koide Y, Murai M (2019) Responses of earliness and lateness genes for heading to different photoperiods, and specific response of a gene or a pair of genes to short day length in rice. Heredias 156(36):1–11

Razaq K, Rauf S, Shahzad M, Ashraf E, Shah F (2017) Genetic analysis of pollen viability: an indicator of heat stress in sunflower (*Helianthus annuus* L.). Int J Innov Approaches Agric Res 1(1):40–50

Reynolds MP, Ortiz R (2010) Adapting crops to climate change: a summary. In: Reynolds MP (eds) Climate change and crop production, CAB International, pp 1–8

Roohi E, Tahmasebi-Sarvestani Z, Modarres Sanavy SAM, Siosemardeh A (2015) Association of some photosynthetic characteristics with canopy temperature in three cereal species under soil water deficit condition. J Agr Sci Tech 17:1233–1244

Sabaghnia N, Karimizadeh R, Mohammadi M (2012) Genotype by environment interaction and stability analysis for grain yield of lentil genotypes. Žemdirbyst 99:305–312

Sahat SA (2015) Effect of climatic changes on yield and quality properties of some Egyptian cotton genotypes. PhD thesis, Department of Agricultural, Faculty of Agriculture, Cairo University, Egypt

Sahito A, Baloch ZA, Mahar A, Otho SA, Kalhoro SA, Ali A, Kalhoro FA, Soomro RN, Ali F (2015) Effect of water stress on the growth and yield of cotton crop (*Gossypium hirsutum* L.). Am J Plant Sci 6:1027–1039

Santhosh B (2014) Studies on drought tolerance in sunflower genotypes. M.Sc. in Agriculture. Deppartment of Crop Physiology College of Agriculture. In: Acharya NG (ed) Ranga Agricultural University Rajendranagar, Hyderabad -500 030 Crop Physiology

Santhosh B, Reddy SN, Prayaga L (2017) Physiological attributes of sunflower (*Helianthus annuus* L.)as influenced by moisture regimes. Green Farming 3:680–683

Shaar-Moshe L, Blumwald E, Peleg Z (2017) Unique physiological and transcriptional shifts under combinations of salinity, drought, and heat. Plant Physiol 174:421–434

Shafii B, Price WJ (1998) Analysis of genotype-by-environment interaction using the additive main effects and multiplicative interaction model and stability estimates. J Agric Biol Environ Stat 3:335–345

Shrestha J, Kushwaha UKS, Maharjan B, Kandel M, Gurung SB, Poudel AP, Karna MKL, Acharya R (2020) Grain yield stability of rice genotypes. Indones J Agric Res 03(02):116–126

Sreedhar S, Reddy TD, Ramesha MS (2011) Genotype x environment interaction and stability for yield and its components in hybrid rice cultivars (*Oryza sativa* L). Int J Plant Breed Genet 5(3):194–208

Stokstad E (2020) Rice genetically engineered to resist heat waves can also produce up to 20% more grain. Science. https://www.sciencemag.org/news/2020/04/rice-genetically-engineered-res ist-heat-waves-can-also-produce-20-more-grain

Swindell WR, Huebner M, Weber AP (2007) Transcriptional profiling of Arabidopsis heat shock proteins and transcription factors reveals extensive overlap between heat and non-heat stress response pathways. BMC Genomics 8:125

Tabrizi HZ (2012) Genotype by environment interaction and oil yield stability analysis of six sunflower cultivars in Khoy Iran. Adv Environ Biol 6(1):227–231

Tahir NA (2015) Identification of genetic variation in some faba bean (*Vicia faba* L.) genotypes grown in Iraq estimated with RAPD and SDS-PAGE of seed proteins. Indian J Biotechnol 14:351–356

Tai GCC (1971) Genotypic stability analysis and it's application to potato regional trials. Crop Sci 11:184–190

Tekalign A (2018) Genotype X environment interaction for grain yield of faba bean (*Vicia faba* L.) varieties in the highlands of Oromia region, Ethiopia. M.Sc. thesis, HaramayaUniversity, Haramaya

Temesgen T, Keneni G, Sefera T, Jarso M (2015) Yield stability and relationships among stability parameters in faba bean (*Vicia faba* L.) genotypes. Crop J 3:258–268

Tolessa TT (2015) Application of AMMI and tai's stability statistics for yield stability analysis in Faba bean (Vicia faba L.) cultivars grown in central highlands of Ethiopia. J Plant Sci 3(4):197–206

Turner MG, Romme WH, Gardner RH, O'Neill RV, Kratz TK (1993) A revised concept of landscape equilibrium: disturbance and stability on scaled landscapes. Landsc Ecol 8:213–227

USDA (2019) World agricultural production. United States department of agriculture, foreign agri-cultural service, circular series WAP 6–19 June 2019, office of global analysis, FAS, USDA, foreign agricultural service/USDA

Xu Y, Zhan C, Huang B (2011) Heat shock proteins in association with heat tolerance in grasses. Int J Proteomics 2011:1–12

Yang S, Vanderbeld B, Wan J, Huang Y (2010) Narrowing down the targets: towards successful genetic engineering of drought tolerant crops. Mol Plant 3:469–490

Yaseen AAM, Wasan MH (2015) Response of sunflower (Helianthus annuus L.) to spraying of Nano silver, organic fertilizer (Algastar (and salicylic acid and their impact on seeds content of fatty acids and vicine. AJEA 9(1):1–12

Part II
Impact of Climate Change on Sustainable Crop Production and the Physiological and Biochemical Basis for Crops Tolerance

Chapter 2
Climate Change and Its Impact on Sustainable Crop Production

Abstract Climate change is currently considered one of the most important challenges facing workers in the field of crop science and plant breeding in the world. This is due to climate changes that have affected all quality of air, water and soil, agriculture, and hence human and animal health. The different climatic elements represented in temperature, humidity, and length of day affect decisively the physiological and biochemical of plants, hence the growth, flowering, fruiting and ripening of crops in all agricultural systems. The negative effects of climate change on agricultural crops can be mitigated by implementing integrated adaptation strategies and understanding the nature of the relationships between different crop traits, and cultivating more adapted genotypes to environmental changes leading to sustained growth even under environmental stress. Therefore, this chapter will focus on critical periods of crop plants to stress conditions and the expected impacts of climate changes mainly on sustainable crop production.

Keywords Climate conditions · Greenhouse gas emissions · Agricultural crops · Critical periods · Adapted genotypes · Sustainability

2.1 Introduction

Climate is defined as the average weather condition in a region in the climatic factors, such as temperature, humidity, air pressure, wind, rain and others over a long period of time (Chen et al. 2017). The change in climate circumstances is mostly attributable to the change in temperature, greenhouse gases for instance carbon dioxide, methane, nitrogen oxides and sulfur oxides in the atmosphere, and others which cause raising the level of sea (Omambia et al. 2017).

The Intergovernmental Panel on Climate Change (IPCC 2018) Special Report on Global Warming of 1.5 °C showed how keeping temperature increases below 2 °C would reduce the risks to human well-being, ecosystems and sustainable development. Limiting warming to 1.5 °C would require an unprecedented response, according to the report's Summary for Policy Makers, which was adopted at meeting in Incheon, the Republic of Korea, on 6 October. The report said that net zero emissions of carbon dioxide must reach zero by 2050. At the current rate of emissions, the world will reach 1.5 °C warming by between 2030 and 2052. The global average temperature is already more than 1 °C higher than the pre-industrial era. Arctic sea is shrinking and sea level rise is accelerating. For instance, by 2100, global sea level rise may be 26 to 77 cm higher than the 1986–2005 baselines under a 1.5 °C temperature increase. This would mean that up to 10 million fewer people would be exposed to related impacts such as saltwater intrusion, flooding and damage to infrastructure. But even with a temperature increase of 1.5 °C, coral reefs are expected to decline by 70–90%, whereas more than 99% would be lost with 2 °C. For the main crops as wheat in tropical and temperate areas, temperature increases of 2 °C or more, induced by climate change, will negatively affect grain yield if there is no adaptation (Porter et al. 2014). It should be noted that crop production is susceptible to changes in climate, hence, this will lead in the future to significant changes in the distribution of crops. Moreover, statistics indicate that climate changes are likely to affect crop production in the twenty-first century in different regions of the world (Osman 2013). Because of the rapid changes in the global climate, abiotic pressures are one of the major restrictions to crop production and food security around the world. In this regard, heat and drought are the two most important pressures effecting negatively on the growth, productivity and quality of agricultural crops.

Environmental pressures negatively affect plant morphology, physiology, and biochemistry, and thus crop yields. This is due to the complex and multilateral effects due to physical damage, physiological disorders, and biochemical changes to these pressures. Halford (2011) stated that water shortage is already a major constraint for crop production in many regions of the world and as major dangers to food security. Water deficit is frequently, accompanied by high temperatures and these two stresses evoke different responses in plants and one stress exacerbates the effects of the another. Several publications have also been delivered concerning the fluctuating climate of Egypt comprising the impact of changing climate on crop production (e.g., El Massah and Omran 2015; El Afandi 2017; Mahmoud 2019).

Egypt is vulnerable to the impacts of climate change as a result of its dependence on the Nile as a primary source of water, its large traditional agricultural base and its long coastline that passes through intensification of development and erosion. Due to poor management of the soil and burning of fossil fuels, greenhouse gas emissions have increased alarmingly. In Egypt, climate change has three risky scopes that represent the risk of drowning the Nile Delta, the water crisis and low crop production. The results of the Agricultural Meteorology and Climate Change Research Unit of the Soil, Water and Environment Research Institute of the Agricultural Research Center, Egypt showed that the results of long-term prediction by simulations and various climate change scenarios, that climate change and the subsequent increase in temperature of the earth surface will damagingly affect the yield of Egyptian agricultural crops. This causing a great deficiency in the productivity of the major food crops in Egypt besides increasing water consumption. The most significant outcomes of studies conducted in this respect revealed that the yield of the wheat crop will reduced by about 9% if the temperature increases 2 °C, and the rate will decline to approximately 18% if the temperature is around 3.5 °C, with an increase in water consumption by about 2.5% rather than those in current weather conditions.

The mitigation of greenhouse gases in the atmosphere of carbon dioxide, methane and nitrous oxide that contribute significantly to global warming is a major task in recent agriculture. Moreover, up to 90% of the methane gas generated in the cultivation of submerged rice is released into the atmosphere. The remaining 10% of CH_4 in soil is often re-oxidized to CO_2 and differs broadly between rice genotypes such as hybrids, lines and conventional varieties. Therefore, the selection of the varieties and the treatments that will reduce the anthropogenic GHG emissions is considered an important goal in this regard (Wu et al. 2018). Consequently, heat stress decreased significantly grain weight but varied among genotypes by 52% in PHB-71 and 15% in IET 22,218 (Karwa et al. 2020).

Sunflower yield will decrease by about 27% and will raise water consumption by approximately 8%. However, climate change will have a beneficial effect on cotton yield, and its productivity will increase by about 17% when the air temperature rises around 2 °C. The rate of rise in this crop will increase to around 31% at 4 °C. Conversely, its water consumption will increase by around 10% rather than its water consumption in current weather conditions (El-Marsafawy 2007). Also, environmental stress harmfully affected growth and yield of faba bean (Abdelmula and Abuanja 2007, Tayel and Sabreen 2011 and Hegab et al. 2014) and productivity and physiological characteristics of sesame, endogenous phytohormones comprising Salicylic Acid and kinetin (Hussein et al. 2015) and yield (Awaad and Ali 2002 and Ayoubizadeh et al. 2019).

The relationship between environmental change and agricultural production is one of the most important problems facing developed and developing countries in the future decades of the twenty-first century. According to Mahmoud (2019) global and regional climate change will lead to the following potential impacts:

1. Impact on the agricultural sector and food security stability in Egypt.
2. Influence on both agricultural production sites and production capacity.
3. Influence on crop distribution and planting dates.
4. Adversely affect crop production with projected increases in temperatures, severe weather factors, drought and plant pests.
5. The pattern of land use change due to floods and rising sea level, and the infiltration of sea water, secondary and salinity.
6. Water resources may be affected by global warming and low rains.
7. Crop water requirements are expected to increase.

So, global climate change could dramatically increase respiration rate, reduced photosynthesis, translocation rate, available assimilates for grain filling and then crop production (Fig. 2.1). This chapter will focus on climate changes in Egypt and the world as well as the expected impacts of these changes mainly on crop production.

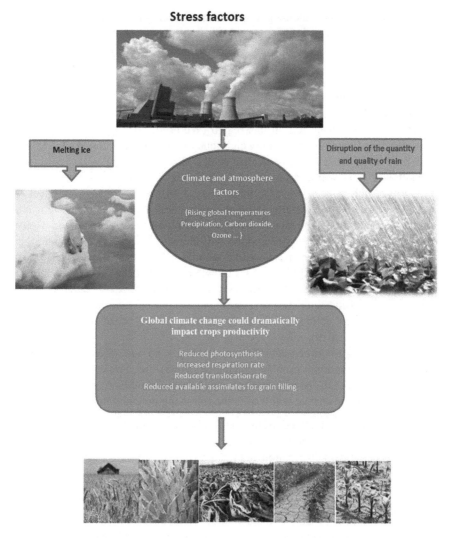

Fig. 2.1 The impact of climate change on the living environment (drawn by the author H A Awaad)

2.2 Critical Periods of Crop Plants to Environmental Stress

Abiotic stresses can significantly decrease crop yields and limit the latitudes and soils on which commercially essential species can be cultivated (Blum 1985). The seriousness of stress depends on its timing, duration and intensity (Serraj et al. 2005). Heat stress causes crops to develop and mature more quickly; it also leads to an increase in respiration and an inhibition of photosynthesis. This is due to a reduction in the activity of ribulose 1, 5-bisphosphatecarboxylase/oxygenase RubisCO and the efficacy of photosystem II. The decrease in RubisCO activity happens as a result of the enzyme responsible for protecting its activity, RubisCO activase, is affected at high temperatures. This makes the genetic manipulation of RubisCO activase to improve its stability at high temperatures is important target.

The effect of environmental stress on crop plants depends on the growth stage, growth habit, crop species and cultivar. Awareness of sensitive periods in plant life is useful in developing breeding programs to deal with current environmental conditions. The following are the critical stages of the impact of environmental stress on some important field crops.

1. Wheat: Crown root initiation, tillering, jointing, booting, flowering, milk and dough stages.
2. Rice: Initial tillering, flowering/anthesis phase.
3. Faba bean: Flowering and podding.
4. Sesame: Flowering and capsule formation.
5. Sunflower: Flowering and achene filling period.
6. Cotton: Flowering and boll formation.

2.3 Impact of Climate Change on Sustainable Crop Production

The adverse effects represent the main challenge for researchers in field crops. And this needs to modification in crop systems to accommodate future requirements, with an increasing population and the threat of climate change. Sustainable crop improvement strategies are an important goal for tackling stress challenges that reduce crop productivity and quality. Improving stress tolerance is a very complex task for field crop improvement scientists. The adaptation of crop varieties to stress conditions with appropriate agricultural managements for crop species is an important means in this regard.

Therefore, the study of the impact of climate change is necessary so that appropriate measures can be taken to adapt with expected changes in the environmental conditions (Awaad et al. 2021). The negative impacts of heat stress in the presence of soil moisture pressure on crop genotypes are illustrated in Fig. 2.2.

Wheat is a cold season crop, and is sensitive to thermal and wet stress during the reproductive phase (Castro et al. 2007). Heat stress throughout the late sown condition reduced total chlorophyll content, leaf area, grain filling duration, plant height, and grain yield. Furthermore, high terminal temperature through flowering increased canopy temperature and leaf and stem rust susceptibility. Chlorophyll content weakened significantly under the late sown condition, due to high temperature damage on chloroplast (Akter and Islam 2017). The incidence and severity of stem and leaf rust were increased with the late sown condition as favorable conditions for both diseases (Draz et al. 2015). The late sown condition reduced grain filling duration, whereas the rise in the temperature during the reproductive stage reduced the quantity of the assimilates (Blanco et al. 2012). The increase in temperature is accompanied by an increase in the grain filling rate, nonetheless without a compensatory effect of the

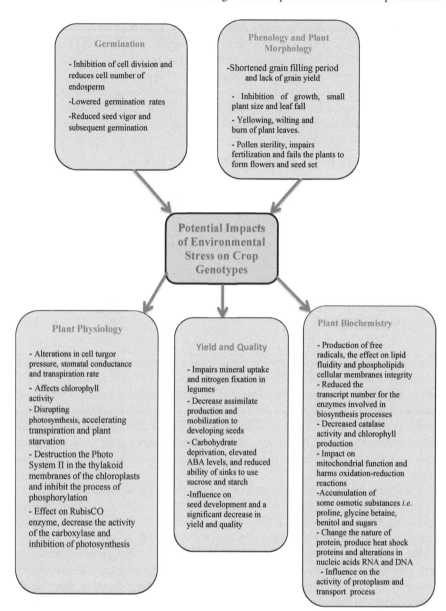

Fig. 2.2 Potential impacts of environmental stress on crop genotypes (drawn by the author H A Awaad)

loss resulting from shortening the grain filling period duration (Ayeneh et al. 2002). ElBasyoni (2018) showed variable response of the tested genotypes to heat stress under different environmental conditions which approves with previous findings. Fatma, Farag (2019) displayed adaptable response of the verified wheat genotypes to water stress under different seasonal conditions. Also, showed that drought stress reduced significantly grain yield. Hence, identifying the most adapted and stable genotypes was definite agronomically as that performs well, as "good yielding" and stable across a wide range of environments. Qaseem et al. (2019) showed that independent and combined drought and heat stress negatively affect plant physiology and wheat yield. Drought stress had greater reduction effects by 45% on grain yield. Harvest index came in the second most severely influenced trait and it was decreased by 37% due to drought. Heat stress affected 53% reduction in grain yield followed by 39% reduction in harvest index. Simultaneously effect of both drought and heat stress caused reduction by 56% in grain yield, and 41% in harvest index. Fatma, Farag (2019) showed that wheat grain yield was progressively decrease from normal irrigation (22.71 and 21.05 ardab/fad.), mild stress (19.77 and 18.31 ardab/fad.), moderate stress (15.88 and 14.85 ardab/fad.) and severe stress (12.82 and 12.26 ardab/fad.) through 1st and 2nd season, respectively.

Photosynthesis is one of the main metabolic processes that govern the **rice** production, is influenced by environmental stress. Photosynthetic capabilities of leaves and water availability to the zone of root system are very important factors that reduce growth and yield of sensitive rice varieties under water stress condition (Zhu et al. 2020). Stomata are closed under water stress environments, decreasing carbon dioxide influx to leaves and driving additional electrons for creation of reactive oxygen species (Mishra et al. 2018). During water stress, inequity is identified between capture and use of light, decrease and impairment in RubisCO activity, pigments and photosynthetic machinery, which are the causes for photosynthesis declines (Farooq et al. 2009). Water stress damages the normal functions of PSI and PSII (Mishra and Panda 2017) and inhibition of electron transport chain and subsequent inactivation of PSII (Mishra et al. 2018). It also affects the both content and activity of chlorophyll (Mishra et al. 2019) and Carotenoids (Ashraf and Harris 2013). However, under stress conditions, higher accumulation of proline is commonly accompanying with water stress tolerance and it helps for conservation of leaf turgor and progress in stomatal conductance (Kumar et al. 2016 and Mishra et al. 2019).

In Egypt, to assess the impact of salinity and water stress on promising **rice** genotypes can be grown under stress environments, Hassan et al. (2013) found that all yield traits i.e. total duration, number of panicles/plant, number of filled grains/panicle, 100-grain weight, sterility %, drought tolerance index and grain yield/plant were affected by the diverse salinity levels and water stress environments during the two seasons. Under both conditions, all characters were reduced significantly by increasing salinity and drought stress levels excluding total duration and sterility %, increased by increasing salinity levels. Whereas, total duration and sterility% were increased from normal to drought stress. This might be due to that water stress causes delay in heading and low yield components led to low grain yield.

Rice genotypes varied in their react to heat stress, in this concern, Karwa et al. (2020) showed that high temperature stress decreased significantly rice grain weight by 45% in rice genotype IET 23,296. Conversely, IET 22,218 had displayed a decline of 15% in grain weight under heat stress. Spikelet fertility was reduced significantly by 10–54% under heat stress in the tested rice genotypes, with highest reduction (> 50%) in rice genotype PHB-71, conversely, IET 22,218, and IET 23,324 exhibited smaller reduction (18%) in spikelet fertility. Furthermore, pollen viability display greater reduction (47%) in PHB-71 and least reduction in both IET 22,218 (16%) and IET 23,324 (14%) under heat stress.

Moreover, worldwide, **faba bean** production morally influenced by water stress as one of the major abiotic stresses. Siddiqui et al. (2015) revealed that drought stress impairs the plants growth, moprpho-physiological characteristics i.e. plant height, fresh and dry weights, leaf area, leaf relative water content and biochemical characters i.e. proline content, total chlorophyll content, and activities of catalase, peroxidase and superoxide dismutase of faba bean genotypes. Drought stress reduced also seed yield. Abid et al. (2017) recorded differential response of faba bean cultivars to water deficit. Water stress reduced significantly faba bean plant leaf area at 30% FC than at 50% FC with reduction fluctuated from 12.30% to 50.21% in 'Hara' and 'VIR 490' genotypes. Also, water deficit decreased shoot and root fresh weights and seed yield than the control.

In contrast, **sesame** as oil crop is moderately tolerant to drought stress (Eskandari et al. 2009). However, severe water stress adversely injuries the plant growth and development (Anjum et al. 2011) as well as the reproduction and yield (Sun et al. 2010). Hussein et al. (2015) found that water stress harmfully affected growth parameters of sesame comprising plant height, fresh weight, dry weight, number of leaves and leaf area. Also, drought stress decreased significantly endogenous phytohormones i.e. salicylic acid and kinetin. Whereas, drought stress caused significant increase in total soluble protein, poly peptide bands and proline. The influence in yield attributes as a result of water stress have been assured by Golestani and Pakniyat (2015) showed that drought reduced all yield relevant traits i.e. number of days to maturity, number of capsules/plant, 1000-seed weight, biological yield and seed yield, and physiological traits i.e. leaf water potential and leaf osmotic potential. It is reasonable to suggest that high yield of sesame plants under water deficit conditions could be attained by selecting breeding materials with lowest decrease in number of days to maturity, number of capsules/plant, 1000-seed weight, biological yield, seed yield, leaf water potential and leaf osmotic potential, and the highest reduction in rate of water loss. Consequently, breeding sesame genotypes with higher tolerance to abiotic stresses is considered unlimited attention from breeders (Dossa et al. 2017).

In another direction, **sunflower** crop could participate to the mitigation solution as a low greenhouse gas emitter compared to cereals and oilseed rape (Debaeke et al. 2017). Under water stress, some sunflower genotypes showed better drought tolerance and the others were sensitive. Santhosh (2014) revealed that stress imposed from 30 to 65 DAS resulted in decreased plant height, leaf number, days to 50% flowering, days to maturity, leaf area index, crop growth rate, leaf chlorophyll content. Also, water stress decreased physiological characters i.e. stomatal conductance, fluorescence, photosynthetic rate, chlorophyll stability index, membrane stability index, while, proline content was increased. Furthermore, water stress imposition has negatively affected yield related attributes like capitulum diameter, capitulum dry weight, total number of seeds/capitulum, test weight seed yield/plant compared to control as indicated by Santhosh et al. (2017).

Finally with regard to the **cotton** crop as originates from hot climates, a negative association was registered between cotton yield and high temperature during early boll development. While cotton is sensitive to high temperature at all stages of growth, it is sensitive to high temperatures through reproductive development, and environmental stress throughout floral development represents a major restraint to crop development and productivity (Oosterhuis 2011). Differential responses between cotton cultivars in water and heat stress tolerance has been recorded. The sensitive cotton cultivar 84-Shad decreased activities of catalase, and peroxidase under stress condition. Conversely, the drought-tolerant cultivar M-503 preserved levels of enzyme activities of superoxide dismutase and ascorbate peroxidase and increased the levels of catalase, peroxidase, and proline content under drought and heat stresses (Sekmen et al. 2014). On the other hand, cotton yield and its attributes as well as fiber traits adversely affected by salinity stress. Moustafa (2006) showed that seed cotton yield was influenced by salinity levels of irrigation water, where increasing salinity levels from 2000 to 8000 ppm of irrigation water led to relative reduction by 82 and 84% in the 2001 and 2002 season, respectively. The decrease in the seed cotton yield might be attributed to the decrease in number of fruiting branches/plant, number of open bolls/plant and boll weight. Furthermore, Manikandan et al. (2019) revealed that response of cotton species and their cultivars differs with different salinity levels. Salt stress at root zone negatively reduced the cotton growth, nutrient uptake and plant physiology. The expansion of leaf area, osmotic potential, leaf water potential and higher osmotic stress, ionic stress affects the photosynthetic rate. Salinity reduced the cotton growth, nutrient discrepancy and seed cotton yield as well as fiber quality. In terms of water stress, Azza, El-Hendawy (2020) indicated that prolonging irrigation intervals to 4 weeks significantly decreased plant height, number of fruiting branches/plant, number of open bolls/plant, boll weight, seed index, seed cotton yield/fed., fiber length and fiber strength, but lint percent was significantly increased in Giza 94 cotton cultivar.

Based on this information, the study of the sensitivity of the crop growth appears important to avoid the possible impacts of climate change.

2.4   Impact of Climate Change on Interrelationships Among Yield Traits of Field Crops

Understanding impact of climate change on the nature of the association between yield and other important traits, breeders can select and develop more stress tolerant varieties that give satisfactory yield levels.

2.4.1   Wheat

The selection in wheat on the basis of number of spikes/m^2, number of spikelets/spike and number of grains/spike would likely to be most useful for increasing grain yield as their direct positive role to grain yield under irrigated condition as revealed by Khan and Naqvi (2012). They also showed that number of spikes/m^2, number of spikelets/spike, spike length and number of grains/spike might be utilized as effective selection criteria for increasing wheat grain yield under different water levels. Consequently, these traits could be selected for the diverse stress environments for supporting grain yield. El-Rawy and Hassan (2014) showed that grain yield/plant was positively associated with grain yield/spike, number of tillers, plant height, flowering time, stomata length, drought sensitivity index, while negatively associated with stomata frequency and drought sensitivity index in normal (clay fertile soil, E_1), 100% (E_2), and 50% (E_3) field water capacity in sandy calcareous soil, respectively. Therefore, greatly heritable traits and closely correlated with grain yield under stress conditions particularly stomata frequency and length might be used as selection traits for identifying high-yielding genotypes tolerant to drought stress. Seyoum (2018) showed that grain yield had positive association with days to heading, days to maturity, 100-grain weight, spike length, number of spikelets/spike at both phenotypic and genotypic levels. While, plant height had highly significant negative association with grain yield at phenotypic one.

Factor analysis was used for understanding the data structure and trait relations. Motavaseel (2014) studied 10 genotypes of bread wheat based on yield and some agronomic traits. According to results of factor analysis, similarity coefficients of most traits are high, indicating that the number of selected factors was suitable and the chosen factors could justify trait variations in a desirable way. In the first factor which includes the most range of data change (43.7%) traits of 1000-grain, days to physiological maturity and grain yield were consisted of high positive and negative coefficients. Therefore, this factor can be recognized as a yield factor. The second factor justified 28.45% of the variations. Traits of plant height, spike length and number of grains/spike were influential on this factor and this factor can be titled plant height factor. Moreover, the third factor accounted for by 16.13% of the variations. Generally, these three factors evaluate a specific trait in contradiction of other traits.

Moetamadipoor et al. (2015) investigated association between morpho-phonological traits and yield using multivariate statistical technique of factor analysis under water stress for 18 phenotypes of bread wheat. Factor analysis based on extraction of latent roots through principal components analysis of seven factors was comprised in the model. Results showed that these factors overall accounted for by 87.7% of variation in traits. Two first factors justified total variation of traits as 17% and 16%, respectively and they were titled as yield and yield components factor and maturity factor. Results attained from factor coefficients show importance of traits affecting yield and traits related to early maturity in selecting of ideal genotypes for drought conditions. Thus, these two factors can be used as criteria for selection in wheat breeding programs under stress conditions. Fatma, Farag (2019) showed that, the factor analysis technique divided the studied variables into four main factors (Tables 2.1 and 2.2) under stress condition. These four factors accounted about 86.309% of the total variability in the dependence structure of wheat grain yield.

Table 2.1 Principal factor matrix after orthogonal rotations for studied characters of wheat in the 2nd season under the fourth water regimes (Fatma, Farag 2019)

Variables	Common factors coefficients				Communality %
	Factor 1	Factor 2	Factor 3	Factor 4	
Plant height (cm)	0.738	0.297	0.374	−0.050	0.776
Flag leaf area (cm^2)	0.158	0.888	0.091	0.246	0.882
Spike length (cm)	0.452	0.494	0.612	0.250	0.886
Number of spikelets/spike	0.395	0.842	−0.26	0.221	0.915
Chlorophyll content	0.002	−0.031	0.925	0.071	0.861
Relative water content	0.476	−0.295	0.083	−0.768	0.910
Transpiration rate	−0.699	−0.628	−0.221	0.081	0.939
Osmotic pressure	0.164	0.645	0.556	−0.171	0.781
Prolien content	0.330	0.060	0.293	0.817	0.867
Number of spikes/m^2	0.913	0.259	−0.092	−0.009	0.910
Number of grains/spike	0.929	0.178	0.245	0.156	0.978
1000-grain weight (gm)	0.425	0.218	0.578	0.301	0.653
Variance ratio	30.497	23.977	18.600	13.234	86.309

Table 2.2 Summary of factor loading for some important traits of wheat in the 2nd season under the fourth water regimes (Fatma, Farag 2019)

Variables	Loading	Percentage of total
Factor 1		30.497
Plant height (cm)	0.738	22.51
Transpiration rate	−0.699	21.32
Number of spikes/m^2	0.913	27.84
Number of grains/spike	0.929	28.33
Factor 2		23.977
Flag leaf area (cm^2)	0.888	37.39
Number of spikelets/spike	0.842	35.45
Osmotic pressure	0.645	27.16
Factor 3		18.600
Spike length (cm)	0.612	28.94
Chlorophyll content	0.925	43.74
1000 grain weight (gm)	0.578	27.33
Factor 4		13.234
Relative water content	−0.768	48.45
Proline content	0.817	51.55
Cumulative variance		86.309

The first factor included four variables and accounted for 30.497%. These variables were plant height (22.51%), transpiration rate (21.32%), number of spikes/m^2 (27.84%) and number of grains/spike (28.33%). It is clear that these variables had high loading coefficients and participate much more on the dependence structure. Most of these variables exhibited positive and significant correlation values with wheat grain yield. The second factor consists of two variables and accounted for 19.034% of the total variability of wheat grain yield. These two variables were proline content (58.44%) and 1000-grain weight (41.56%). The third factor included two variables and accounted for 15.054% of total variance. These variables were flag leaf area (45.72%) and osmotic pressure (54.28%). Two variables were loaded in the fourth factor and accounted for 12.104% of the total variance of the dependence structure. These variables were chlorophyll content (47.3%) and relative water content (52.7%).

2.4.2 Rice

Maintenance of stability in membrane index of the plant under water scarcity has been used to know its correlation with yield of rice under drought stress (Upadhyaya and Panda 2019).

In Egypt, under water stress conditions, El-Khoby et al. (2014) assessed phenotypic correlation coefficients among all possible pairs of the combinations of root characters, yield and its related traits in the F2 generation of the three crosses cross I{Tsuyuake (tolerant) X Sakha 103 (sensitive)}, cross II {Zenith (tolerant) X Sakha 104 (moderate)} and cross III {BL1(moderate) X Sakha 106 (sensitive)}. Grain yield/plant was highly significant and positively correlated with root length, root number/plant, root volume, panicle length, number of panicles/plant and 100-grain weight in the three studied crosses. On the contrary, rice grain yield/plant appeared to be highly significant and negatively associated with plant height in the first two crosses. Moreover, under 18 environments, El-Degwy and Kmara (2015) registered positive and significant association between grain yield/plant with each of; panicle length, number of spikelets/panicle, number of panicles/plant under all environments. Under late sowing (15th May), number of grains/panicle showed the highest positive direct effect on grain yield/plant followed by number of filled grains/panicle and days to 50% heading with values of 0.80, 0.626 and 0.378, respectively. While, number of spikelets/panicle, fertility percentage, number of panicles/plant recorded highest positive direct effect on grain yield under intermediate sowing date (1st May). Moreover, number of spikelets/panicle and panicle length were the most contributors at early sowing date (15th April).

Under rainfed of Bangladesh, Nuruzzaman et al. (2017) registered positive and significant association between grain yield/plant with each of; plant height, total tillers/hill, productive tillers/hill and filled grains/panicle under rainfed conditions at both phenotypic and genotypic levels. Whereas, days to maturity had significantly negative correlation with grain yield. Higher phenotypic association estimates for the traits showed that the environmental effects on traits are great under rainfed conditions. Plant height, total tillers/hill, productive tillers/hill, panicle length, filled grains/panicle and 100-grain weight indicated direct positive effect on grain yield/plant at both phenotypic and genotypic levels. Thus, these traits would be suitable criteria for improving yield.

At the International Rice Research Institute (IRRI), Los Baños, Philippines, Melandri, et al. (2020) assessed correlation analysis to evaluate the influence of flowering time differences on the drought-induced grain yield loss performance in 292 accessions. Flowering under drought significantly and negatively associated with grain yield loss (rs = –0.35), however only 12% of the variation was described by the corresponding linear model. Generally, the association tendency displays that accessions that already flowered before stress burden (<10% of the total) exhibited higher severity of grain yield loss than those that had nearly or already flowered (booting stage for 60% and heading stage for 30% of the total) through stress imposition. Stimulatingly, a significant negative correlation was detected between flowring under control and grain yield loss. This similarity is determined by the almost perfect correlation (rs = 0.96; R2 = 0.94; P-value < 0.001) observed between flow under control and drought. But, drought stress significantly affected the days to 50% flowering with a delay of ~3 day compared with control.

2.4.3 Faba Bean

Environmental pressures directly affect the production of faba bean, especially in semi-arid regions that are a result of the changing global climate and drought conditions in agriculture. The happenstance of water shortage, cold, heat stress is the challenge in faba bean production under stress conditions. Eldardiry et al. (2017) showed that at 60 and 90% ETo under saline and fresh irrigation water salinity, faba bean yield ranged from 1216 to 1590 kg/fed, respectively with the same trend for crude protein content. Also, rise water salinity from 0.68 to 2.45 dS/m lead to reduce seed yield from 1450 to 1319, respectively with reduction percentage 9%. Whereas, Migdadi et al. (2016) showed that severe drought stress caused reduction in faba bean seed yield by 67.9% across genotypes as compared with well irrigated management. The reduction in seed yield owing to drought was varied from 58.3% in genotype Hassawi 2 to 79.4% in genotype ILB 1814.

Under Khatara region as stress environment in Egypt, positive and highly significant correlation was found between seed yield/fed and each of plant height, number of branches/plant, number of pods/plant, number of seeds/pod, seed index, seed yield/plant, harvest index and LAI in both Egyptian cultivars Sakha 1 and Giza Blanca (Table 2.3) as revealed by Mokhtar (2003). Hereby increasing faba bean seed yield might be achieved through increasing the foregoing characters. For Sakha 1 cultivar, path coefficient analysis (Table 2.4), the studied yield contributing characters, accounted for 98.87% from the total variance of faba bean seed yield. Whereas, the residual effect was 1.122% only. The highest direct effect on seed yield/fed was excreted by leaf area index LAI (11.409%), followed by plant height (7.927%), and seed index (4.815), whereas other characters were the lowest ones in their contribution in faba bean seed yield variation. Reinforcing the importance role of physiological character leaf area index, with plant height and seed index in increasing faba bean seed yield of Sakha 1 cultivar. The maximum joint effect on seed yield variation was recorded for, plant height via LAI, seed index via LAI, plant height via seed index, number of pods/plan via LAI, number of seeds/pod via LAI, number of pods/plant via seed index, number of branches/plant via LAI and plant height via number of pods/plant. Whereas, the other indirect effects had a little effect on seed yield variation. Based on the total contribution, the studied characters could be arranged as follows LAI (29.026%), seed index (23.188%), plant height (20.771%), number of pods/plant (11.144%), number of seeds/pod (8.374%) and number of branches/plant (6.375%).

Furthermore, Mokhtar (2003) added that, for Giza Blanca cultivar, (Table 2.4) the studied yield contributing characters, accounted 98.424% from the total variation of faba bean seed yield. Whereas, the residual effect was 1.576% only. The highest direct effect on seed yield was excreted by number of seeds/pod (61.756%), whereas the contribution of plant height was (0.6555%) and number of branches/plant (0.456%) in faba bean seed yield variation. The maximum indirect effect on seed yield variation was recorded for, plant height via number seeds/pod, number of branches/plant via seed index, plant height via number of seeds/pod, number of seeds/pod via LAI,

Table 2.3 Simple correlation between seed yield/fed and its attributes of Sakha 1 (S. 1) and Giza Blanca (G.B) cultivars (Combined) (Mokhtar 2003)

Characters		X_2	X_3	X_4	X_5	X_6	X_7	X_8	Seed yield/fed
Plant hight (X_1)	S.1	0.755**	0.703*	0.620*	0.548	0.661*	0.676*	0.976**	0.673*
	G.B	0977**	0.888***	0.878**	0.800**	0.878**	0.798**	0.860**	0.881**
No. of branches/plant (X_2)	S.1		0.865**	0.916**	0.862**	0.864**	0.920**	0.897**	0.869**
	G.B		0.917**	0.922**	0.800**	0.919**	0.794**	0.860**	0.911**
No. of branches/plant (X_3)	S.1			0.951**	0.949**	0.992**	0.978**	0.800**	0.989**
	G.B			0.981**	0.925**	0.989**	0.929**	0.942**	0.992**
No. of branches/plant (X_4)	S.1				0.976**	0.970**	0.980**	0.795**	0.953**
	G.B				0.905**	0.995*	0.913**	0.927**	0.991**
No. of branches/plant (X_5)	S.1					0.976**	0.973	0.729**	0.958**
	G.B					0.907**	0.976	0.965**	0.938**
No. of branches/plant (X_6)	S.1						0.983	0.785**	0.991**
	G.B						0.904	0.917**	0.994**
No. of branches/plant (X_7)	S.1							0.816**	0.983**
	G.B							0.974	0.936**
No. of branches/plant (X_8)	S.1								0.779**
	G.B								0.952**

*, **Significant at P = 0.05 and P = 0.01, respectively

Table 2.4 Direct and indirect effects of seed yield attributes as percentage of seed yield variation in Sakha 1 and Giza Blanca cultivars (Combined) (Mokhtar 2003)

Character	Sakha 1	Giza Blanca
	Relative importance %	Relative importance %
Plant height	7.827	0.6555
Number of branches/plant	0.442	0.456
Number of pods/plant	1.044	0.086
Number of seeds/pod	0.634	61.756
Seed index	4.815	0.036
Leaf area index	11.409	0.0899
Plant height × No. of branches/plant	2.809	1.0681
Plant height × No. of pods/plant	4.018	0.4215
Plant height × No. of seeds/pod	2.763	11.172
Plant height × seed index	6.729	0.207
Plant height × leaf area index	17.879	0.4176
No of branches/plant × no. of pods/plant	1.176	0.2847
No. of branches/plant × no. of seeds/pod	0.97	9.7841
No. of branches/plant × seed index	2.516	0.1726
No. of branches/ plant × leaf area index	4.303	0.3583
No. of pods/plant × No. of seeds/pod	1.547	4.517
No. of pods/plant × seed index	4.255	0.0866
No. of pods/plant × leaf area index	5.521	0.1656
No. of seeds/pod × seed index	3.411	2.2273
No. of seeds/pod × leaf area index	4.276	4.3698
Seed index × leaf are index	10.807	0.0924
R^2	98.878	98.424
Residual	1.122	1.576
Total contribution		
Plant height	20.771	7.141
Number of branches/plant	6.375	6.158
Number of pods/plant	11.144	2.909
Number of seeds/pod	8.374	77.967
Seed index	23.188	1.454
Leaf are index	29.026	2.795
Total	98.878	98.424

number of seeds/pod via seed index, plant height via number of branches/plant, plant height via number of pods/plant and plant height via LAI. Whereas, the other indirect effects had a little effect on seed yield variation. Generally, the total contribution of the studied characters could be arranged as follows, number of seeds/pod (77.967%), plant height (7.141%), number of branches/plant (6.158%), number of pods/plant (2.909%), LAI (2.795%) and seed index (1.454%). Furthermore, the author reached a very important finding in the variable contribution of root and nodules growth attributes to seed yield variation between the two faba bean cultivars. In Giza Blanca this contribution amounted to 92% compared with 70% in Sakha 1. This refers to a physiological harmony between the root nodule sink and the top sink, particularly in Giza Blanca. Also, the variable contributions of seed yield attributes to seed yield variation are quite interesting. In Giza Blanca, the number of seeds/pod made the greatest contribution (78%), whereas in Sakha 1, LAI, seed index, and plant height all together made this contribution (72%) since the former had large seeds (110–120gm/100 seed), seed yield was governed by seed number/pod rather than seed index. Furthermore, Naglaa Qabil et al. (2018) found that seed yield/plant was positively and significantly associated with each of pod length, number of pods/plant, number of seeds/pod, number of seeds/plant as well as 100-seed weight through the two seasons. Whereas, correlations between seed yield/plant and each of plant height and number of fruiting nodes/plant were negative and significant during the two seasons. Path coefficient analysis showed that number of seeds/plant followed by 100-seed weight had the maximum direct effect on seed yield/plant during the two seasons. Suggest that selection in for these characters will also increase faba bean yield.

2.4.4 Sesame

Stress occurred at flowering stage, had a significant effect on reduction of sesame seed yield (Khammari et al. 2013). Studies show that sesame yield was reduced when the number of irrigations was reduced from seven to five irrigations (Tantawy et al. 2007).

Yield potentiality and yield analysis of seven sesame genotypes grown was computed under newly reclaimed sandy soil. Awaad and Basha (2000) revealed that oil yield of sesame showed positive and significant correlations with each of seed yield/fad (0.987) > 1000-seed weight (0.943) > number of capsules/plant (0.805) > seed oil content (0.801) > fruiting zone length (0.754), in respective order (Table 2.5). These results reflected the importance of these characters for improving oil yield under sandy soil conditions. Also, positive and significant associations were detected between seed yield/fad. and each of 1000-seed weight (0.974) > fruiting zone length (0.769) and number of capsules/plant (0.759), suggesting that the increasing genes for these characters were associated with increasing ones for sesame seed yield.

The results of factor analysis (Table 2.6) showed that eight variables are divided into three main factors as clusters which made up 80.1% of the total variability

Table 2.5 Simple correlation coefficients between oil yield and its contributing characters (combined data of the two seasons) (Awaad and Basha 2000)

Character	2	3	4	5	6	7	8	Y
1- Fruiting zone length	0.236	0.029	0.392	0.570	0.719	0.769*	0.541	0.754*
2- Capsule length		−0.033	0.132	−0.148	−0.058	0.012	0.238	0.071
3- No. of branches/plant			0.689	0.269	0.378	0.341	0.056	0.307
4- No. of capsules/plant				−0.069	0.726	0.759*	0.738	0.805*
5- No. of seeds/capsule					0.726	0.259	−0.259	0.147
6- 1000- seed weight						0.276	0.974**	0.572	0.743**
7- seed yield							0.0.699	0.987**
8- seed oil content								0.801*
Y- oil yield								–

*, **Significant at P = 0.05 and P = 0.01, respectively

Table 2.6 Summary of factor loading for 8 characters of 7 sesame genotypes based on combined data of the two seasons (Awaad and Basha 2000)

Factor	Loading	% of total communality
Factor 1:		40.2
No. of capsules/plant	0.93386	
No. of branches/plant	0.71829	
Seed oil content	0.82192	
1000- seed weight	0.62396	
Factor 2:		24.5
Plant height	0.91646	
Fruiting zone length	0.9074	
No. of seeds/capsule	0.65091	
Factor 3:		15.4
Capsule length	0.84903	
Total contribution	80.1	

in the dependence structure. The first factor included four variables i.e. number of capsules/plant, number of branches/plant, seed oil content and 1000-seed weight which accounted for 40.2% of the total variation. The second factor involved three variables, namely plant height, fruiting zone length and number of seeds/capsule which accounted for 24.5% of the total variability in the dependence structure.

However, the third factor included only one variable i.e. capsule length which made only 15.4% of the total variability.

An update on the results of the previous study, oil yield was positively and significantly correlated with seed yield/fad > 1000-seed weight > number of capsules/plant > seed oil content > fruiting zone length, in respective order. The main sources of oil yield variation based on their total relative importance were number of capsules/plant (37.421%) followed by seed oil content (20.989%), 1000-seed weight (19.233%) and number of branches/plant (16.734), however the two remaining characters i.e. number of seeds/capsule and fruiting zone length recorded lower relative contribution values of 2.791 and 2.687%, respectively. The results of factor analysis coupled with path coefficient analysis, and in most cases with correlation results, indicated that factor 1 which included number of capsules/plant, number of branches/plant, seed oil content and 1000-seed weight was the best one as it accounted for 40.2% of the total variation compared with factor 2 and 3, reinforcing its importance in selection programs to improve oil yield under sandy soil stress conditions. Oil yield showed positive and highly significant association with fruiting zone length, capsule length and seed yield. Whereas, Kuol (2004) computed putative yield-based drought tolerance parameters and pattern of their associations with yield. Sesame seed yield under normal irrigation and seed yield under drought stress were strongly correlated ($r = 0.88**$) in parental genotypes and moderately correlated ($r = 0.36**$) in progenies. Stress tolerance index was positive and highly associated with seed yield under normal ($r = 0.95**$) and with seed yield under drought stress ($r = 0.98**$) in parental genotypes, whereas it was insignificant with yield under both water regimes in progenies. Finally, sesame seed yield had significant and positive association with plant height, first capsule height and capsule number at genotypic and phenotypic, respectively. The tested characters showed positive direct effects on seed yield/plant, except the days to 50% flowering had negative effect. Boureima et al. (2016) revealed that plant height, number of capsules per plant, and the length of the capsules should be considered in selection for obtaining high-yielding sesame cultivars in drought-stressed environments. Mohamed and Bedawy (2019) concluded that the yield traits found to be higher beneficial to genetic diversity in Egyptian sesame populations and genetic improvement of production via plant breeding programs.

2.4.5   Sunflower

Drought is the major environmental factor restrictive sunflower plant growth in a wide range of environments. Sunflower is considered a crop with moderate water requirements (Ky < 1), has the capability to endure a short period of water stress, recover moderately from pressure with less decreases in yield with reduced water use (García-Vila et al. 2012). By its great ability to extract water from the subsoil, The crop has the potential to access deeper water resources (Cabelguenne and Debaeke 1998). Varieties of sunflower have been shown to exhibit contrasting responses to drought (Awaad et al. 2016 and González-Barrios et al. 2017). Rafiei et al. (2013)

stated that the effect of water stress on seed weight, seed yield, oil percentage, harvest index, biological yield was significant so that the maximum amount of seed weight, seed yield, and biological yield was attained in irrigation treatment with 180 mm evaporation from the pan and maximum percentage of oil was obtained by control treatments.

Loose et al. (2019) found that water deficit decreases sunflower oil yield (kg/ha) from 1830.6, 1211.3 to 632.2 under control, Surplus and water stress, respectively. Also, yield components and oil content were decreased, while the oleic/linoleic ratio in oil was increased.

At well-watered condition and stress conditions, Darvishzadeh et al. (2011) and found that genotypic correlations manifest that seed yield/plant was positively and significantly associated with head diameter, plant height and achene traits at well-watered condition and with head diameter and chlorophyll content at water-stressed state. Head diameter and number of achene at both conditions and chlorophyll content at water-stressed condition have positive direct effect on seed yield/plant. Hamza and Safina (2015) showed that seed yield/fed showed close and significantly associated with seed yield/plant (r = 0.87), number of leaves plant (r = 0.83) and less extent to head diameter (r = 0.75). Whereas, Awaad et al. (2016) computed phenotypic, genotypic and environmental correlation coefficients to determine the most effective characters which played an important role in the final yield across the three environments as given in Table 2.7. Positive and significant associations was registered between achene yield/plant and each of leaf water content, transpiration rate, plant height, head diameter and 100-achene weight at both phenotypic and genotypic levels, except for leaf water content which showed significant correlation at genotypic level only. Whereas, increasing values of achene oil content resulted in decreasing achene yield/plant. Both 100-achene weight and achene yield/plant, having the same environmental requirements as found by positive and highly significant environmental correlation between them. Moreover, positive and highly significant interrelationship was recorded between leaf water content with each of leaf chlorophyll content, transpiration rate and 100-achene weight, and between transpiration rate with each of plant height, head diameter and 100-achene weight, as well as, between plant height and both head diameter and 100-achene weight at both genotypic and phenotypic levels in most cases. On the contrary, negative and highly significant association was registered between leaf water content with achene oil content and between leaf chlorophyll content with each of transpiration rate, plant height and head diameter as well as, between transpiration rate, plant height, head diameter on one hand with achene oil content, on the other hand.

They added that, highest direct effect on achene yield/plant was recorded for transpiration rate and plant height with estimates of 12.941% and 12.219%. Whereas, modest direct effects were found by both 100-achene weight and leaf water content with assessments of 7.128 and 7.779%, correspondingly. The maximum indirect effects on achene yield/plant variation were detected for transpiration rate x plant height followed by transpiration rate x 100-achene weight, leaf water content × 100-achene weight, plant height x 100-achen and leaf chlorophyll content x plant height with estimates of 8.442, 5.530, 4.579, 3.181 and 2.858, correspondingly (Table 2.8).

Table 2.7 Genotypic (G), phenotypic (P) and environmental (e) correlation coefficients of various metric traits of sunflower genotypes across three environments (Awaad et al. 2016)

Characters		Leaf chlorophyll content (%)	Transpiration rate	Plant height (cm)	Head diameter	100-achene weight (g)	Achene oil content (%)	Achene yield/plant
Leaf water content (%)	rg	0.452**	0.236*	0.055	−0.070	0.615**	−0.405**	0.381**
	rp	0.166	0.171	0.064	−0.100	0.219*	−0.259*	0.179
	re	−0.063	0.082	0.112	−0.168	−0.139	0.102	−0.002
Leaf chlorophyll content	rg	1	−0.255*	−0.589**	−0.256*	−0.142	−0.144	−0.019
	rp	1	−0.181	−0.414**	−0.190	−0.085	−0.085	0.041
	re	1	−0.090	−0.191	−0.117	−0.035	0.071	0.092
Transpiration rate	rg		1	0.668**	0.601**	0.576**	−0.617**	0.674**
	rp		1	0.584**	0.506**	0.366**	−0.569**	0.476**
	re		1	0.061	0.092	−0.029	−0.097	0.148
Plant height	rg			1	0.679**	0.340**	−0.307**	0.653**
	rp			1	0.566**	0.239*	−0.279**	0.414**
	re			1	0.055	0.065	0.068	−0.044
Head diameter	rg				1	0.097	−0.291**	0.577**
	rp				1	0.101	−0.249*	0.378**
	re				1	0.121	0.047	0.067

(continued)

Table 2.7 (continued)

Characters			Leaf chlorophyll content (%)	Transpiration rate	Plant height (cm)	Head diameter	100-achene weight (g)	Achene oil content (%)	Achene yield/plant
100-achene weight	rg						1	−0.117	0.311**
	rp						1	−0.071	0.298**
	re						1	0.085	0.285**
Achene oil content (%)	rg							1	−0.563**
	rp							1	−0.390**
	re							1	−0.007

*, **Significant at P = 0.05 and P = 0.01, respectively

Table 2.8 Direct and indirect effect of various metric traits of sunflower genotypes on achene yield/plant across three environments (Awaad et al. 2016)

S.O.V	CD	RI%	Total contribution on achene yield/plant
Leaf water content % (X_1)	0.07779	7.779	12.796
Leaf chlorophyll content (X_2)	0.01917	1.917	4.997
Transpiration rate (X_3)	0.12941	12.941	22.778
Plant height (X_4)	0.12219	12.219	20.413
Head diameter (X_5)	0.00081	0.081	0.871
100-achene weight (X_6)	0.07128	7.128	13.939
Achene oil content % (X_7)	0.00438	0.438	2.571
X_1 x X_2	0.01746	1.746	
X_1 x X_3	0.02367	2.367	
X_1 x X_4	0.00537	0.537	
X_1 x X_5	0.00056	0.056	
X_1 x X_6	0.04579	4.579	
X_1 x X_7	0.00748	0.748	
X_2 x X_3	0.01270	1.270	
X_2 x X_4	0.02858	2.858	
X_2 x X_5	0.00101	0.101	
X_2 x X_6	0.00525	0.525	
X_2 x X_7	0.00132	0.132	
X_3 x X_4	0.08422	8.422	
X_3 x X_5	0.00615	0.615	
X_3 x X_6	0.05530	5.530	
X_3 x X_7	0.01470	1.470	
X_4 x X_5	0.00678	0.678	
X_4 x X_6	0.03181	3.181	
X_4 x X_7	0.00713	0.713	
X_5 x X_6	0.00074	0.074	
X_5 x X_7	0.00055	0.055	
X_6 x X_7	0.00207	0.207	
R^2	0.78365	78.365	78.365
Residual	0.21635	21.635	21.635
Total	1.00000	100	100.000

Based on the total contribution of the deliberated characters, might be arranged as follows, transpiration rate (22.778%), plant height (20.413%), 100-achene weight (13.939%), leaf water content (12.796%), leaf chlorophyll content (4.997%), achene oil content (2.571%) and head diameter (0.871%). Commonly, the studied characters accounted for 78.365% of the achene yield/plant variation, whereas, the residual effect was 21.635%. Therefore, the authors stressed the importance of transpiration rate, plant height, 100-achene weight and leaf water content as selection criteria for improving sunflower yield under various environments.

2.4.6   Cotton

Cotton originates from hot climates, and a negative relationship has been recorded between yield and high temperature during early boll development (Oosterhuis 2011). Water stress caused a reduction of 13% in days to first square formation, 14% in days to first flower formation, 19% in plant height, 27% in number of bolls/plant, 14% in boll weight, 4% in ginning out turn and 37% in seed cotton yield (Bakhsh et al. 2019). Seed cotton yield has close relationship with formation and retention of bolls. Significant decreases were detected in plant height, bolls/plant, sympodial branches and seed cotton yield when adequate amount of water was not applied at most sensitive growth stages such as bud formation, flowering and boll formation (Jayalalitha et al. 2015; Zonta et al. 2015; Zhang et al. 2016).

Fiber quality traits comprising fiber length, fiber strength and fiber fineness were likewise reduced under water stress situations (Sahito et al. 2015; Amin et al. 2016). Whereas, Dewdar (2019) showed that most of fiber properties did not affect by water stress situations. Whereas, water stress treatments caused significant reduction in yield and yield components. The interaction between genotypes and stress treatments was significant on most studied traits, with substantial differences between cotton varieties.

Sustainable cotton production in the future will depend on the development of cotton varieties with higher yield potential and quality of seed cotton as well as better tolerance to biotic and abiotic stresses. Abd El-Mohsen and Amein (2016) showed that seed cotton yield/plant was significantly and positively correlated with number of bolls/plant (r = 0.85**), boll weight (r = 0.68**), seed index (r = 0.91**) and lint percentage (r = 0.70**). Regression analysis using step-wise technique revealed that 96.51% of total variation in seed cotton yield was accounted for by number of bolls/plant, boll weight and lint %. Path analysis indicated high positive direct effect of number of bolls/plant (0.57), boll weight (0.39), while lint % had moderate positive direct effect (0.24) on seed cotton yield/plant. Factor analysis indicated that three factors could explain approximately 73.96% of the total variation. The first factor accounted for about 53.21% of the variation was strongly associated with number of bolls/plant, boll weight, seed index and lint percentage, whereas the second factor was strongly related and positive effects on earliness index only, which accounts for about 20.75% of the variation. Hereby high seed cotton yield of Egyptian cotton could be achieved by selecting genotypes with high number of bolls/plant, boll weight and lint percentage.

2.5 Conclusions

Strategies for improving economic crops to tolerate environmental stress are important goal in light of the problems facing agricultural production, represented by climate change, which negatively affect crop productivity. Therefore, identifying the most sensitive periods of plant life to extreme environment conditions is of great importance. It is also considered important to analyze the crop productivity under different environmental conditions. In this chapter, several reference studies showed variation in the amount of reduction in the crop production and quality under different stress environments, which reflects the variance of associations between plant traits.

2.6 Recommendations

Climate change represents the most important challenges facing the world today that have effect on agriculture and human health. The negative effects of climate change on agricultural crops can be mitigated by implementing integrated adaptation strategies and understanding the nature of the relationships between different crop traits, and cultivating more adapted genotypes to environmental changes. Therefore, the breeding program to release varieties that more tolerant to such environmental stresses, in addition to high productivity and high quality, should be essential goals in the national improvement programs in Egypt.

References

Abd El-Mohsen AA, Amein MM (2016) Study the relationships between seed cotton yield and yield component traits by different statistical techniques. IJAAR 8(5):88–104

Abdelmula1 AA, Abuanja IK (2007). Genotypic responses, yield stability, and association between traits among some of sudanese faba bean (*Vicia faba* L.) genotypes under Heat stress: tropentag 2007 university of Kassel-Witzenhausen and university of Göttingen, 9–11 Oct 2007. Conference on international agricultural research for development, pp 1–7

Abid KH, Aouida M, Aroua I, Baudoin J-P, Muhovski Y, Mergeai G, Sassi K, Machraoui M, Souissi F, Jebara M (2017) Agro-physiological and biochemical responses of faba bean (*Vicia faba* L. var. 'minor') genotypes to water deficit stress. Ghassen Biotechnol Agron Soc Environ 21(2):146–159

Akter N, Islam RM (2017) Heat stress effects and management in wheat. Review Agron Sustain Dev 37:1–17

Amin A, Nasim W, Mubeen M, Sarwar S, Urich P, Ahmad A, Wajid A, Khaliq T, Rasul F, Hammad HM, Rehmani MIA, Mubarak H, Mirza N, Wahid A, Ahamd S, Fahad S, Ullah A, Khan MN, Ameen A, Amanullah B, Shahzad S, Saud H, Alharby ST, Ata-Ul-Karim M, Adnan F, Islam QSA (2016) Regional climate assessment of precipitation and temperature in Southern Punjab (Pakistan). Theoret Appl Climatol 131(1–2):121–131

Anjum AH, Xie XY, Wang LC, Saleem MF, Man C, Le W (2011) Morphological, physiological and biochemical responses of plants to drought stress. Afr J Agric Res 6:2026–2032

Ashraf M, Harris PJC (2013) Photosynthesis under stressful environments: an overview. Photosynthetica 51(2):163–190

Awaad HA, A. H. Salem; M. M. A. Ali, K.Y. Kamal (2016) expression of heterosis, gene action and relationship among morpho-physiological and yield characters in sunflower under different levels of water supply. J Plant Prod, Mansoura Univ 7(12):1523–1534

Awaad HA, Abu-hashim M, A Negm (2021) Handbook of mitigating environmental stresses for agricultural sustainability in Egypt. Springer Nature Switzerland AG

Awaad HA, Aly AA (2002) Genotype x environment interaction and interrelationship among some stability statistics in sesame (Sesamum indicum L.). Zagazig J Agric Res 29(2):385–403

Awaad HA, Basha HA (2000) Yield potentiality and yield analysis of some sesame genotypes grown under two plant population densities in newly reclaimed sandy soil. Zagazig J Agric Res 27(2):239–253

Ayeneh A, Van Ginkel M, Reynolds MP, Ammar K (2002) Comparison of leaf, spike, peduncle and canopy temperature depression in wheat under heat stress. Field Crops Res 79:173–184

Ayoubizadeh N, Laei G, Dehaghi MA, Sinaki JM, Rezvan S (2019) Seed yield and fatty acids composition of sesame genotypes as affecet by foliar application of iron Nano0chelate and fulvic acid under drought stress. Appl Ecol Environ Res 16(6):7585–7604

Bakhsh A, Rehman M, Salman S, Ullah R (2019) Evaluation of cotton genotypes for seed cotton yield and fiber quality traits under water stress and non-stress conditions. Sarhad J Agric 35(1):161–170

Blanco A, Mangini G, Giancaspro A, Giove S, Colasuonno P, Simeone R, Signorile A, De Vita P, Mastrangelo AM, Cattivelli L et al (2012) Relationships between grain protein content and grain yield components through quantitative trait locus analyses in a recombinant inbred line population derived from two elite durum wheat cultivars. Mol Breed 30:79–92

Blum A (1985) Breeding crop varieties for stress environment. CRE Rev Plant Sci 2:199–238

Boureima S, Diouf M, Amoukou AI, Van Damme P (2016) Screening for sources of tolerance to drought in sesame induced mutants: assessment of indirect selection criteria for seed yield. Int J Pure App Biosci 4(1):45–60

Cabelguenne M, Debaeke P (1998) Experimental determination and modelling of the soil water extraction capacities of maize, sunflower, soya bean, sorghum and wheat. Plant Soil 202:175–192

Castro M, Peterson CJ, Dalla Rizza M, Díaz Dellavalle P, Vázquez D, Ibañez V, Ross A (2007) Influence of heat stress on wheat grain characteristics and protein molecular weight distribution. In: Wheat production in stressed environments. Springer, Dordrecht, The Netherlands 365–371, ISBN 978-1-4020-5496-9

Chen WY, Suzuki T, Lackner M (2017) Handbook of climate change mitigation and adaptation. Springer International Publishing Switzerland. https://doi.org/10.1007/978-3-319-14409-2

Darvishzadeh R, Hatami Maleki H, Sarrafi A (2011) Path analysis of the relationships between yield and some related traits in diallel population of sunflower (Helianthus annuus L.) under well-watered and water-stressed conditions. Australian J Crop Sci 5:674–680

Debaeke P, Casadebaig1 P, Flenet F, Langlade N (2017) Sunflower crop and climate change: vulnerability, adaptation, and mitigation potential from case-studies in Europe. EDP Sci 24(1):1–15

Dewdar MDH (2019) Productivity; cotton; stress susceptibility index; relative productivity. Int J Plant Soil Sci 27(5):1–7

Dossa K, Diouf D, Wang L, Wei X, Yu J, Niang M, Fonceka D, Yu J, Mmadi MA, Yehouessi LW et al (2017) The emerging oilseed crop Sesamum indicum enters the "Omics" era. Front Plant Sci 8:1154

Draz IS, Abou-Elseoud MS, Kamara AEM, Alaa-Eldein OAE, El-Bebany AF (2015) Screening of wheat genotypes for leaf rust resistance along with grain yield. Ann Agric Sci 60:29–39

El-Hendawy AAM (2020) The effect of spraying with drought tolerance inducers on growth and productivity of cotton plants growing under water stress conditions in Delta Egypt. In: 16th international conference for crop science, Agronomy Deppartment, Faculty of Agriculture, Al-Azhar University, Egypt, Oct 2020 Abstract p 61

El Afandi G (2017) Impact of climate change on crop production. In: Chen WY et al (eds) Handbook of climate change mitigation and adaptation. Springer, Science + Business Media New York, pp 723–748. https://doi.org/10.1007/978-3-319-14409-2_64

El Massah S, Omran G (2015) Would climate change affect the imports of cereals? the case of Egypt. In: Leal Filho W (ed) Handbook of climate change adaptation. Springer, Heidelberg. https://doi.org/10.1007/978-3-642-38670-1_61

ElBasyoni IS (2018) Performance and stability of commercial wheat cultivars under terminal heat stress. Agronomy 8(4):5–29

Eldardiry IE, Abd El-Hady M, Ageeb GW (2017) Maximize faba bean production under water salinity and water deficit conditions. Middle East J Appl Sci 7(4):819–826

El-Degwy IS, Kmara MM (2015) Yield potential, genetic diversity, correlation and path coefficient analysis in rice under variable environments. J Plant Prod 6(5):695–714

El-Khoby WMH, Abd El-lattef ASM, Mikhael BB (2014) inheritance of some rice root characters and productivity under water stress conditions. Egypt J Agric Res 92(2):529–549

El-Marsafawy S (2007) Climate change and its impact on the agriculture sector in Egypt. http://www.radcon.sci.eg/environment2/ArticlsIdeasDetails.aspx?ArticlId=35

El-Rawy MA, Hassan MI (2014) Effectiveness of drought tolerance indices to identify tolerant genotypes in bread wheat (*Triticum aestivum* L.). J Crop Sci Biotechnol 17(4):255–266

Eskandari H, Zehtab-Salmasi S, Ghassemi-Golezani K, Gharineh MH (2009) Effects of water limitation on grain and oil yields of sesame cultivars. J Food Agric Environ 7:339–342

Farooq M, Kobayashi N, Wahid A, Ito O, Basra SMA (2009) Strategies for producing more rice with less water Adv Agron 101:351–388

Fatma FM (2019) Breeding parameters for grain yield and some morpho-pysiological characters related to water stress tolerance in bread wheat. M.Sc. thesis, Agron Department, Faculty of Agriculture Zagazog University, Egypt

García-Vila M, Fereres E, Prieto MH, Ruz C, Soriano MA (2012) Sunflower. In: Crop yield response to water. FAO Irrigation and Drainage, Paper 66

Golestani M, Pakniyat H (2015) Evaluation of traits related to drought stress in sesame (*Sesamum indicum* L.) genotypes. J Asian Sci Res 5(9):465–472

González-Barrios P, Castro M, Pérez O, Vilaró D, Gutiérrez L (2017) Genotype by environment interaction in sunflower (*Helianthus annus* L.) to optimize trial network efficiency. Span J Agric Res 15(4):1–13

Halford NG (2011) The role of plant breeding and biotechnology in meeting the challenge of global warming, planet earth 2011—global warming challenges and opportunities for policy and practice. In: Prof Carayannis E (ed) eBook (PDF) ISBN: 978-953-51-6063-2 Copyright year: 2011

Hamza M, Safina SA (2015) Performance of sunflower cultivated in sandy soils at a wide range of planting dates in Egypt. J Plant Prod, Mansoura Univ 6(6):821–835

Hassan HM, El-Khoby WM, El-Hissewy AA (2013) Performance of some rice genotypes under both salinity and water stress conditions in Egypt. J Plant Prod, Mansoura Univ 4(8):1235–1257

Hegab ASA, Fayed Maha MTB, Hamada MA, Abdrabbo MAA (2014) Productivity and irrigation requirements of faba-bean in North Delta of Egypt in relation to planting dates. Ann Agric Sci 59(2):185–193

Hussein Y, Amin G, Azab A, Gahin H (2015) Induction of drought stress resistance in sesame (*Sesamum indicum* L.) plant by salicylic acid and kinetin. J Plant Sci 10(4):128–141

IPCC (2018) Climate change report is a "wake-up" call on 1.5°C global warming. https://pub lic.wmo.int/en/media/press-release/climate-change-report-%E2%80%9Cwake-%E2%80%9D-call-15%C2%B0c-global-warming

Jayalalitha K, Rani YA, Kumari SR, Rani PP (2015) Effect of water stress on morphological, physiological parameters and seed cotton yield of Bt cotton (*Gossypium hirsutum* L.) hybrids. Int J Food Agric Vet Sci 5(3): 99–112

Karwa S, Bahuguna RN, Chaturvedi AK, Maurya S, Arya SS, Chinnusamy V, Pal M (2020) Pheno-typing and characterization of heat stress tolerance at reproductive stage in rice (*Oryza sativa* L.). Acta Physiol Plant 42(1):1–16

Khammari M, Ghanbari A, Rostami H (2013) Evaluation indicator of drought stress in different cultivars of sesame. Int J Manag Sci Bus Res 2(9):28–39

Khan N, Naqvi FN (2012) Correlation and path coefficient analysis in wheat genotypes under irrigated and non-irrigated conditions. Asian J Agric Sci 4(5):346–351

Kumar A, Basu S, Ramegowda V, Pereira A (2016) Mechanisms of drought tolerance in rice. In: Sasaki T (ed) Achieving sustainable cultivation of rice, 11, Burleigh Dodds Science Publishing

Kuol BG (2004) Breeding for drought tolerance in sesame (*Sesamum indicum* L.) in Sudan. https:// cuvillier.de/uploads/preview/public_file/6165/3865370659

Loose LH, Heldwein AB, da Silva JR, Bortoluzzi MP (2019) Yield and quality of sunflower oil in Ultisol and Oxisol under water regimes. Rev Bras Eng Agríc Ambient 23(7):532–537

Mahmoud MA (2019) Impact of climate change on the agricultural sector in Egypt. Springer International Publishing, Publisher

Manikandan A, Ashu DK, Blaise D, Shukla PK (2019) Cotton response to differential salt stress. Int J Agric Sci 11(6):8059–8065

Melandri G, AbdElgawad H, Riewe D, Hageman JA, Asard H, Beemster GTS, Kadam N, Jagadish K, Altmann T, Ruyter-Spira C, Bouwmeester H (2020) Biomarkers for grain yield stability in rice under drought stress. J Exp Bot 71(2):669–683

Migdadi HM, El-Harty EH, Salamh A, Khan MA (2016) Yield and proline content in faba bean genotypes under water stress treatments. J Anim Plant Sci, 26(6):1772–1779

Mishra SS, Panda D (2017) Leaf traits and antioxidant defense for drought tolerance during early growth stage in some popular traditional rice landraces from Koraput. India Rice Sci 24(4):207–217

Mishra SS, Behera PK, Panda D (2019) Genotypic variability for drought tolerance-related morpho-physiological traits among indigenous rice landraces of Jeypore tract of Odisha, India. J Crop Imp 33:254–278

Mishra SS, Behera PK, Kumar V, Lenka SK, Panda D (2018) Physiological characterization and allelic diversity of selected drought tolerant traditional rice (Oryza sativa L.) landraces of Koraput, India. Phys Mol Biol Plants 24(6):1035–1046

Moetamadipoor, S.A., M.Mohammadi, G. R. B. Khaniki, R.A. Karimizadeh (2015). Relationships between traits of wheat using multivariate analysis. Biological Forum – An International Journal, 7(1): 994–997.

Mohamed NE, Bedawy IMA (2019) Comparative and multivariate anaslysis of genetic diversity of three sesame populations based on phenotypic traits. Egypt J Plant Breed 23(2):369–386

Mokhtar TS (2003) Effect of some agronomic practices on faba bean under newly cultivated lands. M.Sc thesis, Agronomy Department, Faculty of Agriculture Zagazog University, Egypt

Motavaseel H (2014) Study on yield and some agronomic traits of promising genotypes and lines of bread wheat through factor analysis. Indian J Fundam Appl Life Sci 4(3):231–233

Moustafa EA (2006) Gene action and stability of Egyptian cotton yield (*Gossypium barbadense* L.) under Wady Seder conditions. Ph.D. thesis, Department of Agronomy Faculty of Agriculture Zagazig University, Egypt

Nuruzzaman M, Hassan L, Begum SN, Huda MM (2017) Correlation and path coefficient analysis of yield components in NERICA mutant rice lines under rainfed conditions. Nuruzzaman et al. JEAI 16(1):1–8

Omambia AN, Shemsanga C, Hernandez IAS (2017) Climate change impacts, vulnerability, and adaptation in East Africa (EA) and South America (SA). In: Chen WY et al (eds) Handbook of climate change mitigation and adaptation. Springer International Publishing Switzerland, pp 749–799

Oosterhuis, DM (2011) Stress physiology in cotton: the cotton foundation Cordova, Tennessee, USA 2011

Osman KT (2013) Climate change and soil. In: Osman KT (ed) Soils: principles, properties and management. Springer Science + Business Media Dordrecht, pp 253–261. https://doi.org/10. 1007/978-94-007-5663-2_15

Porter JR, Xie L, Challinor V et al (2014) Food security and food production systems. Climate change 2014: impacts, adaptation, and vulnerability. Cambridge University Press, Cambridge, UK/New York, USA, pp 485–533

Qabil N, Helal AA, Rasha YS, El-Khalek A (2018) Evaluation of some new and old faba bean cultivars (Vicia faba L.) for earliness, yield, yield attributes and quality characters. Zagazig J Agric Res 45(3):821–833

Qaseem MF, Qureshi R, Shaheen H (2019) Effects of pre-anthesis drought, heat and their combination on the growth, yield and physiology of diverse wheat (Triticum aestivum L.) genotypes varying in sensitivity to heat and drought stress. Sci Rep 9:6955:1–12

Rafiei F, Darbaghshahi MRN, Rezai A, Nasiri BM (2013) Survey of yield and yield components of sunflower cultivars under drought stress. Int J Adv Biol Biomed Res 1(12):1628–1638

Sahito A, Baloch ZA, Mahar A, Otho SA, Kalhoro SA, Ali A, Kalhoro FA, Soomro RN, Ali F (2015) Effect of water stress on the growth and yield of cotton crop (Gossypium hirsutum L.). Am J Plant Sci 6:1027–1039

Santhosh B (2014) Studies on drought tolerance in sunflower genotypes. M.Sc. In: Agriculture Department of Crop Physiology College of Agriculture Acharya N. G. Ranga Agricultural University Rajendranagar, Hyderabad -500 030 Crop Physiology

Santhosh B, Reddy SN, Prayaga L (2017) Physiological attributes of sunflower (L.) Helianthus annuus as influenced by moisture regimes. Green Farming 3:680–683

Sekmen AH, Ozgur R, Uzilday B, Turkan I (2014) Reactive oxygen species scavenging capacities of cotton (Gossypium hirsutum) cultivars under combined drought and heat induced oxidative stress. Environ Exp Bot 99:141–149

Serraj R, C, T Hash, SMH Rivzi, (2005) Recent advances in marker assisted selection for drought tolerance in pearl millet. Plant Prod Sci 8:334–337

Seyoum EG (2018) Genetic variability and gain in grain yield and associated traits of Ethiopian bread wheat (Triticum aestivium l.) varieties. M.Sc. thesis, the School of Graduate studies, Jimma University College of Agriculture and Veterinary Medicine, Ethiopia

Siddiqui MH, Al-Khaishany MY, Al-Qutami MA, Al-Whaibi MH, Grover A, Ali HM, Al-Wahibi MS, Bukhari NA (2015) Response of different genotypes of faba bean plant to drought stress. Int J Mol Sci 16(5):10214–10227

Sun J, Rao Y, Le M, Yan T, Yan X, Zhou H (2010) Effects of drought stress on sesame growth and yield characteristics and comprehensive evaluation of drought tolerance. Chin J Oil Crop Sci 32:525–533

Tantawy MM, Oudu SA, Khalil FA (2007) Irrigation optimization for different Sesame varieties grown under water stress condition. J Appl Sci Res 3(1):7–12

Tayel MY, Sabreen KhP (2011) Effect of irrigation regimes and phosphorus level on two Vica faba varieties: 1-growth traits. J Appl Sci Res 7(6):1007–1015

Upadhyaya H, Panda SK (2019) Drought stress responses and its management in rice. In: Hasanuzzaman M, Fujita M, Nahar K, Biswas JK (eds) Advances in rice research for abiotic stress Tolerance, Elsevier, UK, pp 177–200

Wu S, Hu Z, Hu T, Chen J, Yu K, Zou J, Liu S (2018) Annual methane and nitrous oxide emissions from rice paddies and inland fish aquaculture wetlands in southeast China. Environ. https://doi.org/10.1016/j.atmosenv.2017.12.008

Zhang D, Luo Z, Liu S, Li W, Dong H (2016) Effects of deficit irrigation and plant density on the growth, yield and fiber quality of irrigated cotton. Field Crops Res 197:1–9

Zhu R, Wu FY, Zhou S, Hu T, Huang J, Gao Y (2020) Cumulative effects of drought-flood abrupt alternation on the photosynthetic characteristics of rice. Environ Exp Bot 169, 103901

Zonta JH, Bezerra JRC, Sofiatti V, Brandao ZN (2015) Yield of cotton cultivars under different irrigation depths in the Brazilian semi-arid region. Revista Brasileira De Engenharia Agrícola e Ambiental 19(8):748–754

Chapter 3
Foundations of Crop Tolerance to Climate Change: Plant Traits Relevant to Stress Tolerance

Abstract Climate change is described via higher temperatures, elevated atmospheric CO_2 concentrations, great climatic threats, and less water available for agriculture. Crop plants show a wide range of responses to tolerance to these stresses, such as morphological, physiological and biochemical responses. Crop physiology and economic yield are adversely affected by climate change due to physical damage, physiological disturbances and biochemical changes. Adaptations through breeding for stress tolerance, crop management and biotechnology and combined to partially cope with the negative impacts of climate change and achieve sustainability. This chapter focused mechanisms of adaptation to environmental stresses via various plant characters, represented in phenological, morphological, physiological and biochemical traits.

Keywords Climate conditions · Morpho-physiolgical traits · Biochemical and molecular base · Agricultural crops · Adapted genotypes

3.1 Introduction

Climate change represents a major warning to agricultural crops in arid and semi-arid regions worldwide. Global climate change and related adverse abiotic stress conditions, for instance drought, salinity, heavy metals, heat stress, etc., significantly affect plant growth and development, ultimately affecting crop yield and quality, as well as agricultural sustainability (Hasanuzzaman et al. 2020). For this purpose, it is necessary to reconsideration the various plant traits associated with the tolerance of crop varieties to stress environmental conditions such as heat, drought, salinity and others in order to improve crop adaptation. The progression achieved in understanding environmental stress tolerance is attributed to evolutions in plant physiology, crop breeding, and gene manipulation. Crop improvement strategies are an important goal to address the challenges of stress that reduce crop productivity and quality. Improving stress tolerance is a very complex mission for field crop improvement workers, and further study is required to understand the stress mechanism more deeply. Also, identifying genes that control the traits responsible for tolerating environmental stress conditions is the key to genetic improvement (Seyoum 2018; Shams 2018; Singh et al. 2018).

Environmental studies point to the vital role of breeding and molecular biology in supporting plant scientists and breeders to release new cultivars that will address the different environmental stresses under the current climate change. It is therefore important to know and understand the various physiological and chemical characteristics associated with these pressures to improve crop tolerance for stress conditions (Moucheshi et al. 2019; Lehari et al. 2019).

Crop plants show a wide range of responses to tolerance to these stresses, such as morphological, physiological and biochemical responses. Crop production is adversely affected by environmental stresses due to physical damage, physiological disturbances and biochemical changes. Both of these pressures have multilateral effects and thus are complex in mechanical work.

Because of the rapid changes in the global climate, abiotic pressures are one of the major restrictions to crop production and food security around the world. In this regard, heat and drought are the two most important pressures effecting negatively on the growth, productivity and quality of agricultural crops. In this concern, Fadia Chairi et al. (2020) practiced two different patterns of selection due the G × E interaction and changes in the ranking of genotypes: in the high yielding environments (grain yield > 5 Mg ha-1), genotypes favor increased water uptake, with higher levels of transpiration and more open stomata (more negative values of carbon isotope composition, δ13C, and higher canopy temperature depression), however, in low yielding environments (grain yield < 5 Mg ha − 1) genotypes close their stomata and favor greater water use efficiency (less negative δ13C values and lower canopy temperature depression values).

Several investigators described the plant response to stress through escape, avoidance, tolerance and resistance mechanisms as illustrated in Fig. 3.1 (Foulkes et al. 2007 and Awaad et al. 2021).

Escape is defined as the ability of a crop cultivar to germinate and grow only during the favorable conditions and complete its life cycle. While, **avoidance** is the ability of a cultivar to maintain (relatively) high tissue water content despite lower water content in the soil. Whereas, **tolerance** is the ability of a crop cultivar to preserve its biomass production during stress conditions**.** However, **resistance** requires the maintenance of economically viable plant production under environmental stress condition.

Plants adopt on several strategies to avoid the effects of heat or drought. For example, if stress is most likely to occur in late summer they may avoid it by suitable sowing date and tolerant genotype. Other 'avoidance' strategies include the expansion of deeper and more widespread root systems, permitting the plant to get more water from the soil to maintain his life during droughts. Crop plants can develop responses that enable them to survive under water pressure. These are called tolerance traits that vary from species to another and between different varieties, developmental stages, organs and tissue types. The plant hormone, abscisic acid, shows a significant role, inducing a network of signaling pathways relating multiple protein kinases

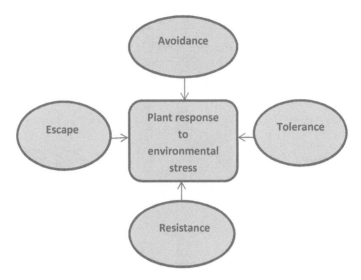

Fig. 3.1 Plant response to environmental stress (Foulkes et al. 2007 and Awaad et al. 2021) (drawn by the author H. A. Awaad)

and transcription dynamics. Transcription factors include responsible responses to drought stress like ABA response element binding proteins, members of the zinc finger homeodomain, myeloblastosis and myelocytomatosis families, dehydration-responsive element binding protein, and the NAC family (Semenov and Halford 2009). Over-expression of another transcription factor, plant nuclear factor has been shown to improve drought tolerance. This chapter focused on plant mechanisms of adaptation to environmental stress conditions, plant characters related to environmental stress tolerance, represented in phenological, morphological, physiological and biochemical traits.

3.2 Plant Traits Relevant to Environmental Stress Tolerance

Crop plants have characteristics that help them to adapt with environmental stress factors like developmental, morpho-physiological and biochemical traits (Fig. 3.2). Therefore, these traits could be used as selection criteria for high yield under stress, and could be discussed as follows:

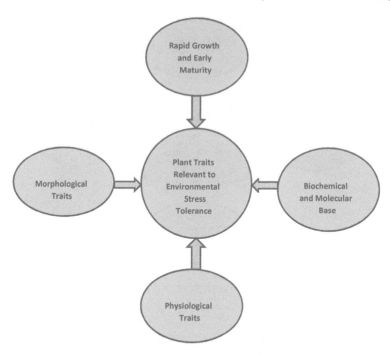

Fig. 3.2 Plant traits relevant to environmental stress tolerance (drawn by the author H. A. Awaad)

3.2.1   Rapid Growth and Early Maturity

Early growth genotypes offer a mechanism for escaping unfavorable environmental conditions (Awaad 2009). The early genotypes in semi-arid condition are capable of escaping the terminal heat stress, which is the major yield reducing factor. Crop varieties vary in the rapidity and germination rate under stress conditions. Islam et al. (1999) recorded significant variances between bread wheat genotypes in germination rate under water stress (−8 bar) conditions.

However, Zaid et al. (2016) registered a decrease in the germination rate in wheat genotypes as a result of water stress. Wheat grain yield was genetically and significantly correlated with days to heading, and could be used as direct selection criteria. The indirect effect of days to heading via flag leaf area (6.54%) and days to heading via flag leaf chlorophyll content (6.19%) could be considered useful pairs for improving grain yield through simultaneous selection (Awaad 2001). Early growth is a rapid measure to high temperature tolerance, Singh and Ahmed (2003) identified the importance of seedling vigor and earliness as indicators of tolerance to heat stress in wheat. In Egyptian wheat cultivar Sakha 93, Gemmeiza 11 and Sids 12, Omnya, El- Moselhy (2015) revealed a suitable level of earliness with desirable values of growing degree days. Early maturation displayed a positive and significant relationship with canopy temperature depression under heat stress situation.

To avoid contamination by heavy metals, copper, lead, cadmium and zinc, Alybayeva et al. (2014) showed that wheat yield is connected with the ability to quickly enter to the tillering stage, successfully overwinter, preserve during the summer vegetation like winter wheat variety Mironovskaya-808. For salinity stress tolerance, El-khamissi (2018) evaluated three Egyptian wheat genotypes namely; Landrace 1, landrace 2 and Sakha 93 cultivar using different levels of NaCl (0, 50, 100 and 150 mM) at seedling stage. They found that germination percentage, growth parameters and photosynthetic pigments were reduced by increasing NaCl concentrations in wheat genotypes. However, Sakha 93 cultivar might be more salt tolerance as it achieved the highest value of germination percentage, growth parameters, photosynthetic pigments rather than Landrace 1, while Landrace 2 was the most sensitive to salinity stress.

A significant however negative association for wheat grain yield/plot with days to maturity was detected (Bhanu et al. 2018). Based on Pearson correlation coefficients under late sowing date, grain filling duration was significant and positively correlated with plant height and grain yield, but it was significant and negatively correlated with leaf and stem rust scores. Furthermore, leaf rust was significant and negatively correlated with the number of days to flowering (ElBasyoni 2018). To explain the variation in grain yield and related traits in diverse wheat germplasm, Qaseem et al. (2019) found that days to anthesis and days to maturity were negatively associated with grain yield under stress showing advantage of early maturity during stress. Traits having a major contribution in the first two principal components under different stress treatments may lead to improved varieties with heat and drought stress tolerance.

For rice and with limited irrigation water, earliness and photo-sensitivity represent an important role in adaptation of varieties and avoiding environmental stress conditions. Several genes and/or alleles control photoperiod sensitivity and length of basic vegetative growth in rice (Rana et al. 2019). Salinity stress caused earliness in heading, rice cultivar Sakha 102 was the earliest in total duration, but produced the lowest yield followed by Sakha101 and Giza 182 in the two seasons of study (Hassan et al. 2013). In Egypt, rice breeder succeeded in developing water-saving early maturity varieties i.e. Giza 178, Giza 179, Sakha 107, Sakha super 300, Giza 182 characterized by tolerant to abiotic stress conditions especially salinity, drought and heat combining with high yielding and good milling quality (Anonymous 2021).

Faba bean genotypes differed significantly for days to flowering and maturity under rainfed and irrigated conditions. Maalouf et al. (2015) found that both days to flowering and days to maturity were positively associated with seed yield. Golestani and Pakniyat (2015) showed that drought reduced number of days to maturity. They suggested that high sesame yield under water stress could be achieved with lowest decrease in number of days to maturity. Donatelli et al. (2015) showed that the length of the biological cycle is important factor for adaptations of different sunflower maturity groups. Growing degree-days was manipulated to get a realistic variation of flowering and physiological maturity. Debaeke et al. (2017) showed that adaptations through breeding for earliness and stress tolerance in sunflower could be developed, to partly cope with these negative environmental impacts. Finally, Arafa et al. (2008)

observed that, the higher estimates of earliness for Giza 45, Giza 87, Giza 70, Giza 85, Giza 86 and Giza 89 were recorded at the maximum heat units DD15's. However, Giza 88 need heat units less than the other cultivars. Devi and Reddy (2018) showed that the conserved water would be useful to cotton genotypes at maturity stage and hence for the yield improvement when late season drought develops.

3.2.2 Morphological Traits

Assessments of genotypic, phenotypic and environmental associations among the characters are useful in planning the selection strategies. Since, the relations between morphological, physiological traits i.e. leaf chlorophyll content, relative water content, transpiration rate, stomatal conductance osmotic adjustment in particular during heat or water stress, are well described in crop plants (Hirasawa et al. 1995; Guidi and Soldatini 1997; Pankovic et al. 1999; Bhanu et al. 2018 and Fatma, Farag 2019).

3.2.2.1 Root System

Root system plays a vital role in meeting the water and nutrient requirements of the plant and in avoiding water deficiency conditions. Deep rooted varieties have the advantage of avoiding drought compared to plants with shallow root. Streda et al. (2011) found significant relationship between root system size and wheat yield under extremely dry year with values of $r^2 = 0.285*$ and $0.284*$, in both heat and drought stress, respectively. Wheat varieties with a 21% superior in root system size had a 420 kg/ha increase in grain yield. Moreover, wheat genotypes TAM 111 and TAM 112 produce more tillers and bigger root mass and shoot mass compared to TAM 304. Under dry management, TAM 112 had 67 and 81% more grain yield rather than TAM 111 and TAM 304, respectively. Water use efficiency for grain and water use efficiency for biomass were also more in TAM 112 compared to the rest cultivars under water stress (Thapa et al. 2018).

Root characteristics are the vital attributes for enhancing rice production under drought stress. Crop function under water stress is determined by the constitution and formation of rice root system. Rice production under water stress can be forecasted by taking root mass and length into account (Comas et al. 2013). Diversified and varied responses are observed on root growth characteristics under water stress. Generally, rice varieties with profound and prolific root system show better adaptability under drought (Mishra et al. 2019 and Kim et al. 2020). Genotypes having profound root system, coarse roots, capacity of producing many branches and high root and shoot ratio are important for drought tolerance. The morpho-physiological characteristics of rice roots play a major role in determining shoot growth and overall grain yield under water stress (Kim et al. 2020). Under the experimental farm of Rice Research and Training Center, Sakha, Kafr El-Sheikh, Egypt, El-khoby et al. (2014) crossed six rice materials with diverse water stress tolerance to produce three crosses. They indicated that grain yield/plant was highly significant and positively correlated with root length, root number/plant and root volume in the three crosses. Also, under water deficit conditions, Gaballah (2018) recorded positive and highly significant correlation among rice grain yield/plant, root length, root volume, number of roots/plant, root thickness and root: shoot ratio.

While, in faba bean, genetic differences in root length and total root dry matter were recorded. Roots of line ILB938/2 were 20% longer than those of most other lines, and those of line L-7 and Apollo/1 were 15% longer. Cultivar Enantia had almost twice the root dry matter of most other lines. Enantia and L-7 showed reasonably good stomatal traits. Lines L-7 and Apollo/1 along with cultivar Enantia are sources of good rooting traits to give drought avoidance. Also, ILB938/2 is a valuable resource of dehydration avoidance through stomatal traits, drought response by osmotic adjustment and drought avoidance by deep rooting (Khan et al. 2007 and Khazaei et al. 2013 and Anonymous 2017). From another point of view, after exposure sesame to drought stress, the contents of soluble sugar, soluble protein, reduced glutathione, reduced ascobate, catalase, phenylalanine ammonialyase activity in sesame roots were increased with water stress level. Also, contents of proline and superoxide dismutase, peroxidase activity in roots revealed a tendency of increase and then decreased with stress degree (Jian et al. 2019).

Whereas, sunflower has a deep and extensive root system that can extract water up to 270 cm (Rachidi et al. 1993). Reports have characterized heritable variation in sunflower for root length and any greater root length associated with longer growth duration (Rauf and Sadqat 2008). With increase in the osmotic stress, the root length was reduced in genotypes. Greater root system of seedling could be beneficial in preserving water availability under inadequate moisture supply (Santhosh 2014).

Cotton crop is characterized by tap roots with ten times the plant length (Larcher 2003). It was observed that the initial or moderate drought was accompanied by increased root length under drought conditions (Luo et al. 2016). Therefore, shoot length, shoot and root fresh and dry weights, and root to shoot ratios are potential traits related to abiotic stress tolerance (Ashraf and Ahmad 2000; Basal et al. 2006; Dewi 2009). Kubure et al. (2016) observed that the local cotton cultivar produced significantly longer tap roots (21.4 cm) rather than in improved genotypes Hachalu (18.9 cm) and Walki (18.3 cm), showing its adaptation to water stress. Abdel-Kader et al. (2015) noticed that cotton genotypes Tamcot C. E. x Deltapine, Giza 90 x (Giza 90 X Australian) and Giza 80 x Deltapine were drought tolerance through maintaining the uppermost values of root length, shoot length, root/shoot ratio. Yield was correlated with each of the previous morphological traits under both normal (100% ETc, 1269 mm/season) and water stress (60% ETc, 761 mm/season) conditions.

3.2.2.2 Leaf Features

Drought-tolerant wheat varieties are characterized by small size and low leaf area, with less transpiration surface and higher stomata resistance, resulting in increased adaptability to water shortage environments, (Golestani-Araghi and Assad 1998). Leaf rolling in cereals, is a common reaction to stress and results in a 50–70% decrease in transpiration (Gusta and Chen 1987). The reduced leaf angle combined with winding the leaf blade under water pressure efficiently reduces radiation load on plant canopy (Boyer 1996). Water stress increased excised leaf water retention which reproduces the water retaining mechanism at stress that may be due to leaf rolling and lessening the exposed leaf surface area, hereafter improve drought tolerance and grain yield (Lonbani and Arzani 2011).

Several leaf traits have been used for the determining of drought tolerant rice genotypes i.e. leaf rolling (Anjum et al. 2017), higher flag leaf area, leaf area index, leaf relative water content (Mishra and Panda 2017).

Leaf waxes characteristic helps the plants to adapt with the conditions of heat and drought by reducing water loss. Significant differences were recorded among wheat genotypes for epicuticular wax content and agronomic traits (Al-Bakry 2007). Epicuticular waxes of epidermal played a main role in controlling water loss and may be an important characteristic in water stress tolerant genotypes, for its role in reducing water loss from leaf surfaces (Ahmed et al. 2012). Leaf cuticular wax was related to water stress tolerance and might be utilized as an operative selection aspect in the improving water stress-tolerant wheat cultivars (Guo et al. 2016). Leaf wax content and composition increased in response to abiotic stress, specifically to water stress in plant species such as sesame (Kim et al. 2017).

In sunflower genotypes, significant differences for morpho-physiological characters among genotypes of at 45, 55, 65 and 75 DAS was recorded by Santhosh (2014). They found significant interaction between moisture stress x genotype on leaf area index at 60, 75 and 90 DAS. Genotype DRSF-113, followed by EC-602063 under control has recorded highest LAI during perevious stages. Lowest LAI has been detected under stress by GMU-337 at 60, 75 and 90 DAS. In cotton, Feng et al. (2016) showed that leaf chlorophyll content and leaf angle were correlated with the photosynthetic potential of various genotypes and with adaptability to light environmental change.

3.2.3 Physiological Traits

3.2.3.1 Leaf Chlorophyll Content

The content of chlorophyll and chlorophyll fluorescence is of paramount importance and is closely related to photosynthesis, as an important physiological criterion in improvement programs to environmental stress tolerance. Leaf chlorophyll content was considered an important determinant of crop yield and unique character as key to tolerate the terminal drought and heat stresses. Awaad (2001) recorded genetically positively and significantly correlated between wheat grain yield and both flag leaf area and flag leaf chlorophyll content. Days to heading, flag leaf area and flag leaf chlorophyll content were the main characters to be used as direct selection criteria. Whereas days to heading via flag leaf area as well as days to heading via flag leaf chlorophyll content might be expressed beneficial pairs for improving wheat grain yield through simultaneous selection. Positive association was noted between chlorophyll fluorescence stability after flowering and grain yield of wheat (Yang et al. 2002), as well as between chlorophyll content and 1000-grain weight across locations (Mondal et al. 2013). Furthermore, a strong correlation was found between chlorophyll content and wheat grain yield under water shortage conditions (Akram et al. 2014).

At thermal pressure, chlorophyll fluorescence has extensively been used in assessing response of genotypes to environmental stress. In this situation, positive and significant relationship has been found between flag leaf area and each of chlorophyll content and 1000-grain weight under different levels of heat stress, (Omnya, El-Moselhy 2015). ElBasyoni (2018) recorded positive and significant correlation between total chlorophyll content with leaf area, grain filling duration, plant height, and grain yield under the late sown condition. Bhanu et al. (2018) showed that chlorophyll content at anthesis and canopy temperature depression CTD at heading and late milking showed significant and positive association with wheat grain yield. However, ElBasyoni (2018) showed that total chlorophyll content was not significantly correlated with grain yield in either the timely-sown or the late-sown plots.

Whereas in rice, Farid et al. (2016) showed that crosses Sakha 102 × Giza 178, Sakha 102 × A22, Giza 178 × WAB 54–125 and Sakha 105 × WAB 54–125 gave the highest mean performances for leaf chlorophyll content, number of panicles/plant and grain yield/plant under drought conditions, revealing the importance of these traits in improving drought tolerance in rice. Furthermore, Gaballah (2018) tried four irrigation managements and found that Sakha 105 gave the best values of chlorophyll content under drought stress compared to the remaining genotypes.

In faba bean, plant leaf area varied significantly between studied cultivars, where under normal situations, 'VIR 490' had the largest leaf area (703.85 ± 19.59 cm^2. plant-1), whereas 'Mahasen' and 'F. 390' exhibited the lowest (201.25 ± 12.86 and 204.55 ± 9.55 cm^2.plant, respectively). 'Giza 3' and 'NEB 482' were capable to maintain leaf surface like to control plants at 50% FC. than the remaining cultivars (Abid et al. 2017). Under moisture stress, sunflower achene yield was found to be more associated with leaf greenness as SPAD chlorophyll meter readings and fluorescence (Fv/ Fm) (Santhosh et al. 2017). Finally, cotton genotypes Tamcot C. E. x Deltapine, Giza 90 x (Giza 90X Australian) and Giza 80 × Deltapine were tolerant to drought by maintaining the uppermost values of Chlorophyll a, b, total chlorophyll, chlorophyll a/b ratio, chlorophyll stability index and lowest values of electrolyte leakage % under drought stress. Yield was correlated with each of the morphological and physiological traits under normal and water stress conditions (Abdel-Kader et al. 2015).

3.2.3.2   Relative Water Content

Relative water content and water retention are important physiological features associated with tolerance of environmental stress such as drought, salinity, nutrient deficiencies beside crop production. A positive significant association was registered between relative water content and wheat grain yield under water stress through reproductive stage in wheat, consequently could be used in improving water stress tolerance in a combination with high yield (Tahara et al. 1990). Saleh (2011) demonstrated the importance of the relative water content and its relation with the grain yield of bread wheat under normal conditions and water stress.

Several leaf traits were utilized for screening cold, high salinity and drought tolerant rice varieties, i.e. water use efficacy, (Li et al. 2011), leaf relative water content and leaf pigment content (Mishra and Panda 2017; Hussain et al. 2018) and water use efficacy (Gaballah et al. 2021) for water and chilling stress tolerance.

Ten Egyptian and various faba bean genotypes were evaluated to water stress tolerance based on physiological characteristics by Siddiqui et al. (2015). Genotype "C5" and "Zafar 1" were found to be relatively tolerant to drought stress owing to its better leaf relative water content and total chlorophyll content. In sesame, relative water content, proline, chlorophyll content and normalized differential vegetation index are powerful tools for studying leaf transpiration and photosynthetic performance. Bayoumi et al. (2017) showed that the highest values of water use efficiency of seed or oil yields (kg/m^3) were gotten by irrigating population 4 under sever water stress. The correlations between normalized differential vegetation index and grain yield differed according to growth stage, moisture availability, and genotypes composition.

The genotypes were accumulated more proline due to stress had the same genotypes which gave the highest relative water content, peroxidase and catalase enzymes. At genetic levels, Awaad et al. (2016) found positive and significant correlations between sunflower achene yield/plant and each of leaf water content and transpiration rate. While, under two irrigation regimes i.e., 100% ETc, 1269 mm/season (normal) and 60% ETc, 761 mm/season (water stress), Abdel-Kader et al. (2015) verified 21 cotton genotypes. They indicated that cotton genotypes Tamcot C. E. x Deltapine, Giza 90 x (Giza 90X Australian) and Giza 80 × Deltapine were drought tolerance by maintaining the uppermost values of relative water content under drought stress. Yield was correlated with each of the morphological and physiological traits under normal and water stress conditions.

3.2.3.3   Stomatal Conductance

Crop plants regulate water loss via controlling stomatal conductivity. In this regard, water supply and temperature affect the opening of the stomatal, and then the transpiration rate. El-Rawy and Hassan (2014) evaluated 50 bread wheat genotypes under three environments; normal (clay fertile soil, E_1), 100% (E_2), and 50% (E_3) field water capacity in sandy calcareous soil. Grain yield/plant was strongly positively correlated with flowering time, stomata length, stress tolerance index, while negatively correlated with stomata frequency and drought sensitivity index in E_2 and E_3, respectively. Thus, highly heritable traits strongly correlated with grain yield under stress conditions especially stomata frequency and length could be used as reliable indices for selecting high-yielding genotypes tolerant to drought stress. At the same time, Akram et al. (2014) and El-Rawy and Hassan (2014) emphasized the strong relationship between stomatal conductivity and wheat grain yield under water stress.

Stomata are closed in environmental conditions of limited water, reducing carbon dioxide influx to rice leaves and driving extra electrons for formation of reactive oxygen species (Farooq et al. 2009a; Mishra et al. 2018). The regulation of stomatal closure and improved osmoregulation in transgenic of rice plants are important for drought tolerance, and the OsCPK9 gene plays a significant role in this respect (Wei et al. 2014).

Under environments of Mediterranean region, from the point of view drought stress as the major abiotic factor limiting faba bean yield, both cultivar Mélodie and inbred line ILB 938/2 appeared to be good drought-tolerance attributes as revealed by stomatal features i.e. stomatal conductance, water use and carbon isotope discrimination along with high leaf temperature and transpiration efficiency. These traits maintain water in plant tissues through reducing water loss under water stress, as a mechanism of dehydration avoidance. ILB938/2 showed evidence of osmotic adjustment in response to moisture stress (Khan et al. 2007; Khazaei et al. 2013, Anonymous 2017 and Fadia Chairi et al. 2020).

To determine the effect of different leaf hydraulic traits on yield traits of sunflower under irrigated and drought regimes. Rauf (2008) found that stomatal conductance showed strong relationship with achene yield under irrigated condition. Whereas, under drought regime all genotypes showed lower stomatal conductance. Ghobadi et al. (2013) revealed that the stomatal conductance was significantly decreased with drought stress intensity. Moderate drought stress decreased stomatal conductance at flowering and mid-grain filling stages as much as 33.5 and 28.2%, respectively. Santhosh (2014) found that the uppermost stomatal conductance was recorded in genotype DRSF-113 (193 mmol [H2O] m^2/s) at 75 DAS. The treatment combination of DRSF-113 and EC-602063 under control has verified highest stomatal conductance. Lowest stomatal conductance was obtained at stress by GMU-337 through different stages 55 and 65 and 75 DAS. In cotton, Karademir et al. (2018) indicated that genotypes with warmer canopy temperature, implying more-closed stomata, gave the greatest yields.

3.2.3.4 Transpiration Rate

Transpiration represents for most water loss by the plant leaves (http://www.knowpia.com/pages/Transpiration). To better understand the basis of water stress tolerance, Thapa et al. (2018) were subjected wheat genotypes TAM 111, TAM 112 and TAM 304 under two water treatments i.e. adequate water and water-limited. They showed that wheat cultivars TAM 111 and TAM 112 used additional water for cumulative evapotranspiration and produce more tillers compared to TAM 304. Wheat genotype TAM 112 exhibited lower stomatal conductance and transpiration rate compared to TAM 111 and TAM 304 under water stress. A negative and highly significant relationship was registered between transpiration rate and both 1000-grain weight and grain yield under sever water stress rather than normal conditions. This means that drought tolerant varieties have less water loss through transpiration than sensitive ones (Fatma, Farag 2019).

Meanwhile, Cabuslay et al. (2002) showed that rice cultivars tolerant to mild water stress had a high relative transpiration under stress compared with that under non-stressed conditions, low initial leaf area, high carbon isotope discrimination in the leaf, and low specific leaf weight. These features assisted tolerant cultivars to maintain high moisture in the leaf and to have high values of leaf area, shoot dry matter, and sugar and starch in tissues in stressed cultivars comparative to the control. Hirayama et al. (2006) reported that rice genotypes with lower leaf temperature during water deficit showed higher rates of transpiration, photosynthesis and grain yield.

Previous studies have shown a lower transpiration rate due to greater cuticular wax deposition, which is repeatedly associated with improved water stress tolerance. In sesame, Bayoumi et al. (2017) showed that proline and normalized differential vegetation index are considered strong indicators for determining leaf transpiration rate. The ability of genes controlling increase transpiration efficiency and photosynthetic pigments plays an important role in improving water use efficiency.

Positive and significant correlations at both phenotypic and genotypic levels between sunflower achene yield/plant and transpiration rate were found by Awaad et al. (2016). The highest indirect effects on achene yield/plant variation were detected for transpiration rate via plant height followed by transpiration rate via 100-achene weight; leaf water content via 100-achene weight with values of 8.442%, 5.530% and 4.579%, respectively across three environments.

In cotton genotypes, transpiration rate under water scarcities is a characteristic of water conservation under high water stress environments. Devi and Reddy (2018) showed that the conserved water would be useful to cotton genotypes at maturity stage and hence for the yield improvement when late season drought develops. Water stress tolerance in genotypes can be improved with limited transpiration rate optimized for use at water stress environments.

3.2.3.5 Canopy Temperature

In crop genotypes, heat stress increased average temperature of the canopy. A low canopy temperature shows the capability of crop variety to tolerate heat stress. Rane (2003) indicated that wheat varieties with low canopy temperature through grain filling period were more tolerant to heat stress. Genotypes with cooler canopies produce higher grain yield under stress environments (Mondal et al. 2013).

Under heat stress conditions, canopy temperature, stomatal conductance, leaf and chlorophyll content were highly associated with yield (Lu et al. 1998 and Roohi et al. 2015). The correlation between canopy temperature and grain yield was negative and significant and approves with previously described results (Asthir 2015). Genotypes with high yield under well water conditions were also characterized by low canopy temperature and high chlorophyll content. The genotypes of wheat with a low canopy temperature can maintain a high transpiration and photosynthetic rate as well as give a high yield under moisture-stressed environments. Significant positive relationship between canopy temperature and leaf chlorophyll content with wheat grain yield was verified by Talebi et al. (2011). Under thermal stress environment, Omnya, El- Moselhy (2015) indicated that most of wheat genotypes tended to increase their canopy temperature depression CTD rather than the optimum or moderate late sowing date. Under Kafr Al-Hamam location, at late sowing date, Misr 2 displayed the highest CTD (7.99), however, at Sids location, Gemmeiza 11 presented the highest CTD (7.62). Quiet variances in correlation coefficients were observed from sowing date to another. Wheat grain yield showed positive and significant association with canopy temperature depression and grain protein content, which indicates that the importance of these characteristics in wheat breeding programs. ElBasyoni (2018) revealed that canopy temperature was positively and significantly related with leaf and stem rust scores, however negatively associated with grain filling duration, wheat grain yield and number of days to flowering.

Karwa et al. (2020) screening collection of rice genotypes for tolerant or sensitive to heat stress. Genotype IET 22,218 showed higher canopy temperature depression and accumulation of endogenous level of polyamines under both optimum and heat stress environments and gave higher head rice yield (>85%) under heat stress. Hence, IET 22,218 was recognized as the unique donor for heat stress tolerance.

Beneficial associations were found between crop water stress index estimated from single leaf temperatures and stomatal resistance, leaf area index and available water in the root zone of sunflower (Orta et al. 2002). Significant and positive association between seed cotton yield and canopy temperature at various cotton growing stage was found by Karademir et al. (2018). However, strong relationships with canopy temperature were detected at peak flowering stage of cotton development period.

3.2.3.6 Osmotic Potential

Osmotic adjustment is promising, because it resists the effects of a rapid shortage of leaf water potential. There is a genotypic difference in osmotic potential as physiological character related to prolonged droughts. Farag, Fatma (2019) displayed adaptable response of the verified wheat genotypes to water stress through osmotic pressure under different seasonal conditions. A positive significant association was registered between osmotic pressure and both 1000-grain weight and wheat grain yield under sever water stress with coefficients (0.572) and (0.698), respectively.

Physiological research suggests that drought resistance in rice mainly relies on water use efficiency that allows minimum water usage for maximum production and osmotic adjustment that enables plants to maintain turgor and protect the meristem (Nguyen et al. 2004). Screening four-week-old rice seedlings against drought was done on set of rice cultivars by Choudhary et al. (2009). They demonstrated a uniform rise in RWC around 48–72 h due to osmotic adjustment. Accumulation of osmopro-tectants i.e. proline, glycinebetaine and soluble sugar provides osmotic adjustments in rice plants (Kumar et al. 2016; Upadhyaya and Panda 2019).

Regarding salinity stress, the ability to exclude salt along with organic solvents is an important component in salt t tolerant genotypes. The exclusion of Na^+ and Cl^- is an important basic mechanism for salt tolerance. In faba bean, Tavakkoli et al. (2012) found significant variations in the level of salt tolerance in eleven faba bean genotypes. Concentrations of Na^+ and Cl^- at 75 mM NaCl were significantly associated with biomass production under normal environments ($r = -0.97$ and -0.95). However, salt tolerance was more strongly related to the osmotic potential of the leaves. The reduction in salt tolerance differed between genotypes and associated with the capability to retain a great water content of the tissue. Abid et al. (2017) showed that 'Hara' cultivar was found to be less affected by water stress and produced the maximum value of higher antioxidant enzyme activities (catalase, Glutathione peroxidase and ascorbate peroxidase), osmoprotectant accumulations, Chlorophyll b and relative water content. Hence 'Hara' cultivar was found to be more tolerant to water deficit stress compared to the other cultivars.

In sesame, drought-tolerance genotypes displayed comprehensive defense in several issues on physiological and biochemical traits. Jian et al. (2019) indi-cated that drought-tolerant varieties exhibited more accumulation of osmotic regu-latory substances, and more content of antioxidant substances than those in drought-sensitive ones.

Whereas when cotton cultivars exposed to water-deficit stress at peak flowering (70 days after planting) Osmotic adjustment happened in reproductive tissues and their subtending leaves by different primary mechanisms. Pistils accumulated greater sucrose levels, maintaining cell turgor in plants subjected to drought at like levels to those in well-watered plants. However, subtending leaves lowered osmotic potential and maintained cell turgor by accumulating more proline (Pilon et al. 2019).

3.2.3.7 Photosynthesis Rate

Photosynthesis process is one of the key physiological phenomena affected by the drought and heat stress in plants (Farooq et al. 2009b). Photosynthesis is the main driver of plant growth and grain yield. Henceforth role of photosynthesis in appre-ciative the physiological foundation of a plant's response to drought is important. Variation in photosynthetic efficiency is the vital to determine the extent of photosyn-thesis in genotypes under water stress conditions. In this connection, flag leaves in wheat genotype TAM 112 at mid-grain filling stage exhibited lower photosynthetic rate, but higher photosynthetic water use efficiency compared to TAM 111 and TAM 304 at water stress (Thapa et al. 2018). Under different levels of NaCl (0, 50, 100 and 150 mM) at seedling stage, El-khamissi (2018) showed that Sakha 93 cultivar was found to be more salt tolerance as it attained uppermost value of photosynthetic pigments rather than Landrace 1, while Landrace 2 was the most sensitive to salinity stress.

Higher enzyme activity and improved photosynthesis are identified in transgenic rice by interrogation of some genes i.e. OsTPS1, enhances abiotic stress tolerance (Li et al. 2011). Rice genotype IET 22,218 verified higher photosynthesis, under both optimum and heat stress environments with higher head rice yield (>85%) under heat stress (Karwa et al. 2020).

While, Kubure et al. (2016) showed that photosynthesis in faba bean, is a necessary physiological trait to improve faba bean productivity; mainly at rain-fed environments where often soil moisture is a restrictive element. genotypes Hachalu and Walki exhibited significantly larger leaf area index than the local cultivar. In seame, the higher LAI in developed genotypes lead to higher Normalized differential vegetation index is powerful tools for influential photosynthetic performance. Though, when stress is occurred, the ability of genotypes to continue a relatively high rate of photosynthetic activity might contribute well to seed yield (Bayoumi et al. 2017).

Photosynthetic rates of the tolerant sunflower genotypes were significantly superior over other less tolerant ones. The genotype DRSF-113 (26 μ moles $CO_2 m^2/S$) followed by EC-602063 (25 μ moles CO_2 m^2/s) attained highest photosynthetic rate at 75 DAS. At 55 and 65 DAS, genotype DRSF-113 under control, the same genotype followed by EC-602063 under moisture stress showed higher photosynthetic rates than other genotypes at 75 DAS. Genotype GMU-337 have shown lowest photosynthetic rate under stress (Santhosh 2014).

Photosynthate accumulation of various cotton genotypes increased primarily with the growth stage as shown by Feng et al. (2016). The accumulation of total photosynthate in cotton genotype Xinluzao 43 during boll opening stage fluctuated from 27,792.4 to 28,087.8 kg/ha and was greater compared to the rest three varieties. Total photosynthate accumulation rate of Xinluzao 43 was higher by 20.5%–70.2% than Shiza 2; 21.5%–92.7% greater than Xinluzao 33, as well as higher by 37.0%–154.1% than Xinluzao 13 from the initial boll setting stage to the boll opening stage. Leaf chlorophyll content was related with the photosynthetic potential of different cotton genotypes and with adaptability to light environmental change. It was reported that leaf wax content and constitutes increased in response to abiotic stress (Xue et al. 2017), specifically to drought stress in plant species such as Arabidopsis (Kosma and Jenks 2007), alfalfa (Ni et al. 2012), Populus euphratica (Xu et al. 2016) and sesame (Kim et al. 2007).

3.2.4 Biochemical and Molecular Base

3.2.4.1 Proline Content

Proline accumulation is an adaptive behavior of crop plants when they are exposed to various environmental pressures (Oncel et al. 2000). Proline, an amino acid (Fig. 3.3), plays a greatly useful role in plants exposed to different stress conditions i.e. drought,

Fig. 3.3 Proline structure

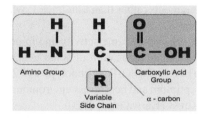

salinity, cold and heat stress. Further acting as an excellent osmolyte, proline shows three major roles during stress, as a metal chelator, an antioxidative defense molecule and a signaling molecule. Review of the literature point out that a stressful environment led to an overproduction of proline in plants which in turn imparts stress tolerance by maintaining cell turgor or osmotic balance, stabilizing membranes thus avoiding electrolyte leakage and bringing concentrations of reactive oxygen species within normal ranges, consequently stopping oxidative burst in plants. Paleg et al. (1981) previously stressed the importance role of proline in osmoregulation and serves as a source of energy, carbon and nitrogen in addition to protecting various plant enzymes from heat and water deficit. It is therefore a selection indicator in crop breeding programs to drought tolerance.

Initial effects of heat stress, are on plasmalemma, which appearances in the form of more fluidity of lipid bilayer. Proline content has been related with performance of irrigated wheat under high temperature level which might also be used as selection indicator to recognize wheat tolerant cultivars (Hasan et al. 2007). Wheat cultivars with high proline content at heat stress tended to show greater thermotolerance. Seedling proline content at 35 °C and membrane injury (%) maintained a significant negative correlation (r = -0.818**) across six Egyptian wheat cultivars (Abou Gabal and Tabl 2014). A positive and significant association was recorded between proline content and 1000-grain weight under sever water stress with coefficient 0.501* (Farag, Fatma 2019). Qaseem et al. (2019) exposed one hundred and eight elite different wheat germplasm to drought, heat and combined heat + drought treatments. Multivariate analysis displayed a strong correlation of water-soluble carbohydrates, proline content and all other studies agronomic and physiological traits with grain yield.

In rice, higher accumulation of proline is usually associated with drought tolerance and it helps for maintenance of leaf turgor and progress in stomatal conductance (Kumar et al. 2016). Choudhary et al. (2009) screened some rice genotypes against water stress and demonstrated an increase in osmotic adjustment as result of increased proline accumulation. Mishra et al. (2019) stated that proline content can act as a biochemical marker under drought screening of rice genotypes.

In faba bean, Migdadi et al. (2016) evaluated nine genotypes for seed yield and its components and free proline content in leaves under three water regimes (well-watered, mild and severe drought)). They recorded negative correlation between proline content and seed yield/plant (r = −0. 0.650*) over all treatments and was insignificant under both well-watered (r = 0.62) and high drought stress (r = 0.43). Molaei et al. (2012) found that the sesame cultivar Oltan recognized as a drought tolerance cultivar in compare with two other cultivars Hendi and Hendi 14 had the highest increase in amount of proline content under water stress condition and showed not significant decrease in its leaf water content. Recently, Jian et al. (2019) showed that after water stress, the contents of proline, soluble sugar, soluble protein in sesame leaves, showed an increase with the increase in the level of stress.

Furthermore, significant genetic variability between genotypes has been detected in sunflower for osmotic adjustment. There are many molecules which may be considered as candidates of compatible solutes such as K^+, Ca^{++}, proline, HSPs, and dehydrins. Though, in sunflower great number of investigations represented that proline played an important osmolyte under water stress condition. Where, Rauf and Sadaqat (2008) compared high and low osmotic adjustment genotypes and found advantage of high osmotic adjustment genotypes owing to higher harvest index. The highest amount of proline in leaf and root of sunflower is accumulated in drought stress after 48 h. Najafi et al. (2018) showed that under abiotic of drought, salinity, cold and heat stresses, the amount of proline increased in leaves and roots by increasing the levels of stress. Proline content tended to be increased in young leaves as a result of drought stress increased (Badr et al. 2004; Cechin et al. 2006 and Oraki et al. 2012). Iqbal et al. (2005) reported that shortage of water caused an increase in leaf proline concentration by 28% at vegetative stage and by 24% at the reproductive stage, than that of the non-stressed ones.

While, Ghobadi et al. (2013) stated that the extent of the changes in proline accumulation was related to the intensity of the stress. Proline content in genotypes subjected to moderate drought stress was 42% more than the control. In six cotton genotypes, with decreasing water content there was a progressive increase in free proline. Maximum accumulation of free proline in drought stressed cotton occurred under severe water stress. The combination of heat and drought stress showed an increase in proline content in five cultivars. Diverse proline profiles were detected between various treatments and diverse mechanisms for heat and drought (de Ronde et al. 2000). The accumulation of proline, soluble proteins, soluble sugars, amplified significantly in TM-1.TM-1 as more tolerance than Zhongmian-16 and Pima4-S. The expression of drought-reactive genes covering coding for transcription regulatory proteins or enzymes controlling genes were greater in TM-1 at stress, conferring a more tolerant than other both genotypes (Hasan et al. 2018).

3.2.4.2 Abscisic Acid

Abscisic acid (Fig. 3.4) is a plant hormone. ABA functions in many plant developmental processes, of them stomatal closure. It is especially important for plants in the response to environmental stresses, including drought, soil salinity, cold tolerance, freezing tolerance, heat stress, heavy metal ion tolerance and plant pathogens. Abscisic acid plays an significant role in plant signaling system which helps the plant to complete function normally under water and heat stress conditions (Finkelstein et al. 2002; Hussain et al. 2010).

Fig. 3.4 Abscisic acid

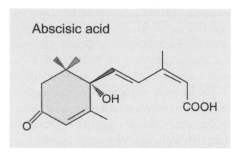

Abscisic acid is also produced in the roots in response to decreased soil water potential and other situations in which the plant may be under stress. Abscisic acid then translocates to the leaves, where it rapidly alters the osmotic potential of stomatal guard cells, causing them to shrink and stomata to close. The ABA-induced stomatal closure reduces transpiration. Thus preventing further water loss from the leaves in times of low water availability. A close linear correlation was found between the abscisic acid content of the leaves and their conductance represents stomatal resistance on a leaf area basis (Barbara et al. 1988). The plant hormone, abscisic acid, shows a significant role, inducing a network of signalling pathways relating multiple protein kinases and transcription dynamics (Semenov and Halford 2009).

Under water stress, abscisic acid significantly increased the activities of super-oxide dismutase and peroxidase, showing a significant decrease on rewatering. The relative water content was significantly improved by abscisic acid priming under water stress in the two wheat cultivars Chakwal-97 (drought tolerant) and Punjab-96 (drought sensitive). The sensitive cultivar Punjab-96 showing lower endogenous abscisic acid content was more responsive to abscisic acid priming. Abscisic acid was highly effective in improving grain weight of tolerant cultivar Chakwal-97 under drought stress (Bano et al. 2012). Pál et al. (2018) showed a relationship between polyamine metabolism and abscisic acid signalling leads to the controlled regulation and maintenance of polyamine and proline levels under osmotic stress situations in wheat seedlings.

ABA can also improve osmotic regulation potentially leading to sustained growth under drought (Tardieu et al. 2010). Liu et al. (2014) showed that OsbZIP71, a Bzip transcription factor, confers salinity and drought tolerance in rice. Abscisic acid, brassinosteroids and ethylene phytohormone pathways improves drought response without depressing yield (Gupta et al. 2020).

Abscisic acid and proline helped the plant to restore the altered physiological process under water stress conditions. Ali et al. (2013) showed that abscisic acid and proline alone or in combination improved all growth characteristics of faba bean by improving photosynthetic pigments, catalase and peroxidase activities. While, in sunflower, Unyayar et al. (2004) exposing genotypes to drought and water logging. Significant differences in proline accumulation were revealed under stress conditions. Hussain et al. (2015) added that abscisic acid, was helpful in improving drought stress tolerance by enhancing water relations and achene yield of sunflower hybrids under drought at bud or flower initiation. Sunflower hybrid DK-4040 displayed better enhancement of drought tolerance by abscisic acid under drought compared to SF-187 and S-278 because it displayed more improvement in water potential, osmotic potential, turgor pressure, relative leaf water contents and achene yield.

Finally, Li et al. (2010) showed endogenous molecules such as jasmonic acid, ethylene, and abscisic acid play an important role in abiotic stress tolerance. The two cotton Di19 proteins i.e., GhDi19-1 and GhDi19-2 were involved in response to salt stress and abscisic acid signaling.

3.2.4.3 Antioxidants

Antioxidants play a regulatory role in the work of controlling genes in the production of various enzymes related to tolerance of stress environmental conditions, scavenging and decreasing destructive effect of reactive oxygen species produced during drought stress on crop plants might correlate with water stress tolerance. Research studies showed that several enzymatic (ascorbate peroxidase, catalase, superoxide dismutase, peroxidase, monodehydroascorbate reductase, dehydroascorbate reductase, glutathione reductase) as well as non-enzymatic (ascorbate, phenolic compounds, carotenoids, glutathione, glycine betaine, proline, sugar and polyamines) antioxidants present in tissues and played significant roles in environmental stress tolerance (Gill and Tuteja 2010; Karuppanapandian et al. 2011). So, to alleviate contrary effects of reactive oxygen species, genotypes have evolved an antioxidant defense mechanism represents enzymes like superoxide dismutase, peroxidase and catalase etc. (Agarwal and Pandey 2004). The activities of superoxide dismutase and peroxidase and endogenous abscisic acid in plants were measured under water stress in wheat cv. Chakwal-97 (drought tolerant) and cv. Punjab-96 (drought sensitive). Drought tolerant cultivar exhibited more efficient mechanism to scavenge reactive oxygen species as revealed by a significant increase in the activity of antioxidant enzyme superoxide dismutase. Under drought stress, abscisic acid significantly increased the activities of superoxide dismutase and peroxidase, displaying a significant decline on rewatering.

Higher antioxidant and enzyme activity are detected in transgenic rice plants, enhances abiotic stress (Li et al. 2011). Three popular traditional rice landraces, namely Kalajeera, Machakanta and Haladichudi, from Koraput, India were tested by Mishra and Panda (2017) for drought tolerance by exposing to different levels of drought stress induced by varying concentrations of polyethylene glycol (PEG) 6000. Activities of antioxidative enzymes increased under drought stress. The traditional rice landraces displayed higher ratios of various parameters compared to the sensitive IR64. Karwa et al. (2020) showed that polyamines and antioxidant enzymes activity in rice genotype IET 22,218 under heat stress were associated with lowering oxidative stress and conserved higher pollen viability and spikelet fertility.

In faba bean, Siddiqui et al. (2015) subjected faba bean genotypes to different levels of drought stress i.e., normal irrigation, mild stress, moderate stress and severe stress. They showed that water stress reduced all growth parameters and total Chl content of all genotypes. Conversely, the effect of water stress on both genotypes "C5" and "Zafar 1" were relatively low as a result its better antioxidant enzymes activities (catalase, superoxide dismutase and peroxidase), and accumulation of Chlorophyll, and leaf relative water content. So, genotype "C5" and "Zafar 1" were appeared to be relatively tolerant to drought stress, while genotypes "G853" and "C4" were sensitive to drought stress.

Whereas, Bayoumi et al. (2017) showed that he activities of peroxidase and catalase antioxidant enzymes were changed when sesame plants were exposed to water stress. Peroxidase and catalase activities increased significantly in all populations under drought stress Levels. Catalase activity increased in Population 4 more than 20 folds due to water stress. Jian et al. (2019) found that after water deficit stress, the activities of superoxide dismutase, peroxidase, catalase, phenylalanine ammonialyase in the leaves of sesame were increased with increasing the level of stress.

In cotton, Ratnayaka et al. (2003) reported that, reactive oxygen species (ROS) under drought stress were produced however, the ascorbate peroxidase and its activities also increased and plants maintained the ROS scavenging process until the plant recovered from the stress conditions. Sarwar et al. (2020) indicated that subjected cotton plants to heat stress showed increased SOD and CAT activity in comparison with non-stressed plants.

3.2.4.4 Heat Shock Proteins

Heat shock proteins (HSP) are one of the imperative classes of molecular chaperones, which act in response to numerous stresses i.e. extreme temperature, water stress, salinity, oxidative, heavy metals, and high intensity irradiations (Swindell et al. 2007; Al-Whaibi 2011; Xu et al. 2011). The roles of HSP are apparently more in response to high temperature stress when compared with other stresses such as Hsp90 in wheat to heat stress (Majoul et al. 2004); Hsp70 in wheat to cold stress (Vítámvás et al. 2012; Kosová et al. 2013); Hsp100/clpB2, B4 & D2 in wheat to salt stress (Muthusamy et al. 2017); Hsp90 in sunflower to cold stress (Balbuena et al. 2011); as well as in cotton to heat stress (Burke et al. 1985 and Zhang et al. 2016).

Transcription factors that are specifically associated with heat stress include heat shock factors (HSFs) reviewed by Semenov and Halford (2009) and over-expression experiments with HSFs have resulted in increased thermo-tolerance in transgenic plants. Molecular, chaperones', such as heat-shock proteins (HSPs), play a vital role in keeping proteins and RNA in their correct conformation, mitigate the effect of stress on crop plants. Many researches showed the different functions of HSP and their homologues through their expression under the tight regulation of cell cycle, cell growth and development. Al-Whaibi (2011) has confirmed the role of HSP in plant growth and development largely in embryogenesis and seed development.

The effect of heat stress on the synthesis of heat shock proteins (HSP) was identified with the seven wheat parents, Badr Asmaa (2017) revealed that the seedling exposed to heat stress formed heat shock proteins in significant quantity after exposed to 45 °C. The most stable cultivars were Sakha 93 where its thermoplastic reactions did not change under heat shock circumstances compared to the control. Sids1 presented high gene expression with gene copy number 3.61 compared to Gemmeiza 10 which had little gene expression with gene copy number 0.05. The detected expression of hsp22 (mitochondrial) might contribute to heat stress tolerance.

In rice, OsHSP50.2, an HSP90 family gene up-regulated by heat and osmotic stress managements, positively regulates drought stress tolerance. Overexpression of OsHSP50.2 in transgenic rice plants decreased water loss and chlorophyll and enhanced tolerance to drought and osmotic stresses than wild-type plants (Xiang et al. 2018).

Nieden et al. (1995) illustrious species-specific variants in HSP17 expression between faba bean, maize and tomato plants. They indicated that HSP17 was mainly contained in protein bodies in mature seeds of faba bean and tomato but not in maize. Therefore, only two ESTs, one for HSP17.9 (GenBank KC249973.1) and other for HSP70.1 (GenBank EU884304.1) of faba bean, are present in National Centre for Biotechnology Information dbEST database.

3.2.4.5   Cell Membrane Stability

Cell membrane thermostability (CMS) is considered to be one of the major selection criteria for diverse abiotic stresses comprising water and high temperature in wheat. Abou Gabal and Tabl (2014) based on membrane thermostability assessment, cultivar which showed equal or more than 50% membrane injury was Misr1, Sids12, Misr 2 and Giza168 were expressed as heat sensitive cultivars. The other two wheat cultivar Sakha 93 and Gemmeiza 9 displayed less than 50% membrane injury in membrane thermostability test and were grouped as heat tolerant cultivars. The higher relative value of proline content was found in heat tolerant cultivars compared to that in heat sensitive cultivar like Giza168 but it showed high relative value of proline content. Ciulca et al. (2017) showed that the degree of stability of the cell membrane shows the ability of cell tissues to retain electrolytes under conditions of water stress. Among 21 F1 hybrids of seven winter wheat varieties, the over dominance is acting in the inheritance of this indicator in case of nine hybrids, being generally related with a reduction of membrane stability, excluding for Fundulea 4 × Turda 2000 and GKKapos x Apullum that cause this indicator to increase. For six crosses, incomplete dominance leads to decrease membrane stability, whereas at four crosses it was related with an increase of membrane stability.

Maintenance of stability in membrane index of the plant under water scarcity has been used to know its correlation with yield of rice under drought stress (Upadhyaya and Panda 2019). Also, Karwa et al. (2020) found that rice variety IET 22,218 registered lower H_2O_2 accumulation, cell membrane damage under high temperature and designated as the unique donor for heat tolerance.

The drought-tolerance varieties of sesame exhibited less cell membrane lipid peroxidation damage, and greater content of antioxidant substances compared to the drought-sensitive varieties Jian et al. (2019). While in sunflower, Santhosh (2014) observed significant differences for physiological characters among genotypes at 45, 55, 65 and 75 DAS. At 55 DAS, the interaction between genotypes and moisture stress was significant. Among genotypes, EC-602063 under control and in stress exhibited greater membrane stability index. While genotype GMU—37 gave lowest membrane stability index both in control and moisture stress treatments. In cotton, Abdel-Kader et al. (2015) indicated that cotton genotypes Tamcot C. E. x Deltapine, Giza 90 x (Giza 90X Australian) and Giza 80 × Deltapine were drought tolerance as uppermost values of membrane stability index, with lowest values of electrolyte leakage % under drought stress.

Therefore, understanding the nature of the association between stress tolerance and other important traits, breeders can select and develop more stress tolerant and adapted varieties that give satisfactory yield levels.

3.3   Conclusions

Improving field crops to environmental stress tolerance is the most important problem facing agricultural production. So, it is very important to determine the plant proper-ties associated with environmental stress conditions, to understand the relevant phys-iological, biochemical and molecular foundations and to tolerate climate changes in order to produce more adaptive varieties. Therefore, this chapter focused on identi-fying the most important characteristics of crop plants related to crop tolerance to environmental stresses.

3.4   Recommendations

Based on the studies discussed in this work, it is recommended to deepen the research work on the nature of inheritance, genetic system and molecular base controlling the inheritance of traits associated with tolerance to environmental stress factors. Understanding the nature of the association between important traits, breeders can select and develop more stress tolerant varieties that give satisfactory yield levels. Also the use of genetic engineering techniques in the genetic improvement process seems beneficial in this regard. This facilitates the possibility of transferring tolerant genes from genetic resources to commercial varieties to cope with climate change and sustain the productivity in the face of expected changes in environmental conditions.

References

Abdel-Kader MA, Esmail AM, El Shouny KA, Ahmed MF (2015) Evaluation of the drought stress effects on cotton genotypes by using physiological and morphological traits. Int J Sci Res 4(11):1358–1362

Abid KH, Aouida M, Aroua I, Baudoin J-P, Muhovski Y, Mergeai G, Sassi K, Machraoui M, Souissi F, Jebara M (2017) Agro-physiological and biochemical responses of faba bean (Vicia faba L. var. 'minor') genotypes to water deficit stress. Ghassen Biotechnol Agron Soc Environ 21(2):146–159

Abou Gabal AA, Tabl KM (2014) Heat tolerance in some Egyptian wheat cultivars as measured by membrane thermal stability and proline content. Middle East J Agric Res 3(2):186–193

Agarwal S, Pandey V (2004) Antioxidant enzyme responses to NaCl stress in cassia angustifolia. Biol Plant 48:555–560

Ahmed M, Asif M, Goyal A (2012) Silicon the non-essential beneficial plant nutrient to enhanced drought tolerance in wheat. In: Goyal A (ed) IntechOpen, London, UK

Akram M, Iqbal RM, Jamil M (2014) The response of wheat (Triticum aestivum L.) to integration effects of drought stress and nitrogen management. Bulg J Agric Sci 20(2):275–286

Al-Bakry MRI (2007) Glaucous wheat mutants: I, agronomic performance and epicuticular wax content. Egypt J Plant Breed 11(1):1–9

Ali HM, Siddiqui MH, Al-Whabi MH, Basalah MO, Sakran AM, El-Zaidy M (2013) Effect of proline and Abscisic acid on the growth and physiological performance of faba bean under water stress. Pak J Bot 45(3):933–940

Al-Whaibi MH (2011) Plant heat-shock proteins: a mini review. J King Saud Univ–Sci 23:139–150

Alybayeva RA, Kenzhebayeva SS, Atabayeva SD (2014) Resistance of winter wheat genotypes to heavy metals. IERI Procedia 8:41–45

Anjum SA, Xie XY, Wang LC, Saleem MF, Man C, Lei W (2017) Morphological, physiological and biochemical responses of plants to drought stress. Afr J Agric Res 6(9):2026–2032

Anonymous (2017) Identification of sources of improved drought tolerance. https://cordis.europa.eu/result/rcn/44796/en?format=xml)

Anonymous (2021) Technical recommendations for rice crop. Ministry of Agriculture and Land Reclamation, Agricultural Research Center, Field Crops Research Institute, Rice Research and Training Center, Egypt.

Arafa AS, Nour ODM, Hassan ISM (2008) Impact of accumulated heat units (DD15 'S) on the performance behavior of some Egyptian cotton cultivars. Egypt J Agric Res 86(5):1945–1956

Ashraf M, Ahmad S (2000) Influence of sodium chloride on ion accumulation, yield components and fiber characteristics in salt-tolerant and salt-sensitive lines of cotton (Gossypium hirsutum L.). Field Crops Res 66:115–127

Asthir B (2015) Protective mechanisms of heat tolerance in crop plants. J Plant Interact 10:202–210

Awaad HA (2001) The relative importance and inheritance of grain filling rate and period and some related characters to grain yield of bread wheat (Triticum aestivum L.) proceed. The second Pl Breed Conf, Oct 2, (Assiut Univ) pp 181–198

Awaad HA, Salem AH, Ali MMA, Kamal KY (2016) Expression of heterosis, gene action and relationship among morpho-physiological and yield characters in sunflower under different levels of water supply. J Plant Prod, Mansoura Univ 7(12):1523–1534

Awaad HA (2009) Genetics and breeding crops for environmental stress tolerance, I: drought, heat stress and environmental pollutants. Egyptian Library, Egypt

Awaad HA (2021) Performance and genetic diversity in water stress tolerance and relation to wheat productivity under rural regions. In: Awaad HA, Abu-hashim M, Negm A (eds) Handbook of mitigating environmental stresses for agricultural sustainability in Egypt. Springer Nature Switzerland AG, pp 63–103

Badr Asmaa MSH (2017) Breeding wheat for heat tolerance. M.Sc. thesis, Department of Agronomy, Faculty of Agriculture, Ain Shams University

Badr NM, Thalooth AT, Mohamed MH (2004) Effect of foliar spraying with the nutrient compound "Streen" on the growth and yield of sunflower plants subjected to water stress during various stages of growth. Bull Natl Res Cent Cairo 29(4):427–439

Balbuena TS, Salas JJ, Martínez-Force E, Garcés R, Thelen JJ (2011) Proteome analysis of cold acclimation in sunflower. J Proteome Res 10:2330–2346

Bano A, Ullah F, Nosheen A (2012) Role of abscisic acid and drought stress on the activities of antioxidant enzymes in wheat. AGRIS 58(4):181–185

Barbara B, Stuhlfauth T, Fock HP (1988) The efficiency of water use in water stressed plants is increased due to ABA induced stomatal closure. Photosynth Res 18(3):327–336

Basal H, Hemphill JK, Smith CW (2006) Shoot and root characteristics of converted race stocks accessions of upland cotton (*Gossypium hirsutum* L.) grown under salt stress conditions. Am J Plant Physiol 1:99–106

Bayoumi TY, Abdou ET, Elshakhess SAM, Mahmoud SA (2017) Impact of antioxidant enzymes and physiological indices as selection criteria for drought tolerance in sesame populations. J Middle East North Afr Sci 3(2):25–34

Bhanu A, Arun B, Mishra VK (2018) Genetic variability, heritability and correlation study of physiological and yield traits in relation to heat tolerance in wheat (*Triticum aestivum* L.). Biomed J Sci Tech Res 2(1):212–216

Boyer SJ (1996) Advances in drought tolerance in plants. Adv Agron 56:187–218

Burke JJ, Hatfield JL, Klein RR, Mullet JE (1985) Accumulation of heat shock proteins in field-grown cotton. Plant Physiol 78(2):394–398

Cabuslay GS, Ito O, Alejar AA (2002) Physiological evaluation of responses of rice (*Oryza sativa* L.) to water deficit. Plant Sci 163(4):815–827

Cechin I, Rossi SC, Oliveira VC, Fumis TF (2006) Photosynthetic responses and proline content of mature and young leaves of sunflower plants under water deficit. Photosynthetica 44(1):143–146

Choudhary MK, Basu D, Datta A, Chakraborty N, Chakraborty S (2009) Dehydration-responsive nuclear proteome of rice (*Oryza sativa* L.) illustrates protein network, novel regulators of cellular adaptation, and evolutionary perspective. Mol Cell Proteom 8(7):1579–1598

Ciulca S, Madosa E, Ciulca A, Sumalan R, Lugojan C (2017) The assessment of cell membrane stability as an indicator of drought tolerance in wheat. In: Conference: 17th international multidisciplinary scientific geo conference SGEM 2017

Comas LH, Becker SR, Cruz VMV, Byrne PF, Dierig DA (2013) Root traits contributing to plant productivity under drought. Front Plant Sci 4:442

de Ronde JA, van der Mescht A, Steyn HSF (2000) Proline accumulation in response to drought and heat stress in cotton. Afr Crop Sci J 8(1):85–91

Debaeke P, Casadebaig1 P, Flenet F, Langlade N (2017) Sunflower crop and climate change: vulnerability, adaptation, and mitigation potential from case-studies in Europe. EDP Sci 24(1):1–15

Devi JD, Reddy VR (2018) Transpiration Response of cotton to vapor pressure deficit and its relationship with stomatal traits. Front Plant Sci 29:1572

Dewi ES (2009) Root morphology of drought resistance in cotton (*Gossypium hirsutum* L.). M.Sc. thesis. Texas A & M University, College Station, TX

Donatelli M, Srivastava AK, Duveiller G, Niemeyer S, Fumagalli D (2015) Climate change impact and potential adaptation strategies under alternate realizations of climate scenarios for three major crops in Europe. Environ Res Lett 10: 075005

ElBasyoni IS (2018) Performance and stability of commercial wheat cultivars under terminal heat stress. Agronomy 8(4):5–29

El-khamissi H (2018) Morphological and biochemical evaluation of some Egyptian wheat genotypes under salinity stress conditions

El-Khoby WMH, Abd El-lattef ASM, Mikhael BB (2014) inheritance of some rice root characters and productivity under water stress conditions. Egypt J Agric Res 92(2):529–549

El-Rawy MA, Hassan MI (2014) Effectiveness of drought tolerance indices to identify tolerant genotypes in bread wheat (*Triticum aestivum* L.). J Crop Sci. Biotechnol 17(4):255–266

Chairi F, Aparicio N, Serret MD, Araus JL (2020) Breeding effects on the genotype × environment interaction for yield of durum wheat grown after the green revolution: the case of spain. Crop J 8(4):623–634

Farid MA, Abou Shousha AA, Negm MEA, Shehata SM (2016) Genetical and molecular studies on salinity and drought tolerance in rice (Oryza sativa L). J Agric Res Kafer El-Sheikh Univ 42(2):1–23

Farooq M, Kobayashi N, Wahid A, Ito O, Basra SMA (2009) Strategies for producing more rice with less water. Adv Agron, 101, 351–388

Farooq M, Wahid A, Basra SMA, Din IU (2009) Improving water relations and gas exchange with Brassinosteroids in rice under drought stress. J Agron Crop Sci 195:262–269

Fatma FM (2019) Breeding parameters for grain yield and some morpho-pysiological characters related to water stress tolerance in bread wheat. M.Sc. thesis, Agronomy Department, Faculty of Agriculture Zagazig University, Egypt

Feng G, H Luo, Y Zhang, L Gou, Y Yao, Y Lin, W Zhang (2016). Relationship between plant canopy characteristics and photosynthetic productivity in diverse cultivars of cotton (Gossypium hirsutum L.). The Crop Journal 4 (6):499–508

Finkelstein R, Gampala S, Rock C (2002) Abscisic acid signaling in seeds and seedlings. Plant Cell 14:15–45

Foulkes MJ, Sylvester-Bradley R, Weightman R, Snape JW (2007) Identifying physiological traits associated with improved drought resistance in winter wheat. Field Crop Res 103(1):11–24

Gaballah MM (2018) Kinetn application on some rice genotypes performance under water deficit. In: The Seventh Field Crops Conference, 18–19 Dec 2018, Absrract No 23

Gaballah MM, Metwally AM, Skalicky M, Hassan MM, Brestic M, EL Sabagh A, Fayed AM (2021) Genetic diversity of selected rice genotypes under water stress conditions. Plants 1(27):1–19

Ghobadi M, Taherabadi S, Ghobadi ME, Mohammadi GR, Jalali-Honarmand S (2013) Antioxidant capacity: photosynthetic characteristics and water relations of sunf lower (Helianthus annuus L.) cultivars in response to drought stress. Ind Crop Prod 50:29–38

Gill SS, Tuteja N (2010) Reactive oxygen species and antioxidant machinery in abiotic stress tolerance in crop plants. Plant Physiol Biochem 48:909–930

Golestani M, Pakniyat H (2015) Evaluation of traits related to drought stress in sesame (Sesamum indicum L.) genotypes. J Asian Sci Res 5(9):465–472

Golestani-Araghi S, Assad MT (1998) Evaluation of four screening techniques for drought resistance and their relationship to yield reduction ratio in wheat. Euphytica 103:293–299

Guidi L, Soldatini GF (1997) Chlorophyll fluorescence and gas exchanges in flooded soybean and sunflower plants. Plant Physiol Biochem 35:713–717

Guo J, Xu W, Yu X, Shen H, Li H, Cheng D, Liu A, Liu J, Liu C, Zhao S, Song J (2016) Cuticular wax accumulation is associated with drought tolerance in wheat near-isogenic lines. Front Plant Sci 7:1809

Gupta A, Rico-Medina A, Caño-Delgado AI (2020) The physiology of plant responses to drought. Science 368:266–269

Gusta LV, Chen TH (1987) The physiology of water andtemperature stress, pp 115–150. In: Heyne EG (ed) Wheat and wheat improvement, 2nd ed. American Society of Agronomy, Madison, WI.

Hasan MA, Ahmed JU, Bahadur MM, Haque MM, Sikder S (2007) Effect of late planting heat stress on membrane thermostability, proline content and heat susceptibility index of different wheat cultivars. J Natn Sci Found Sri Lanka 35(2):109–117

Hasan MdM, Ma F, Prodhan ZH, Li F, Shen H, Chen Y, Wang X (2018) Molecular and physio-biochemical characterization of cotton species for assessing drought stress tolerance. Int J Mol Sci 19(9):2636

Hasanuzzaman M, Bhuyan MHMB, Zulfiqar F, Raza A, Mohsin SM, Mahmud JA, Fujita M, Fotopoulos V (2020) Reactive oxygen species and antioxidant defense in plants under abiotic stress: revisiting the crucial role of a universal defense regulator. Antioxidant 9(681):1–52

Hassan HM, El-Khoby WM, El-Hissewy AA (2013) Performance of some rice genotypes under both salinity and water stress conditions in Egypt. J Plant Prod, Mansoura Univ 4(8):1235–1257

Hirasawa T, Wakabayashi K, Touya S, Ishihara K (1995) Stomatal responses to water deficits and abscisic acid in leaves of sunflower plants (*Helianthus annuus* L.) grown under different conditions. Plant Cell Physiol 36:955–964

Hirayama M, Wada Y, Nemoto H (2006) Estimation of drought tolerance based on leaf temperature in upland rice breeding. Breed Sci 56:47–54

Hussain HA, Hussain S, Khaliq A, Ashraf U, Anjum SA, Men SN, Wang LC (2018) Chilling and drought stresses in crop plants: Implications cross talk, and potential management opportunities. Front Plant Sci 9:393

Hussain S, Saleem MF, Ashraf MY, Cheema MA, Haq MA (2010) Abscisic acid, a stress hormone helps in improving water relations and yield of sunflower (*Helianthus annuus* L.) hybrids under drought. Pak J Bot 42:2177–2218

Hussein Y, Amin G, Azab A, Gahin H (2015) Induction of drought stress resistance in sesame (*Sesamum indicum* L.) plant by salicylic acid and kinetin. J Plant Sci 10(4):128–141

Iqbal N, Ashraf M, Ashraf MY, Azam F (2005) Effect of exogenous application of glycinebetaine on capitulum size and achene number of sunflower under water stress. Int J Biol Biotech 2:765–771

Islam MS, Srivastava PSL, Deshmukh PS (1999) Genetic studies on drought tolerance in wheat. II. Early seedling growth and vigour. Ann Agric Res 20(2):190–194

Jian S, XiaoWen Y, MeiWang LE, YueLiang RAO, TingXian YAN, YanYing YE, HongYing ZHOU (2019) Physiological response mechanism of drought stress in different drought-tolerance genotypes of sesame during flowering period. Sci Agric Sin 52(7):1215–1226

Karademir E, Karademir C, Sevilmis U, Basal H (2018) Correlation between canopy temperature, chlorophyll content and yield in heat tolerant cotton (*Gossypium hisutum* L.) genotypes. PSP 29(8):5230–5237

Karuppanapandian T, Moon JH, Kim C, Manoharan K, Kim W (2011) Reactive oxygen species in plants: their generation, signal transduction and scavenging mechanisms. Aust J Crop Sci 5:709–725

Karwa S, Bahuguna RN, Chaturvedi AK, Maurya S, Arya SS, Chinnusamy V, Pal M (2020) Phenotyping and characterization of heat stress tolerance at reproductive stage in rice (*Oryza sativa* L.). Acta Physiol Plant 42(1):1–16

Khan HR, Link W, Hocking TJ, Stoddard FL (2007) Evaluation of physiological traits for improving drought tolerance in faba bean (*Vicia faba* L.). Plant Soil 292, 205–217

Khazaei H, Street K, Bari A, Mackay M, Stoddard FL (2013) The FIGS (Focused Identification of Germplasm Strategy) approach identifies traits related to drought adaptation in *Vicia faba* genetic resources. PLoS ONE 2013(8). https://doi.org/10.1371/journal.pone.0063107

Kim KS, Park SH, Jenks MA (2007) Changes in leaf cuticular waxes of sesame (*Sesamum indicum* L.) plants exposed to water deficit. J Plant Physiol 164:1134–1143

Kim KS, Park SH, Jenks MA (2017) Changes in leaf cuticular waxes of sesame (*Sesamum indicum* L.) plants exposed to water deficit. J Plant Physiol 164:1134–1143

Kim Y, Chung YS, Lee E, Tripathi P, Heo S, Kim KH (2020) Root response to drought stress in rice (*Oryza sativa* L.). Int J Mol Sci 21(4):1513

Kosma DK, Jenks MA (2007) Eco-physiological and molecular-genetic determinants of plant cuticle function in drought and salt stress tolerance. In: Jenks MA, Hasegawa PM, Jain SM (eds) Advances in molecular breeding toward drought and salt tolerant crops. Springer, Netherlands, Dordrecht, pp 91–120

Kosová K, Vítámvás P, Planchon S, Renaut J, Vanková R, Prášil IT (2013) Proteome analysis of cold response in spring and winter wheat (Triticum aestivum) crowns reveals similarities in stress adaptation and differences in regulatory processes between the growth habits. J Proteome Res 12:4830–4845

Kubure TE, Raghavaiah CV, Hamza I (2016) Production potential of faba bean (*Vicia faba* L.) genotypes in relation to plant densities and phosphorus nutrition on vertisols of central highlands of west showa zone, Ethiopia, East Africa. Adv Crop Sci Tech 4, 214. https://doi.org/10.4172/2329-8863.1000214

Kumar A, Basu S, Ramegowda V, Pereira A (2016) Mechanisms of drought tolerance in rice. In: Sasaki T (ed) Achieving sustainable cultivation of rice, 11, Burleigh Dodds Science Publishing

Larcher W (2003) Physiological plant ecology: ecophysiology and stress physiology of functional groups. Springer, Netherlands, p 514

Lehari K, Kumar M, Burman V, Aastha V, Kumar V, Chand P, Singh R (2019). Morphological, physiological and biochemical analysis of wheat genotypes under drought stress. J Pharmacogn Phytochem SP2, 1026–1030

Li G, Tai FJ, Zheng Y, Luo J, Gong SY, Zhang ZT et al (2010) Two cotton Cys2/His2-type zinc-finger proteins, *GhDi19-1* and *GhDi19-2*, are involved in plant response to salt/drought stress and abscisic acid signaling. Plant Mol Biol 74:437–452

Li HW, Zang BS, Deng XW, Wang XP (2011) Overexpression of the trehalose-6-phosphate synthase gene *OsTPS1* enhances abiotic stress tolerance in rice. Planta 234(5):1007–1018

Liu C, Mao B, Ou S, Wang W, Liu L, Wu Y, Chu C, Wang X (2014) OsbZIP71, a Bzip transcription factor, confers salinity and drought tolerance in rice. Plant Mol Biol 84:19–36

Lonbani M, Arzani A (2011) Morpho-physiological traits associated with terminal drought- stress tolerance in triticale and wheat. Agron Res 9:315–329

Lu Z, Percy RG, Qualset CO, Zeiger E (1998) Stomatal conductance predicts yields in irrigated Pima cotton and bread wheat grown at high temperatures. J Exp Bot 49:453–460

Luo HH, Zhang YL, Zhang WF (2016) Effects of water stress and rewatering on photosynthesis, root activity, and yield of cotton with drip irrigation under mulch. Photosynthetica 54:65–73

Maalouf F, Miloudi N, Ghanem ME, Murari S (2015) Evaluation of faba bean breeding lines for spectral indices, yield traits and yield stability under diverse environments. Crop Pasture Sci 66(10):1012–1023

Majoul T, Bancel E, Triboï E, Ben Hamida J, Branlard G (2004) Proteomic analysis of the effect of heat stress on hexaploid wheat grain: characterization of heat-responsive proteins from non-prolamins fraction. Proteomics 4:505–513

Migdadi HM, El-Harty EH, Salamh A, Khan MA (2016) Yield and proline content in faba bean genotypes under water vstress treatments. J Anim Plant Sci 26(6):1772–1779

Mishra SS, Behera PK, Kumar V, Lenka SK, Panda D (2018) Physiological characterization and allelic diversity of selected drought tolerant traditional rice (*Oryza sativa* L.) landraces of Koraput, India. Physiol Mol Biol Plants 24(6):1035–1046

Mishra SS, Panda D (2017) Leaf traits and antioxidant defense for drought tolerance during early growth stage in some popular traditional rice landraces from Koraput, India. Rice Sci 24(4):207–217

Mishra SS, Behera PK, Panda D (2019) Genotypic variability for drought tolerance-related morpho-physiological traits among indigenous rice landraces of Jeypore tract of Odisha, India. J Crop Imp 33:254–278

Molaei P, Ebadi A, Namvar A, Bejandi TK (2012) Water relation, solute accumulation and cell membrane injury in sesame (*Sesamum indicum* L.) cultivars subjected to water stress. Ann Biol Res 3(4):1833–1838

Mondal S, Singh RP, Crossa J, Huerta-Espino J, Sharma I, Chatrath R, Singh GP, Sohu VS, Mavi GS, Sukaru VSP, Kalappanavarge IK, Mishra VK, Hussain M, Gautam NR, Uddin J, Barma NCD, Hakim A, Joshi AK (2013) Earliness in wheat: a key to adaptation under terminal and continual high temperature stress in South Asia. Field Crop Res 151:19–26

Moucheshi AS, Razi H, Dadkhodaie A, Ghodsi M, Dastfal M (2019) Association of biochemical traits with grain yield in triticale genotypes under normal irrigation and drought stress conditions. Aust J Crop Sci 13(02):272–281

Muthusamy SK, Dalala M, Chinnusamy V, Bansal KC (2017) Genome-wide identification and analysis of biotic and abiotic stress regulation of small heat shock protein (HSP20) family genes in bread wheat. J Plant Physiol 211:100–113

Najafi S, Sorkheh K, Nasernakhaei F (2018) Characterization of the APETALA2/Ethylene-responsive factor (AP2/ERF) transcription factor family in sunflower. Sci Rep 8(11576):1–16

Nguyen TT, Klueva N, Chamareck V, Aarti A, Magpanta G, Millena AC, Pathan MS, Nguyen HT (2004) Saturation mapping of QTL regions and identification of putative candidate genes for drought tolerance in rice. Mol Genet Genomics 272(1):35–46

Ni Y, Guo YJ, Han L, Tang H, Conyers M (2012) Leaf cuticular waxes and physiological parameters in alfalfa leaves as influenced by drought. Photosynthetica 50:458–466

Nieden UZ, Neumann D, Bucka A, Nover L (1995) Tissue-specific localization of heat-stress proteins during embryo development. Planta 196:530–538

El- Moselhy OM (2015) Gene action and stability for some characters related to heat stress in bread wheat. Ph.D. thesis, Agronomy Department, Faculty of Agriculture Zagazig University, Egypt

Oncel I, Keles Y, Ustun AS (2000) Interactive effects of temperature and heavy metal stress on the growth and some biochemical compounds in wheat seedlings. Environ Pollut 107:315–320

Oraki H, Parhizkar KF, Aghaalikhna M (2012) Effect of water deficit stress on proline contents, soluble sugars, chlorophyll and grain yield of sunflower (*Helianthus annuus* L.) hybrids. African J Biotechnol 11:164–168

Orta AH, Erdem T, Erdem Y (2002) Determination of water stress index in sunflower. Helia 25(37):27–38

Pál M, Tajti J, Szalai G, Peeva V, Végh B, Janda T (2018) Interaction of polyamines, abscisic acid and proline under osmotic stress in the leaves of wheat plants. Sci Rep 8:1–12

Paleg LG, Douglas TJ, van Daal A, Keech DB (1981) Proline, betaine and other organic solutes protect enzymes against heat inactivation. Aust J Plant Physiol 8:107–114

Pankovic D, Sakac Z, Kevresan S, Plesnicar M (1999) Acclimation to long-term water deficit in the leaves of two sunflower hybrids: photosynthesis, electron transport and carbon metabolism. J Exp Bot 50:127–138

Pilon C, Loka D, Snider JL, Oosterhuis DM (2019) Drought-induced osmotic adjustment and changes in carbohydrate distribution in leaves and flowers of cotton (*Gossypium hirsutum* L.). J Agron Crop Sci 205(2):168–178

Qaseem MF, Qureshi R, Shaheen H (2019) Effects of pre-anthesis drought, heat and their combination on the growth, yield and physiology of diverse wheat (*Triticum aestivum* L.) genotypes varying in sensitivity to heat and drought stress. Sci Rep 9:1–12

Rachidi F, Kirkham MB, Stone LR, Kanemasu ET (1993) Soil water depletion by sunflower and sorghum under rainfed conditions. Agric Water Manag 24(1:(9–62

Rana BB, Kamimukai M, Bhattarai M, Koide Y, Murai M (2019) Responses of earliness and lateness genes for heading to different photoperiods, and specific response of a gene or a pair of genes to short day length in rice. Hereditas 36(156):1–11

Rane J (2003) Evaluation of advanced wheat accessions high of temperature tolerance. In: Abstract 2nd international congress of plant physiology. New Delhi, India, p 254. 8–12 Jan 2003

Ratnayaka HH, Molin WT, Sterling TM (2003) Physiological and antioxidant responses of cotton and spurred anoda under interference and mild drought. J Exp Bot 54:2293–2305

Rauf S (2008) Breeding sunflower (*Helianthus annuus* L.) for drought tolerance. Commun Biometry Crop Sci 3(1):29–44

Rauf S, Sadaqat HA (2008) Effect of osmotic adjustmenton root length and dry matter partitioning in sunflower (*Helianthus annuus* L.) under drought stress. Acta Agric Scand, Sect B-Soil Plant Sci 58(3):252–260

Roohi E, Tahmasebi-Sarvestani Z, Modarres Sanavy SAM, Siosemardeh A (2015) Association of some photosynthetic characteristics with canopy temperature in three cereal species under soil water deficit condition. J Agr Sci Tech 17:1233–1244

Saleh SH (2011) Performance, correlation and path coefficient analysis for grain yield and its related traits in diallel crosses of bread wheat under normal irrigated and drought conditions. World J Agric Sci 7:270–279

Santhosh B (2014) Studies on drought tolerance in sunflower genotypes. M.Sc. in Agriculture Department of Crop Physiology College of Agriculture Acharya N. G. Ranga Agricultural University Rajendranagar, Hyderabad—500 030 Crop Physiology

Santhosh B, Reddy SN, Prayaga L (2017) Physiological attributes of sunflower (L.) *Helianthus annuus* as influenced by moisture regimes. Green Farming 3:680–683

Sarwar M, Saleem MF, Ullah N, Rizwan M, Ali S, Shahid MR, Alamri SA, Alyemeni MN, Ahmad P (2020) Exogenously applied growth regulators protect the cotton crop from heat-induced injury by modulating plant defense mechanism. Sci Rep 2018(8):1–15

Semenov MA, Halford NG (2009) Identifying target traits and molecular mechanisms for wheat breeding under a changing climate. J Exp Bot 60:2791–2804. https://doi.org/10.1093/jxb/erp164

Seyoum EG (2018) Genetic variability and gain in grain yield and associated traits of Ethiopian bread wheat (*Triticum aestivium* l.) varieties. M.Sc. thesis, the School of Graduate studies, Jimma University College of Agriculture and Veterinary Medicine, Ethiopia

Shams K (2018) Physiological, biochemical and yield responses of wheat cultivars to deficient water stress. Mod Phytomorphology 12:12–130

Siddiqui MH, Al-Khaishany MY, Al-Qutami MA, Al-Whaibi MH, Grover A, Ali HM, Al-Wahibi MS, Bukhari NA (2015) Response of different genotypes of faba bean plant to drought stress. Int J Mol Sci 16(5):10214–10227

Singh G, Kumar P, Kumar R, Gangwar LK (2018) Genetic diversity analysis for various morphological and quality traits in bread wheat (*Triticum aestivum* L.). J Appl Nat Sci 10(1):24–29

Singh NB, Ahmed Z (2003) Seedling vigour as an index for assessing terminal heat tolerance in wheat under irrigated late sown condition. In: Abstract: 2nd international congress of plant physiology. New Delhi, India, p 173. 8–12 Jan 2003

Streda T, Dostál V, Horáková V, Chloupek O (2011) Drought and root system size of barley and wheat. 1. Tagung der Österreichischen Gesellschaft für Wurzelforschung, 65–66

Swindell WR, Huebner M, Weber AP (2007) Transcriptional profiling of Arabidopsis heat shock proteins and transcription factors reveals extensive overlap between heat and non-heat stress response pathways. BMC Genomics 8:125

Tahara M, Carver BF, Johnson RC, Smith EL (1990) Relationship between relative water content during reproductive development and winter wheat grain yield. Euphytica 49:255–262

Talebi R (2011) Evaluation of chlorophyll content and canopy temperature as indicators for drought tolerance in durum wheat (Triticum durum Desf.). Aust J Basic Appl Sci 5(11):1457–1462

Tardieu F, Parent B, Simonneau T (2010) Control of leaf growth by abscisic acid hydraulic or non-hydraulic processes? Plant Cell Environ 33(4):636–647

Tavakkoli E, Paull J, Rengasamy P, McDonald GK (2012) Comparing genotypic in faba bean (*Vicia faba* L.) in response to salinity in hydroponic and field experiments. Field Crops Res 127:99–108

Thapa S, Reddy SK, Fuentealba MP, Xue Q, Rudd JC, Jessup KE, Devkota RN (2018) Physiological responses to water stress and yield of winter wheat cultivars differing in drought tolerance. J Agron Crop Sci 204(5):347–358

Unyayar S, Keles Y, Unal E (2004) Proline and ABA levels in two sunflower genotypes subjected to water stress. Bulg J Plant Physiol 30:34–47

Upadhyaya H, Panda SK (2019) Drought stress responses and its management in rice. In: Hasanuzzaman M, Fujita M, Nahar K, Biswas JK (eds) Advances in rice research for abiotic stress tolerance. Elsevier, UK, pp 177–200

Vítámvás P, Prášil IT, Kosová K, Planchon S, Renaut J (2012) Analysis of proteome and frost tolerance in chromosome 5A and 5B reciprocal substitution lines between two winter wheats during long-term cold acclimation. Proteomics 12:68–85

Wei S, Hu W, Deng X, Zhang Y, Liu X, Zhao X, et al. (2014). A rice calcium-dependent protein kinase OsCPK9 positively regulates drought stress tolerance and spikelet fertility. BMC Plant Biol 14(133):1–13

Xiang J, Chen X, Hu W, Xiang Y, Yan M, Wang J (2018) Overexpressing heat-shock protein OsHSP50.2 improves drought tolerance in rice. Plant Cell Rep 37(11):1585–1595

Xu X, Xiao L, Feng J, Chen N, Chen Y, Song B, Xue K, Shi S, Zhou Y, Jenks MA (2016) Cuticle lipids on heteromorphic leaves of Populus euphratica Oliv. growing in riparian habitats differing in available soil moisture. Physiol Plant 158:318–330

Xu Y, Zhan C, Huang B (2011) Heat shock proteins in association with heat tolerance in grasses. Int J Proteomics 2011:1–12

Xue D, Zhang X, Lu X, Chen G, Chen Z-H (2017) Molecular and evolutionary mechanisms of cuticular wax for plant drought tolerance. Front Plant Sci 8:621

Yang J, Sears RG, Gill BS, Paulsen GM (2002) Genotypic differences in utilization of assimilate sources during maturation of wheat under chronic heat and heat shock stresses. Euphytica 125(179):188

Zaid Chachar NA, Chachar QI, Chachar SMM, Chachar GA, Chachar S (2016) Identification of drought tolerant wheat genotypes under water deficit conditions. Int J Res 4(2):206–214

Zhang J, Srivastava V, McD J, Stewart JU (2016) Heat-tolerance in cotton is correlated with induced overexpression of heat-shock factors, heat-shock proteins, and general stress response genes. J Cotton Sci 20:253–262

Part III
Improve Crop Adaptability and Stability to Climate Change and Modern Technology

Chapter 4
Approaches in Wheat to Mitigate Impact of Climate Change

Abstract This chapter highlights on approaches in wheat to mitigate impact of climate change on crop production and to enhance sustainable agriculture in Egypt, include performance of wheat genotypes in response to environmental changes, models to measure adaptability and yield stability. Approaches should be advanced to manage with the climate change for alleviating the harmful effects of environmental stress on wheat production through several practical options such as releasing new cultivars more adapted and stable. Particular attention is given to role of biotechnology in improving heat, water stress and salinity tolerance by using molecular markers and gene transfer technologies. Also, how can measure sensitivity of wheat genotypes to abiotic stresses, beside the suggested mitigation agricultural techniques and include the use of tolerant cultivars, suitable planting dates, fertilizer and manure managements, irrigation systems and others to mitigate environmental stress on wheat crop.

Keywords Wheat · Climate change · Genotype × environment interaction · Performance · Adaptability and stability · Biotechnology · Gene action · Sensitivity measurements · Mitigation

4.1 Introduction

Wheat is the main food grain crop, grown in a wide range of environmental conditions over an area of 218.28 million hectares worldwide with average production 3.49 metric tons/ha. gave total production 761.88 million metric tons. Despite of this significant growth, the world population in several parts is still suffering from food shortages due to inadequate availability of food grains. To encounter the future food demands imposed by overwhelming increasing population which is projected to reach nine billions in 2050, the world wheat production must continue to increase by 2% annually. So, increasing production per unit area appears to be one of the important factors for narrowing the gap between wheat production and consumption especially with the continuous increase in the world population as shown in Fig. 4.1, which will be accompanied by increased demand for wheat. In the meantime, in Egypt, the area of wheat cultivation is 1.37 million hectares gave total production 9.0 million metric

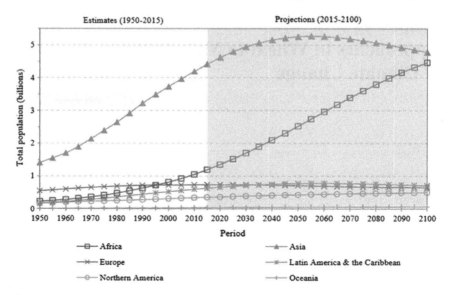

Fig. 4.1 Population by region estimates: 1950–2015 and medium variant projection, 2015–2100. *Source* United Nations, department of economic and social affairs, population division (2017). World population prospects: the 2017 revision, New York: Unite Nations

tons, while the Egyptian consumption of wheat grains is about 19 million tons. This wide gap between consumption and actual production forces the country to import10 million metric tons. Figure 4.2 shows production/yield quantities of wheat worldwide and in Egypt (FAOSTAT 2020).

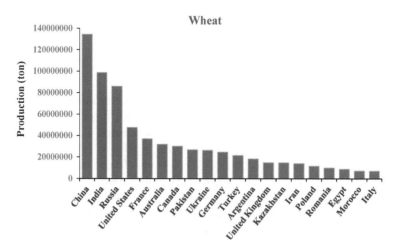

Fig. 4.2 Production/yield quantities of wheat worldwide (FAOSTAT 2020)

Wheat is the most widely grown crop in the world especially in the developing countries. Recently, the climatological extremes is predicted to have a general adverse effect on wheat production due to the damaging effect on plant development especially during anthesis stage. Therefore, the task of breeder is to screen out genotypes planted at different environments to enable selection of those genotypes, which are suitable for wider range of environments.

Environmental stresses are of the most environmental limitations to wheat yield. Drought and high temperature continues to be major challenges to agricultural scientists and plant breeders. By 2025, an assessed 1.8 billion people will live in regions plagued by water scarcity, with two-thirds of the world's population alive in water-stressed areas (Anonymous 2021). Also, productivity of wheat across the globale is influenced by salinity. Egypt is one of the countries that suffer salinity problems (Al-Naggar et al. 2015), and 33% of the cultivated land in Egypt is already salt affected (Yassin et al. 2019). Egypt has been facing climate change, which has impacted on agricultural production. In the future it is expected that climate change will affect Egypt in many ways, in particular with regard to water resources and agricultural production. Simulations of crop production under different climatic scenarios have shown decreasing wheat and maize yields (Mohsen 1999). Reports indicated to serious climate changes occurred in Egypt by detecting variability of land surface temperature (LST) over the last decade, with varied geomorphological characteristics and human stressors. Time series of LSTs were retrieved from satellite images acquired between 2003 and 2014, with a total of 276 images. Hereher (2016) showed that trend analysis suggests a temporal increase in LST over Egypt on the order of 0.3–1.06 °C/decade. Spatially, changes are more pronounced over southern areas of Egypt. Also, in some over-populated regions, LST increased by 1.54 °C/decade for the same study period. So, climate change is real and conceivable in Egypt. Accordingly, increasing wheat production under drought and heat stresses as abiotic stress circumstances has become inevitable through recent years. Meanwhile wheat production with optimal growth circumstances does not meet the requirements of over increasing Egyptian population. Heat tolerance is a complex trait and influence by different components. Also, the drought phenomenon is strong problems in developing countries of the world, where nearly 37% of wheat growing areas are semi-arid possess low moisture as a decisive factor for higher yield (Rajaram 2001). Qaseem et al. (2019) found that grain yield was decreased by 56.47%, 53.05% and 44.66% under combined of both heat + drought, heat and drought treatment, respectively. The combined of heat + drought treatment affects the grain yield by decreasing metabolism and mobilization of reserves to improving grains and leaves. Fatma, Farag (2019) showed that wheat grain yield was progressively decreased from normal irrigation, mild stress, moderate stress to severe stress with relative reduction 12.95, 30.07 and 43.55% through 1st season, as well as 13.02, 29.45 and 41.75% in the 2nd one respectively. Furthermore, El-khamissi (2018) showed that salinity stress reduced significantly germination percentage, growth parameters and photosynthetic pigments in the wheat genotypes.

So, great challenge for Egyptian wheat breeding programs is increasing yield without sacrificing its stability are the main objectives of the plant breeders and crop producer under the prevailing climatic changes (Hamada and Ibrahim 2016).

Therefore, this chapter focuses on the interaction between genotype × environment for economic traits, the performance, adaptability and stability of wheat genotypes in response to various environmental conditions. Also, the different methods for measuring the tolerance in wheat genotypes, genetic behavior of tolerance traits and the role of biotechnologies in this field. It also highlights on agricultural managements to mitigate environmental stress on wheat production.

4.2 Genotype × Environment Interaction and Its Relation to Climate Change

Genotype × environment "G × E" interaction could be defined as the differential response of the genotypes for a particular character under various environments (Campbell and Jones 2005). Application of G × E interaction is of major importance to identify stable genotypes for contradictory environments. G × E interactions are important in the improvement and valuation of wheat cultivars. Allard and Bradshaw (1964) revealed that the magnitude of G × E interaction increases with increasing differences between genotypes in different environments or changes in relative order of genotypes.

Phenotypic differences result from the combined influence of the genotype, the environment, and their interaction. Therefore, these variations are reflected in differences in the sensitivity of the genotypes to environmental changes, influencing the behavior and performance of the genotypes (Allard 1999). G × E interaction is known to be simple or complex nature (Cruz and Castoldi 1991) and that its size might be reduced the correlation between the phenotype and the genotype, affecting genetic variance; subsequently, affecting breeding parameters i.e. heritability and genetic gain from selection. In the light of such information and interpretations, Vencovsky and Barriga (1992) emphasize that it is not enough just to discover the presence of G × E interaction, but it is also important to consider their nature. Xie and Mosjidis (1996) showed that the nature of the interaction simple, or scalar, is due to the amounts of the differences of variability between the genotypes, and complex interaction which depends on the correlation of the genotypes in the environments.

Simple interaction does not cause changes in the ordering of the genotypes amongst environments, demonstrating the adaptation of genotypes to the broad range of environments. Thus, this simple pattern of interaction will allow genotypes to be recommended in a general way (Romagosa and Fox 1993). Conversely, complex interaction changes the ordering of the genotypes amongst environments, showing the presence of genotypes adapted to specific environments, i.e. limiting the recommendation to specific environments (Vencovsky et al. 2012).

The significant differential response of the genotypes to distinct environments obstructs selection process of high-productive and stable cultivars (Cooper and De Lacy 1994), which limits the ability of plant breeders to recommend the cultivation of genotype in specific environments. Therefore, the breeder must conduct the experiments in multi-locations and multi-years as possible, to describe the G × E interaction (Farias et al. 1996). Thus, selection based on the average yield of the genotypes in a particular environment is not operational (Hopkins et al. 1995).

So, the G × E interaction can be minimized by growing particular cultivars for each environment or cultivars with wide adaptability and good stability. Whereas, the G × E interaction can be reduced also by stratification of the production site into areas with similar environmental features so that the interaction becomes unimportant (Cruz and Carneiro 2006). The second alternative is to grow diverse crops (Ramalho et al. 1993). Thus, this information is necessary to guide breeders of the crop and allow estimates of more reliable genetic parameters.

Genotype × environment interaction indicates the impact of environments on the expression of grain yield in wheat genotypes. Genotype × environment interaction analysis presented a certain degree of variation among genotypes; some genotypes displayed wide adaptation while other exhibited specific adaptation either to favorable or unfavorable environments. Significant genotype vs. environment interaction (G × E) is predictable as a result of geographic diversity and differences in agricultural practices in Egypt. Due to the extension and geographical variations of the growing areas in Egypt, the edaphic and climatic diversity and the variations in agricultural managements, significant genotype × environment (G × E) interaction is expected. Studies on Egypt have shown the presence of the G × E interaction in experiments for genotypes or cultivars evaluation for different traits. Previous reports of Aly and Awaad (2002), Tawfelis (2006), Anwar et al. (2007), Hamam and Khaled (2009), Al-Otayk (2010), Arain et al. (2011), El-Ameen (2012), Abd El-Shafi et al. (2014), ElBasyoni (2018) and Awaad (2021) detected significant G × E interaction on grain yield and its contributing traits.

In light of these considerations, with the assumption that the greatness of the genotype × environment interaction has a significant influence on phenotype expression, obstructing the effort of the breeder in the selection it is of importance to measure G × E interaction and partitioning to its portions. Studies of genotype × environment (G × E) interaction makes genotypic evaluation a complicated process because of the different response of genotypes under different locations or years. Variable response of the cultivars to heat or drought stress across all traits varied under different environmental situations. Obviously, grain yield is the most important trait that might solely determine the success of a plant breeder. Studies have shown that stability analyses according to various measures can result in better identification of stable genotypes (Akter and Islam 2017 and Iqbal et al. 2017 and Fadia Chairi et al. 2020). Meanwhile, Fadia Chairi et al. (2020) showed that the most important environmental factors affecting the G × E interaction of wheat yield were the maximum and the mean temperature during the whole crop cycle. An improvement in genetic yield was detected in warm environments and under optimal water conditions. In Egypt, under eighteen diverse environments, Aly and Awaad (2002) evaluated fifteen local and imported bread wheat genotypes for some economic characters. Pooled analyses of variance indicated highly significant differences among wheat genotypes (G), seasons (S), locations (L) and sowing dates (D), as well as their first-order interactions between genotypes and the environmental factors for grain yield/m^2, grain yield/main spike, spike length and grain protein content with a few exceptions, and only second-order (G $\times$ L $\times$ D) interaction for grain protein content. Location effect accounted for most part of the total variation on the studied characters, followed by seasonal and the genotype effects, however, sowing dates had little effect in this respect. Since the contribution of these items were 50.59% for locations, 25.80% for seasons, 12.24% for genotypes and 11.37% for sowing dates from the total variance of wheat grain yield/m^2 (Table 4.1).

The integration with the previous study with the same genetic materials under the eighteen diverse environments, Awaad and Aly (2002) found that combined analyses of variance indicated highly significant differences among seasons, locations, sowing dates and wheat genotypes as well as their first, second and third-order interactions for days to 50% heading, flag leaf area and grain yield/fed. in most cases. This result provide evidence for the necessity of testing studied genotypes in multiple environments to identify the most adapted and stable genotypes to environmental stresses (Table 4.2).

Moreover, stability analysis of variance revealed highly significant G $\times$ E-"linear" for all the studied characters. The G $\times$ E-"linear" interaction also was significant when tested against the pooled deviation for days to 50% heading, flag leaf area and grain yield/fed. (Table 4.3). This result designates that differences in linear response among genotypes across environments had occurred, and the linear regression and the deviation from linearity were the main components for the existing differences.

Under Egyptian conditions, for evaluation heat stress and other environments, twelve genotypes of bread wheat (*Triticum aestivum* L.) were evaluated under twelve environments. The twelve environments were the combinations of three sowing dates, i.e., 20th November (recommended sowing date), 10th December (moderate late sowing date) and 30th December (late sowing date "heat stress"), two locations at two Agriculture Research Stations, 2015 i.e., Kafr Al-Hamam (El-Sharkia) and Sids (Bani-Sweif) during two seasons 2010/2011 and 2011/2012 by Omnya Elmoselhy (). Results showed that mean squares due to seasons (S), locations (L), sowing dates (D), genotypes (G) and their first, second and third interactions were highly significant for earliness characters i.e. days to heading, growing degree days, days to maturity as well as morph- physiological characters *i.e.* flag leaf area, flag leaf chlorophyll content and canopy temperature depression (Table 4.4) in most cases. Highly significant differences were recorded for genotypes (G) and their first-order interaction of (S $\times$ G), (L $\times$ G) and (D $\times$ G). Also, highly significant second (S $\times$ L $\times$ G), (L $\times$ D $\times$ G) and (L $\times$ D $\times$ G) as well as third (S $\times$ L $\times$ D $\times$ G) order interactions have been observed for earliness and physiological characters, implying different response of genotypes over seasons, locations and sowing dates.

Moreover, highly significant mean squares due to wheat genotypes for grain yield and its components were registered by Omnya Elmoselhy et al. (2015), revealing

Table 4.1 Pooled analyses and partitioning of variance for grain yield/m^2, grain yield/main spike, spike length and grain protein content of fifteen wheat genotypes under three sowing dates during two seasons in three locations (Aly and Awaad 2002)

Source of variation S.O.V	d.f	Grain yield/m^2 (kg)	Grain yield/main spike (g)	Spike length (cm)	Grain protein content (%)
Seasons (S)	1	0.3324**	2.068***	100.337***	8.954**
Location (L)	2	0.6402**	15.333***	17.786***	47.963**
S × L	2	0.0195*	0.046	10.507***	0.108
Reps in (S × L) combined	12	0.0390	1.954	0.605	1.762
Sowing dates (D)	2	0.1217**	0.352**	13.453**	1.848*
S × D	4	0.0040	0.148	0.685	1.056
L × D	4	0.0350**	0.106	1.004	1.673*
S × L × D	4	0.0032	0.212	2.308*	0.123
Genotypes (G)	14	0.138**	1.878***	13.784***	5.884**
G × S	14	0.0026	0.829**	4.855***	2.879***
G × L	28	0.0265***	0.679***	2.586***	2.928***
G × D	28	0.0177**	0.298***	3.502**	4.423**
G × S × L	28	0.0044	0.087	0.492	0.055
G × S × D	28	0.0023	0.039	0.574	0.088
G × L × D	56	0.0041	0.172	0.726	0.797*
G × S × L × D	56	0.0028	0.055	0.527	0.059
Error	528	0.0054	0.159	0.886	0.537

(continued)

Table 4.1 (continued)

Source of variation S.O.V	d.f	Grain yield/m² (kg)	Grain yield/main spike (g)	Spike length (cm)	Grain protein content (%)
Contribution of the factors (%)					
Seasons	25.80		12.13	63.32	14.05
Locations	50.59		71.38	15.09	64.25
Sowing dates	11.37		3.51	10.04	7.33
Genotypes	12.24		12.98	11.55	14.37

*, **Denote significant at 5 and 1% levels of probability, respectively

Table 4.2 Pooled analyses of variance for grain yield and its contributing characters of fifteen wheat genotypes under three sow dates during two seasons in three locations (Awaad and Aly 2002)

Source of variation	d.f	Days to 50% heading	Flag leaf area (cm^2)	Grain yield (ard° / fed.')
S.O.V.				
Seasons (S)	1	954.529**	8122.126**	400.900**
Locations (L)	2	6426.265**	39634.430**	1542.325**
Seasons × locations (S × L)	2	2789.609**	2453.070**	221.067**
Reps in (S × L) combined	12	5.556	21.209	1.396
(D)	2	7364.429**	4721.335**	658.695**
S × D	2	753.169**	36.494	4.205**
L × D	4	1521.965**	155.281**	26.789**
S × L × D	4	624.809**	72.845**	15.089**
Genotypes (G)	14	473.248**	432.137**	42.293**
G × S	14	65.899**	112.397**	6.807**
G × L	28	136.893**	250.832**	27.019**
G × D	28	91.449**	75.570**	14.126**
G × S × L	28	26.183**	93.505**	3.098**
G × S × D	28	26.499**	58.045**	6.111**

(continued)

Table 4.2 (continued)

Source of variation S.O.V.	d.f	Days to 50% heading	Flag leaf area (cm^2)	Grain yield (ard^o / fed.$^.$)
G × L × D	56	38.037**	66.162**	13.958**
G × S × L × D	56	21.328**	30.906**	6.348**
Error	528	3.021	12.343	0.688

*, **Denote significant at 5 and 1% levels of probability, respectively
o Ardab = 150 kg ∎ Feddan = 4200 m^2

Table 4.3 Mean squares of stability analysis for wheat grain yield and its contributing characters (Awaad and Aly 2002)

S.O.V.	d.f	Days to 50% heading	Flag leaf area (cm^2)	Grain yield (ard^o/fed.$^.$)
Genotypes	14	473.248**	432.137**	42.293**
E + (G x E)	255	74.267**	165.488***	10.537**
E-"Linear"	1	14920.057**	34914.891**	1805.549**
G x E –"Linear"	14	26.554**	64.421**	9.963**
Pooled deviation	240	15.193**	26.594**	3.091**
Pooled error	504	1.021	4.09	0.229

**Denote significant at 1% level of probability, o Ardab = 150 kg ■ Feddan = 4200 m^2

Table 4.4 Combined analyses of variance over seasons (S), locations (L), sowing dates and genotypes (G) for earliness components and morphophsiological characters (Omnya Elmoselhy 2015)

S.O.V	d.f	Days to heading	Growing degree days	Days to maturity	Flag leaf area (cm^2)	Flag leaf chlorophyll content	Canopy temperature depression
Season (S)	1	29,601.33**	110,787.93**	1993.48**	315.65**	23.79	8.72*
Error	4	1.84	3692.37	10.22	7.81	11.97	0.691
Location (L)	1	25.04*	1688.55	855.70**	0.01	3.49	127.38**
S × L	1	0.10	54.74	592.68**	185.04**	19.61	23.79**
Error	4	1.65	4034.312	6.52	0.27	6.13	0.57
Sowing dates (D)	2	2402.40**	338,270.06**	10,955.78**	35.28.53**	410.56**	23.76**
S × D	2	244.36**	128,509.49**	78.27**	22.3	24.33*	2.09
L × D	2	58.12**	34,581.13**	110.36**	252.49**	68.09**	9.95**
S × L × D	2	79.37**	11,389.15*	83.82**	55.51*	25.99*	0.26
Error	16	0.84	2810.55	2.15	10.04	5.73	0.69
Genotypes (G)	11	168.76**	30,749.74**	88.23**	22,175.21**	129.38**	1.50**
S × G	11	31.29**	7151.26**	18.84**	40.81**	5.35*	0.19
L × G	11	2.99**	2965.88	4.89**	131.85**	4.2	0.34
S × L × G	11	5.12**	3506.11	9.44**	54.06**	5.18*	0.24
D × G	22	20.23**	4737.16**	2.19**	34.33**	3.03	0.49*
S × D × G	22	12.54**	4964.54**	2.09**	44.64**	4.47	0.12
L × D × G	22	1.19*	2952.71	3.00**	38.16**	2.72	0.40
S × L × D × G	22	2.82**	2370.38	1.54*	28.16**	3.13	0.32
Error	264	0.74	2453.93	0.87	7.45	2.42	0.28

*and**Significant at 0.05 and 0.01 probability, respectively

that wheat genotypes were genetically different for genes controlling yield char-acters. The G × E interaction was further partitioned into linear and non-linear (pooled deviation) components. Highly significant environment + (genotype × envi-ronment) component and environment "linear" mean squares were recorded for all grain yield and its components, indicating that the studied characters were highly influenced by the combination of environmental components (seasons, locations and sowing dates). Highly significant genotype × environment "linear" interactions were shown for number of spikes/m^2, number of grains/spike, 1000-grain weight and grain yield/fed. Suggesting that wheat genotypes differed in their response to the environmental variation. The (G × E linear) was highly significant when tested against pooled deviation for number of spikes/m^2, 1000-grain weight and grain yield (ard./fed.), advising that the linear regression and the deviation from linearity were the main components for the variances. The non-linear responses as measured by pooled deviations from regressions were highly significant for number of spikes/m^2, number of grains/spike and 1000-grain weight, indicating that differences in linear response among genotypes across environments did not account for all the G × E interaction effects. Therefore, the fluctuation in performance of genotypes grown in various environments was not fully predictable. Results provide evidence for the necessity of evaluating wheat genotypes under multi-environments in order to iden-tify the best genetic make-up to be grown under particular environments (Table 4.5).

Under El-Kattara (stress environment) and Ghazalla (favorable environment) of Egypt, Ali and Abdul-Hamid (2017) conducted several field experiments to screen 38 wheat genotypes. The combined analyses of variance for grain yield (ard./fed.) showed highly significant differences among environments, years, irrigation systems and genotypes, reflecting the wide genetic diversity between genotypes in their response to tested environments. Significant differences for the first order inter-action of genotypes × environments, genotypes × years, years × irrigation items were detected in each of the drip and surface irrigation systems. The second order (genotypes × years × irrigation) interaction was highly significant for grain yield (ard. /fed.). These results reflected the importance of environmental factors of each year and water irrigation levels on the performance of the tested wheat genotypes.

Climate change and global warming cause a rise of drought effects on wheat yield all over the world. Therefore, improving wheat drought tolerant genotypes is vital, mainly under current water shortage. In this connection, Naglaa Qabil (2017) involved five diverse parental wheat genotypes (Misr 1, Gemmeiza 9, Gemmeiza 11, Line 1 and Line 2) in a half diallel cross fashion under normal irrigation and drought stress. The analysis of variance showed significant differences among genotypes, parents and their F$_1$ crosses for days to heading, flag leaf area, chloro-phyll content (%), number of spikelets/spike, spike length, spike density, number of spikes/plant, number of grains/spike, 1000-grain weight and grain yield/ plant under both conditions, indicating the attendance of sufficient genetic variability among genotypes.

Under two locations, i.e., Elbostan and Elkhazan, Egypt represent newly reclaimed sandy soil and the Nile Delta soil (clay), respectively during three successive seasons

Table 4.5 Joint regression analysis of variance for grain yield and its components of twelve bread wheat genotypes over twelve environments (Omnya Elmoselhy et al. 2015)

S.O.V	d.f	Number of spikes/m²	Number of grains/spike	1000- grain weight	Grain yield (ard°/fed.˙)
Genotypes	11	7363.67**	115.75**	82.12**	11.86**
Environment + (Genotype × environment)	132	1677.44**	40.57**	11.65**	12.32**
Environment (linear)	1	136,633.08**	4022.63**	884.23**	1376.33**
Genotype × environment (linear)	11	1685.35**	10.75**	10.40*	4.02**
Pooled deviation	120	552.08**	10.12**	4.49**	1.71
Pooled error	264	114.71	2.61	1.24	1.92

* and **Significant at 0.05 and 0.01 probability respectively ○ Ardab = 150 kg ■ Feddan = 4200 m²

Table 4.6 The combined analyses of variance over two years, two locations, three planting densities and eleven wheat genotypes for studied traits (Awaad 2021)

S.O.V	d.f	Days to 50% heading		Grain protein content %		Grain yield (ard/fed.)	
		SS	MS	SS	MS	SS	MS
Environments (E)	11	930.78	84.62**	88.30	8.03**	3458.42	314.40**
Reps/Env	24	22.46	0.94**	7.02	0.29	399.99	16.67**
Years (Y)	1	7.17	7.17**	0.68	0.68**	8.60	8.60
Y × L	1	7.66	7.66**	1.41	1.41**	2.55	2.55
Loc. (L)	1	780.45	780.45**	54.57	54.57**	3086.51	3086.51**
Planting density (D)	2	64.38	32.19**	8.25	4.13**	253.21	126.61**
Y × D	2	8.01	4.00**	0.66	0.33	36.16	18.08
L × D	2	50.04	25.02**	21.44	10.72**	69.55	34.78**
Y × L × D	2	13.08	6.54**	1.29	0.65**	1.83	0.92
Genotypes (G)	10	776.60	77.66**	23.27	2.33**	523.78	52.38**
G × E	110	219.33	1.99**	43.11	0.39**	1874.16	17.04**
G × Y	10	9.04	0.90	3.33	0.33**	132.48	13.25**
G × L	10	41.07	4.11**	8.90	0.89**	444.84	44.48**
G × D	20	61.95	3.10**	13.64	0.68**	358.68	17.93**
G × Y × L	10	18.97	1.90**	1.20	0.12	142.31	14.23*
G × Y × D	20	24.56	1.23	5.04	0.25*	197.24	9.86
G × L × D	20	45.76	2.29**	5.63	0.28*	317.98	15.90**
G × Y × L × D	20	17.97	0.90	5.37	0.27**	280.63	14.03**
Error	240	195.09	0.81	32.86	0.14	1775.58	7.40

*, **Significant at 0.05 and 0.01 levels of probability, respectively, Ardab = 150 kg, Fddan = 4200 m^2

2014/2015, 2015/2016, and 2016/2017 under recommended and late sowing dates, ElBasyoni (2018) verified ten Egyptian commercial and broadly cultivated wheat cultivars. Days to flowering, leaf chlorophyll content, canopy temperature, leaf area, grain filling duration, plant height, grain yield showed significant two-way and three-way interactions (environments × cultivars, environments × sowing dates, sowing dates × cultivars and environments × sowing dates × cultivars) effects. Concerning leaf and stem rust scores, environments × cultivars and sowing dates × cultivars interactions were found to be highly significant. This result indicates that the genotypes have varied response to environmental variables. However, environments × sowing dates × cultivars interaction was not statistically significant.

To determine whether it is necessary to test genotypes in multiple environments to ascertain the most adapted and stable genotypes to environmental stresses, Awaad (2021) computed the combined analyses of variance for days to 50% heading, grain protein content and grain yield (ard./fed.) as given in Table (6). Results revealed highly significant variances among environments for the abovementioned traits, hereby the environments were diverse. Furthermore, highly significant effect owing to years (Y) was achieved for days to 50% heading and grain protein content, reflects the extensive differences in climatic circumstances prevailing throughout the growing seasons. The main effect of locations (L) and planting densities (D) was highly significant for all studied traits. Genotypes (G) had highly significant variances of all traits, suggesting that the evaluated genotypes were genetically different for genes governing these traits. Also, highly significant G × E interactions were recorded for the studied traits, provide evidence that the deliberate bread wheat genotypes are varied in their reaction to the environmental circumstances. So, it is necessary to determine the grade of stability for each genotype. The first order interaction of (Y × L) varied significantly for days to 50% heading and grain protein content, demonstrating the different impacts of climatic conditions on locations. Also, significant interactions between (L × D) were found for all traits, while only (Y × D) had significant interaction for days to 50% heading. The interaction components (G × L) and (G × D) were accounted for the most of total G × E interaction of the studied traits, indicating that locations and planting densities had the main effect on the relative genotypic potential of the traits. Moreover, significant interactions between genotypes and years (G × Y) were detected for both grain protein content and grain yield (ard./fed.). The second order (Y × L × D) interaction had highly significant effect for days to 50% heading and grain protein content; (G × Y × L) for days to 50% heading and grain yield (ard./fed.), (G × Y × D) for grain protein content and (G × L × D) for days to 50% heading, grain protein content and grain yield (ard. /fed.). Moreover, the third order (G × Y × L × D) interaction was significant regarding grain protein content and grain yield (ard./fed.). These results reflected the importance of environmental factors of each year, location and planting density on the performance expression of wheat genotypes concerning the foregoing traits.

Continuously, joint regression analyses of variance was measured by Awaad and discovered highly significant changes among genotypes (G), environments (E) and the G × E interaction for days to 50% heading, protein content % and grain yield (ard./fed.) (Table 4.7). This showed the existence of genetic and environmental differences regarding wheat genotypes for the studied traits. Also, (E + G × E) had highly significant effects on all traits. Mean squares attributable to environment (linear) were highly significant for all traits, representing that variances existed between environments and indicated predictable component shared G × E interaction with unpredictable. The linear interaction (G × E linear) was highly significant when verified against pooled deviation for all traits, displaying genetic variances among genotypes for their regression on the environmental index, so it could be continued in the stability analysis (Eberhart and Russell 1966) for these traits. The non-linear responses as estimated by pooled deviations from regressions were significant, showing that differences in linear response among genotypes across environments did not account for all the G × E interaction effects. Therefore, the variability in the performance of genotypes evaluated under different environments was fully predictable.

Table 4.7 Joint regression analysis of variance over twelve environments for eleven wheat genotypes for the studied traits (Awaad 2021)

S.O.V	d.f	Days to 50% heading		Grain protein content %		Grain yield (ardab/fed.)	
		SS	MS	SS	MS	SS	MS
Model	131	642.23	4.90**	51.56	0.39**	1952.12	14.90**
Genotype (G)	10	258.87	25.89**	7.76	0.78**	174.59	17.46**
Environment (E)	11	310.26	28.21**	29.43	2.68**	1152.81	104.80**
G × E	110	73.11	0.66**	14.37	0.13**	624.72	5.68**
E + G × E	121	383.37	3.17**	43.80	0.36**	1777.53	14.69**
Environment (linear)	1	310.26	310.26**	29.43	29.43**	1152.81	1152.81**
G × E (linear)	10	16.17	1.62**	1.47	0.15**	132.65	13.26**
Pooled deviation	110	56.94	0.52*	12.91	0.12**	492.07	4.47**
Pooled Error	240	65.03	0.27	10.95	0.05	591.86	2.47

*, **Significant at 0.05 and 0.01 levels of probability, respectively

o Ardab = 150 kg ∎ Feddan = 4200 m^2

Measuring independent effects of drought, heat and combined heat + drought during reproductive stage on wheat yield and their relevant traits were deliberated on collection of wheat germplasm. Qaseem et al. (2019) showed that the analysis of variance revealed highly significant mean squares due to genotype, treatment, environment, genotype × treatment, genotype × environment, treatment × environment and genotype × treatment × environment on awn length, leaf area, peduncle extrusion, plant height, spike length, spikelets/spike, tillers/plant, grains/spike, biomass and grain yield. On light of these information's, it is of essential to assessment the performance of wheat genotypes, and their adaptation and stability under different environments to determine which ones are more suitable to be grown under certain environmental conditions.

4.3 Performance of Wheat Genotypes in Response to Environmental Changes

It should be noted that, some genotypes performed well under optimum conditions but not under stress and vice-versa, while some of the wheat accessions performed well under wide range of environmental conditions. Previous results indicate that the genotypes used in various studies contain different combinations of genes governing their response to sowing dates and tolerance to environmental stresses. Mean performance is very important in order to determine the differences between genotypes and comparative performance of wheat genotypes for earliness, grain yield and quality characters under various environmental conditions. The ultimate goal of this part is to demonstrate the performance of wheat varieties and strains on a large scale in Egypt and the effect of global warming and drought stress on wheat grain yield. Thus, gave decision-makers in this region an idea of the impact of stress on wheat production and how to avoid it.

4.3.1 Earliness Characteristics

Earliness characteristics represent an important role in adaptation of varieties and avoiding biotic and abiotic stress conditions. Mean performance of wheat genotypes is an important aspect for obtaining genetic progress on phenotypic selection. In this concern, genetic diversity among wheat genotypes under different environments has been recorded in respect to earliness characters and showed that delaying sowing date reduced number of days to heading and maturity as recorded by Irfaq et al. (2005), Sial et al. (2005), Tawfelis (2006), Omnya Elmoselhy (2015) and Qaseem et al. (2019).

Four wheat genotypes were verified under four sowing dates (10th Nov., 21st Nov., 8th Dec. and 22nd Dec.) to examination their performance against heat stress.

Irfaq et al. (2005) indicated highly significant differences in the mean values for earliness traits. They observed decrease in days to heading for all the genotypes with delay in sowing. The mean values for days to heading were fluctuated between 90.8 to118.5, 91.8 to 119, 91.5 to 120.3 and 91.3 to 120.3 for genotypes CT-00231, CT-99187, B-92 and Salem 2000, respectively. A gradual decrease in days to maturity was observed when sowing was delayed. The average between the mean values for days to maturity was 156.8 to128.8, 155.5 to 128.5, 158.5 to 128.8 and 157.5 to 129 for genotypes CT-00231, CT-99187, B-92 and Salem 2000, respectively. Whereas, the percent decrease in days to maturity for the genotypes was observed to be 18, 17, 19 and 18%, respectively. Highly significant differences in earliness traits were due to sowing dates as well as genotypes. Moreover, twelve wheat genotypes were evaluated at two levels of sowing dates i.e. normal sowing date (8th November) and late sowing (11th December) by Sial et al. (2005). Significant decline in the mean values was observed for days to heading and days to maturity under late planting. Number of days necessary for heading under normal sowing varied from 68 and 90. This span was decreased to 80 days (upper limit) under late planting. Wheat genotype Khan 95 was recognized as possessed inherent earliness genes under normal and late planting and appears to have tolerance ability. This decrease was more pronounced in genotypes that needed a greater number of days to heading under normal cultivation. Naturally, the number of days until heading is shortened due to late thermal stress. Tawfelis (2006) detected highly significant differences between years and locations for days to heading and days to physiological maturity. Wheat genotypes showed different responses to environments. Delaying sowing date reduced number of days to heading and days to physiological maturity by an average 13.65 and 13.47%, respectively. Moreover, to select genotypes with tolerance to possible terminal heat stress caused by late sowing, Khan et al. (2007) grown nineteen wheat genotypes under two sowing dates 17th Nov. and 20th Dec. Highly significant differences in mean values were observed for all the genotypes under the influence of different sowing dates for days to heading and days to maturity under normal sowing date. The average values of days to heading of both normal and late sowing dates revealed that minimum days to heading were recorded for wheat genotypes CT-01222, CT-01217 and CT-01008 valued 110, 111 day and 111 days, respectively, representing earliness of the genotypes. Whereas, maximum days to heading were recorded for the genotypes CT-01250, CT-01030 and CT-01079 reached 118, 117 and 116 day, respectively. Regarding days to maturity, minimum days to maturity were recorded for CT-01264, CT-01008 and CT-01382 with 147, 149 and 150 day, respectively, representing earliness of the genotypes. Whereas, maximum days to maturity were recorded for the genotypes CT-01250, CT-01079 and CT-01183 valued 153, 153 and 152 day, respectively.

Assessment the heat tolerance of twelve wheat genotypes under several environments was performed under Egyptian conditions by Hamam and Khaled (2009). Wheat genotypes were sown under two locations (Sohag and Assiut) at two sowing dates 14th Nov. (favorable) and 26th Dec. (heat stress) during winter seasons of 2005/2006, 2006/2007 and 2007/2008. The average number of days to heading at late sowing date was reduced by 9 days. Days to heading of the different genotypes ranged from 81.88 to 104.52 days for Johara 19 and TRI 18,686, respectively, with an overall average 90.47 days. The earliest genotypes were Sahel 1, Bocro-4//kauz"S", TRI 3399 and TRI 2586 about (72.67 days) at Sohag location in the second sowing date. Under Sohag location, early wheat meet high temperature stress at germination stage, however late sown is influenced by high temperature at both reproductive and grain filling stages (>21.37 until 39.55 °C), which leads to a lack of grain yield.

Four heat stress tolerant and three sensitive spring wheat parental genotypes were crossed to produce F_1 progenies of 7 × 7 diallel fashion evaluated under normal and heat stress conditions by Irshad et al. (2012). Genotypic differences were found to be highly significant for days to heading under both temperature circumstances. Days to heading showed a reduction of 18.26% under heat stress conditions. Among the parents, maximum reduction of 23.59% was occurred in the parent Punjab-96. Similarity, in case of F_1 hybrids the reduction ranged from 9.59% (93 T347 × V00183) to 22.46% (Punjab 96 × V 00,183). Moreover, twenty-eight CIMMYT wheat lines and two cheeks at 13 locations across South Asia and two environments in Mexico were evaluated by Mondal et al. (2013). Each location was classified by mega environments (ME): ME1 being the temperate irrigated locations with terminal high temperature stress and ME5 as warm, tropical, irrigated locations. Number of days to heading ranged from 76 to 83 days for all entries across locations with a mean of 79 days. Number of days to heading in the ME5 locations ranged from 57 to 76 days, whereas in the ME1 locations it varied from 81 to 106 days. Under Egyptian conditions, Omnya El-Moselhy (2015) showed that the temperatures from November to March are relatively suitable for wheat crop planted in November while low and high temperature during emergence and grain formation stages respectively are not suitable for crop planted in late December. Days to heading tended to delay from D1 (20th November as optimum sowing date), to 10th December as moderate late sowing date) in almost wheat genotypes under two locations (Kafr Al-Hamam, El-Sharkia and Sids, Bani-Sweif). Whereas, another trend was observed from D2 (10th December as moderate late sowing date) to D3 (30th December as late sowing date) which reduced by delaying sowing date by about 4.1, 8.4 and 6.26 days for first, second and overall seasons, respectively resulted from fewer growing degree days. It is interest to note that, days to heading of bread wheat genotypes in 2nd season was later in heading by 16.56 days than in 1st season at two locations overall three sowing dates resulted in low temperatures degrees in the 2nd season from Nov. up to March causing slow in early vegetative periods of growth which resulted in delaying heading dates of most genotypes rather than 1st season. The Egyptian cultivars Sakha 93, Gemmeiza 11 and Sids 12 exhibited good levels of earliness valued 90.86, 91.17 and 91.75 days and accumulated 957.53, 961.22 and 963.06 growing degree days GDD, respectively, of them Gemmeiza 11 and Sids 12 had high yield potentiality, whereas, genotype Line 16 was the latest one (97.63 days) with GDD (1034.37) despite that it has high yield potential. Delaying sowing date from 10th Dec. to 30th Dec. significantly shortened time to heading and maturity of the studied wheat genotypes. Early in heading and maturity might be due to high temperature degree which reached about 35°C in 1st season and 37 °C in the 2nd season. Delaying sowing date reduced number of days to maturity by about 7.14, 10.07 and 8.61 days for first, second and overall seasons, respectively. Genotypes Gemmeiza 11, Sids 12 were the earliest in maturity with good level of grain yield overall environments.

Under normal irrigation and drought stress conditions, mean performance of parental genotypes and their crosses for days to heading were assessed by Naglaa Qabil (2017). Performance for days to heading under normal irrigation showed an increase compared to drought stress. The imported Line 1 was the earliest one under both situations. The good level of earliness obvious in Line 1 was reflected in the performance of their F1 progenies (Line 1 × Misr 1) and (Line 1 × Line 2) under both conditions. While, the local wheat genotype Gemmeiza 9 was the latest and their F1 crosses (Gemmeiza 9 × Gemmeiza 11), (Line 2 × Gemmeiza 9) and (Misr 1 × Gemmeiza 9) under normal and drought conditions. Therefore, genes controlling early heading have been transferred from the parents to their F1 progeny. At various environments, ElBasyoni (2018) tested ten wheat genotypes for earliness. Heat stress had a significant adverse impact on grain filling duration and number of days to flowering, while it raised the prevalence and severity of leaf and stem rust which contributed to overall yield losses of about 40%. Cultivars Gemmeiza10 and Gemmeiza12 flowered earlier than the other ones under the recommended and late sown conditions. Recommended sowing date prolonged the number of days to flowering for all cultivars across environments. Also, highly significant variance was perceived amongst the environments and deliberated cultivars, and the amount of differences amongst cultivars was adequate to offer a scope to describe the effect of terminal heat stress. The late sowing condition effect the resistance to leaf rust (leaf rust scores increased from 0.34 to 0.73) and stem rust (stem rust scores increased from 0.36 to 0.79). Furthermore, the late sowing condition shortened the grain filling period from 32.5 days to 25.5 days. Cultivar "Gemmeiza12" produced the highest grain yield (8.8 ton/hectare) under the recommended sowing date and flowered earlier than other cultivars under the recommended and late sown conditions. Awaad (2021) assessment means performance of days to heading for eleven genotypes under different environments (Table 4.8). Number of days to 50% heading was reduced from D1 (350 seeds/m^2), D2 (400 seeds/m^2) to D3 (450 seeds/m^2) under Ghazaleh and El-Khattara regions in the first and second seasons. Bread wheat genotypes Line 1, Line 4, Giza 168, Sakha 94 and Misr 1 were the earliest, among them Line 1, Line 4 and Giza 168 exhibited a respectable level of grain yield overall environments compared with the other genotypes. The most earliness wheat genotypes were Line 1 with the value of 79.5 days in the 1st season and 77.33 days in D3 in the 2nd one; Giza 168 with values of 79.33 and 77.40 days in D1 in the 2nd season, under Ghazaleh and El-Khattara regions, respectively. Under El-Khattara region as sandy soil stress environment, Line 4 was the early with 78.53 days in D3 in the 1st season and the 2nd season also in D1, and D3 valued 78.67 day.

Qaseem et al. (2019) showed that at drought stress, days to anthesis and days to maturity were reduced by 10% and 14%, whereas at heat stress they were reduced by 16% and 20%, respectively. Combined drought + heat stress caused 25% reduction in days to anthesis and 31% reduction in days to maturity. At drought stress, genotype SAWSN_3134 was the longest in the number of days to anthesis, however EBWYT_523 and SRSN_6017 had maximum value for days to anthesis at

Table 4.8 Mean performance for eleven wheat genotypes under different environments for days to 50% heading (Awaad 2021)

Genotype	Ghazaleh						El-Khattara						Combined
	2014/2015			2015/2016			2014/2015			2015/2016			
	D1	D2	D3	D1	D2	D3	D1	D2	D3	D1	D2	D3	
Line 1	82.53	80.17	79.50	82.33	80.21	78.87	79.00	77.70	77.47	78.07	77.67	77.33	79.24
Line 2	83.53	82.67	82.00	83.00	82.40	81.73	80.77	79.83	78.77	79.70	81.37	79.67	81.29
Line 3	86.07	85.03	83.40	85.33	83.40	83.70	81.77	81.03	80.50	80.67	83.00	81.33	82.94
Line 4	83.07	82.10	81.07	83.10	81.17	80.43	80.07	79.80	78.53	78.67	80.73	78.67	80.62
Line 5	85.30	84.67	83.70	84.00	84.37	83.87	81.40	81.23	80.47	78.33	81.67	81.40	82.53
Line 6	84.47	82.33	82.03	83.67	80.40	80.33	79.07	82.17	78.70	79.00	80.73	79.07	81.00
Line 7	87.73	88.67	85.43	88.60	86.00	84.90	82.93	83.17	81.60	83.68	84.67	82.90	85.00
Line 8	83.80	83.67	83.10	83.67	82.77	82.83	80.77	79.40	80.67	80.17	81.00	80.70	81.88
Sakha 94	81.27	82.37	80.50	80.83	81.33	82.13	78.83	80.33	79.00	78.66	80.67	79.27	80.43
Giza 168	80.27	80.80	79.67	79.33	80.00	80.43	77.67	79.00	78.03	77.40	78.00	78.17	79.23
Misr 1	83.40	82.97	82.60	83.87	82.70	81.93	80.40	79.13	78.73	78.80	78.83	77.27	80.89
Mean	83.77	83.22	82.09	83.43	82.25	81.98	80.24	80.25	79.31	79.56	80.76	79.53	81.26
L.S.D. 0.05	1.59	1.15	1.18	2.06	1.90	1.28	1.66	1.38	1.11	1.80	1.55	1.43	0.42

D1, D2 and D3 refer to 350 seeds/m^2, 400 seeds/m^2 and 450 seeds/m^2

heat and combined heat + drought stress. Genotype SRSN_6008 at drought stress, ESWYT_110 at heat stress and CHAKWAL-50 at combined drought + heat stress were early maturing ones. These genotypes were shortest in days to maturity among the tested germplasm. Furthermore, in respect to salinity stress, El-khamissi (2018) evaluated three Egyptian wheat genotypes namely; Landrace (LR1), landrace (LR2) and Sakha 93 (S93) cultivar to salinity tolerance using different levels of NaCl (0, 50, 100 and 150 mM) at seedling stage. He concluded that Sakha 93 cultivar show a high degree of tolerance to salinity followed by LR1, while LR2 was the most sensitive one.

Lastly, Yassin et al. (2019) found differential response of wheat genotypes to salinity for earliness attributes. Based on combined analysis across normal and salinity conditions, Line 13 and Sids 12 were the earliest genotypes for days to heading in the first and second seasons with value 84.17 and 88.5 days, respectively. While Line 10 in the first season and Sids 12 in the second season were the earliest genotypes for days to maturity under the normal and salinity conditions valued (131.0 and 135.33 days), respectively. Whereas, Line 8 (94.0 and 99.0 days) and Misr 2 (96.17 and 98.83 days) were the latest for days to heading during both seasons, however Line 16 (152.83 days) in the first season and Misr 2 (140.83 days) in the second one were lately in maturity. The lengthiest GFP was found in Line 16 (59.17 days) and Misr 1 (52.0 days), but Line 5 (43.33 days) and Line 10 (41.83 days) had the shortest GFP under both environments during both seasons, respectively. The highest GFR was observed by Misr2 (40.67 and 34.07 g/m^{-1}/day^{-1}) in the two seasons, while Line 9 (27.47 g/m-1/day-1) in the first season and Line 9 and Sids 12 (17.30 g/m-1/day-1 and 16.95 g/m-1/day-1) in the second season recorded the slowest GFR. It is interest to mention that the differential response of wheat genotypes to environmental conditions indicates their difference in adaptation genes in response to the environments.

4.3.2 Morpho-physiological Characters

Morpho-physiological characters of wheat genotypes represent an important role in acclimatization and avoiding environmental stress conditions. Heat tolerance was assessed by using twelve wheat genotypes under several environments viz. two locations (Sohag and Assiut) and two sowing dates 14th Nov. (favorable) and 26th Dec (heat stress) during three winter seasons of 2005/2006, 2006/2007 and 2007/2008. Hamam and Khaled (2009) found that flag leaf area was significantly influenced by years, locations, sowing dates, nitrogen fertilizer levels and genotypes. The flag leaf area averages ranged from 25.33 to 34.27 cm^2 for Sahel 1 and Johara 19 genotypes, respectively with an overall average 30.27 cm^2. Flag leaf area decreased by 13.29% by delaying sowing date. Whereas, Sikder and Paul (2010) used six wheat cultivars to assess heat tolerance of wheat through physiological approaches under two sowing dates i.e. 30th November as normal sowing condition and 30th December as late sowing heat stress condition. They exhibited variations in canopy temperature depression (CTD) among the cultivars and sowing conditions. Under normal sowing conditions, heat tolerance cultivars Shatabdi and Gourab maintained the highest CTD (4.66), whereas heat sensitive stress cultivar Sonora had the lowest CTD (2.33). Under late sowing heat stress condition, all the cultivars increased their CTD compared to normal sowing condition. At heat stress condition, heat tolerant cultivar Shatabdi showed the highest canopy temperature depression (6.33) which was followed by Gourab, Sourav and Kanchan.

Moreover, under three sowing dates "first November as early sowing date (D1), middle November as favorable sowing date (D2) and middle December as late sowing (D3)", heat stress tolerance of fifty-eight spring wheat genotypes were tested by Hamam (2013). The averages of flag leaf area under D1, D2 and D3 were 15.37, 16.17 and 14.23 cm^2, respectively. The averages of genotypes over the three sowing dates during the three years ranged from 8.31 cm^2 for genotype No. 31 to 27.13 cm^2 for genotype No. 34 with an average of 15.26 cm^2 over all genotypes. Reduction average in flag leaf area under D1 and D3 was low for genotypes No. 8, 10, 14, 23, 32, 34, 38, 47, 55 and 57 as compared with favorable sowing date D2. At the same time, Mondal et al. (2013) showed that canopy temperature (CT) showed significant genotypic differences and significant association with the mean grain yield in Obregon1 and Obregon 2 environments. Entries with cooler canopies had higher grain yield in Obregon1 and Obregon 2 environments. Total chlorophyll content measured in Varanasi and India ranged from 46 to 53 SPAD unites. Chlorophyll content had a strong association with 1000-grain weight across locations but not with grain yield at individual location or across locations. Omnya Elmoselhy (2015) observed significant differences among wheat genotypes in different environments for flag leaf area, flag leaf chlorophyll content and canopy temperature depression. Wheat genotypes Line 16 and Gemmeiza 11 had broader flag leaf area overall environments with values of 65.95 cm^2 and 55.34 cm^2, respectively. Whereas, genotypes Sids 13 and Line 27 were the lowest one and valued 39.11 cm^2 and 39.60 cm^2 among the studied twelve wheat genotypes. Line 16, exhibited high concentration of flag leaf chlorophyll content (< 50.00 SPAD unites), in 1st and 2nd seasons in both locations. While, cultivars Giza 168 in the 1st season, and Line 27 in the 2nd one in D3 under Sids location were the lowest one with values of 41.27 and 40.39 SPAD unites, respectively. Wheat genotypes Line 16, Misr 1 and Gemmeiza 11 exhibited high concentration of flag leaf chlorophyll content overall environments with values of 50.77, 46.37 and 46.18 SPAD unites, respectively. While, cultivars Giza 168 and Sakha 93 were the lowest one among the studied twelve wheat genotypes with values of 43.47 and 42.77 SPAD unites, respectively (Tables 4.9 and 4.10). Under late sowing date (D3) as heat stress environment, most of wheat genotypes tended to increase their CTD compared to D1 and D2. In the 1st season, at Kafr Al-Hamam location, at D3, the highest CTD (<7.00) was recorded by cultivars Misr 2, Misr 1, Sids 1 and Shandweel 1; Gemmeiza 11, Giza 168, Sids 12, Sids 13, Line 14 and Misr 2 at Sids location. In the 2nd season, at Kafr Al-Hamam location, the uppermost CTD (<7.00) was recorded by cultivars Misr 2, Shandweel 1, Line 16, Line 27, Giza 168, Sids 12 and Line 14. Whereas, at Sids location, wheat genotypes

Table 4.9 Mean performance of twelve bread wheat genotypes for flag leaf chlorophyll content in two locations under three sowing dates of 2010/2011 season (Omnya Elmoselhy 2015)

Genotypes	Flag leaf chlorophyll content 2010/2011											
	Kafr Al-Hamam				Sids				Over all			
	D1	D2	D3	Mean	D1	D2	D3	Mean	D1	D2	D3	Mean
Misr 1	46.92	46.18	44.70	45.93	47.26	48.75	43.10	46.37	47.09	47.46	43.90	46.15
Misr 2	47.66	44.82	43.87	45.45	44.27	45.31	41.43	43.67	45.97	45.07	42.65	44.56
Sakha 93	46.79	44.58	41.46	44.28	45.62	45.78	42.17	44.52	46.21	45.18	41.82	44.40
Giza 168	47.39	44.61	43.40	45.13	44.66	44.67	41.27	43.53	46.02	44.64	42.34	44.33
Gemmeiza 11	48.48	45.99	45.64	46.70	45.84	48.41	44.00	46.08	47.16	47.20	44.80	46.39
Shandweel 1	45.28	44.45	42.21	43.98	46.83	49.26	43.10	46.40	6.05	46.86	42.66	45.19
Sids 1	45.18	45.04	42.50	44.24	46.06	47.27	44.40	45.91	45.62	46.16	43.45	45.08
Sids 12	47.39	45.11	44.70	45.73	45.62	49.63	43.40	46.22	46.51	47.37	44.05	45.98
Sids 13	47.82	45.09	42.27	45.06	46.13	46.58	42.63	45.11	46.98	45.83	42.45	45.09
Line 27	45.02	45.67	43.13	44.61	44.21	49.55	42.97	45.58	44.62	47.61	43.05	45.09
Line 14	46.67	45.87	44.68	45.74	46.98	48.57	41.37	45.64	46.82	47.22	43.02	45.69
Line 16	52.31	50.06	48.57	50.31	50.84	53.33	49.13	51.10	51.57	51.69	48.85	50.70
Mean	47.24	45.62	43.93	45.60	46.19	48.09	43.24	45.84	46.82	46.86	43.59	45.72
L.S.D. 0.05	2.57	2.04	2.57		2.25	3.06	2.15					

D1: 20th November as optimum sowing date, D2: 10th December as moderate late sowing date and D3: 30th December as late sowing date "heat stress"

Table 4.10 Mean performance of twelve bread wheat genotypes for flag leaf chlorophyll content in two locations under three sowing dates of 2011/2012 season (Omnya Elmoselhy 2015)

Genotypes	Flag leaf chlorophyll content 2011/2012											
	Kafr Al-Hamam				Sids				Over all			
	D1	D2	D3	Mean	D1	D2	D3	Mean	D1	D2	D3	Mean
Misr 1	49.17	47.04	46.48	47.56	49.08	47.01	45.49	47.19	49.12	47.03	45.98	47.37
Misr 2	44.59	44.26	44.15	44.33	45.44	45.15	40.91	43.83	45.02	44.71	42.53	44.09
Sakha 93	43.74	43.76	40.51	42.67	44.39	44.89	39.31	42.86	44.06	44.33	39.91	42.77
Giza 168	44.45	42.84	42.03	43.11	46.06	44.38	41.06	43.83	45.26	43.61	41.54	43.47
Gemmeiza 11	49.72	46.17	44.23	46.71	47.38	45.01	44.56	45.65	48.55	45.59	44.39	46.18
Shandweel 1	46.07	45.81	45.00	45.63	46.04	44.34	46.18	45.52	46.06	45.07	45.59	45.57
Sids 1	45.24	45.00	45.77	45.34	45.61	44.70	41.58	43.96	45.43	44.85	43.68	44.65
Sids 12	47.02	45.38	45.09	45.83	48.38	45.00	41.03	44.80	47.70	45.19	43.06	45.32
Sids 13	45.48	44.71	42.78	44.32	45.58	45.10	39.03	43.24	45.53	44.91	40.90	43.78
Line 27	46.88	42.66	43.78	44.44	46.68	45.80	40.39	44.29	46.78	44.23	42.09	44.37
Line 14	47.02	45.31	45.09	45.81	45.37	44.17	41.20	43.58	46.19	44.74	43.14	44.69
Line 16	51.93	51.11	49.70	50.91	51.13	50.97	49.77	50.62	51.53	51.04	49.74	50.77
Mean	46.77	45.34	44.55	45.55	46.76	45.54	42.54	44.95	46.77	45.44	43.54	45.25
L.S.D. 0.05	2.56	3.08	4.38		1.83	1.08	4.00					

D1: 20th November as optimum sowing date, D2: 10th December as moderate late sowing date and D3: 30th December as late sowing date "heat stress"

Misr 2, Giza 168, Gemmeiza 11, Shandweel 1, Sids 12 and Sids 1 gave CTD values varied from 6.22 to 6.89 (Tables 4.11 and 4.12).

Mean performance under normal irrigation showed an increase when compared with drought stress. Naglaa Qabil (2017) showed that Local wheat cultivar Gemmeiza 9 and their F1 crosses (Misr 1 × Gemmeiza 9) and (Gemmeiza 9 × Gemmeiza 11) recorded the highest mean performance of flag leaf area under both conditions as well as (Line 1 × Gemmeiz 9) under normal irrigation. The local wheat cultivar Gemmeiza 11 and their relevant cross (Line 1 × Gemmeiza 11) had high mean values of leaf chlorophyll content under both conditions, reinforcing their importance in applied breeding programs. Furthermore, the local cultivars, Misr 1 and Gemmeiza 9 as well as their F1 cross (Misr 1 × Gemmeiza 9) exhibited the highest productivity under both environments. Thus, these genotypes were more tolerant to drought stress. On the other hand, wheat parents Line 1 and Line 2 and their F1 crosses (Line 1 × Misr 1), (Line 1 × Line 2) and (Line 2 × Gemmeiza 9) displayed the lowest productivity under both normal irrigation and drought stress environments. Consequently, these genotypes were more sensitive to drought stress as given in Table 4.13.

ElBasyoni (2018) revealed that heat stress had a significant adverse impact on morphophysiological traits. The analysis of variance showed highly significant effect for six environments on total chlorophyll content, canopy temperature, leaf area and plant height. Sowing dates had a highly significant effect on altogether traits. Highly significant variance was also perceived amongst the considered cultivars and the amount of differences amongst cultivars was adequate to offer a scope to describe the effect of terminal heat stress. Late sown condition demonstrated a significant adverse effect on total chlorophyll content, leaf area and plant height. Total chlorophyll content for all cultivars averaged 31.9 and 28.5 under recommended and late sown condition, respectively. The late sowing condition decreased leaf area from 35.15 to 23.5 cm^2, but increased leaf rust scores from 0.34 to 0.73 and stem rust scores from 0.36 to 0.79, increased canopy temperature from 32.76 °C to 56.16 °C. Under late sowing date, Gemmeiza 9, Gemmeiza11, and Sids12 showed moderate resistance to leaf rust across all environments. Moreover, five wheat cultivars were resistant to stem rust resistant, i.e., Sids12, Sakha 94, Gemmeiza10, Gemmeiza11, and Gemmeiza12, across all environments. Cultivar Giza168 exhibited the highest values for total chlorophyll content (33.5 SPAD units) while "Gemmeiza11" (30.7 SPAD) was the lowest one under the recommended sowing date, but, under the late sown condition, cultivar "Sids13" has the highest total chlorophyll content (29.5 SPAD). Cultivar Gemmeiza 9 had the highest canopy temperature (37.1) under the recommended sowing date, whereas, cultivar Misr 2 had the highest canopy temperature (60.4) under the late sown condition. Leaf areas under the recommended sowing date valued 38.6 to 29.9 cm^2 for cultivars Gemmeiza 7 and Sakha 94, respectively, however, it reduced from 25.2 to 20.1 cm^2 for the same cultivars under the late sown one.

Variation amongst seven wheat varieties and their half diallel cross for physiological drought traits was deliberated by El-Gammaal (2018). Two field experiments were conducted, the first was ordinarily irrigated four times at different stages, and the second one irrigated only once at tillering stage. Water stress treatment declined

Table 4.11 Mean performance of twelve bread wheat genotypes for canopy temperature depression at two locations under three sowing dates of 2010/2011 season (Omnya Elmoselhy 2015)

Genotypes	Canopy temperature depression 2010/2011											
	Kafr Al-Hamam				Sids				Over all			
	D1	D2	D3	Mean	D1	D2	D3	Mean	D1	D2	D3	Mean
Misr 1	6.93	6.66	7.39	6.99	6.06	5.72	6.65	6.14	6.49	6.19	7.02	6.57
Misr 2	7.22	676	7.99	7.33	6.49	5.81	7.02	6.44	6.86	6.29	7.51	6.88
Sakha 93	7.02	6.06	6.18	6.42	5.60	5.29	6.70	5.87	6.31	5.67	6.44	6.14
Giza 168	7.01	6.18	7.50	6.90	6.07	5.46	7.34	6.29	6.54	5.82	7.42	6.59
Gemmeiza 11	7.97	6.63	6.61	6.73	6.28	5.86	7.62	6.59	6.62	6.24	7.11	6.66
Shandweel 1	7.07	6.79	7.22	7.03	5.81	5.68	6.72	6.07	6.44	6.24	6.97	6.55
Sids 1	6.84	6.49	7.27	6.86	5.79	5.91	6.82	6.17	6.32	6.20	7.04	6.52
Sids 12	7.19	6.49	6.71	6.80	6.50	5.47	7.31	6.42	6.84	5.98	7.01	6.61
Sids 13	6.80	6.17	6.01	6.35	5.45	5.42	7.24	6.04	6.13	5.80	6.65	6.19
Line 27	7.34	6.69	6.64	6.89	6.24	5.36	6.99	6.20	6.79	6.02	6.82	6.54
Line 14	7.15	6.84	6.52	6.84	5.99	5.54	7.04	6.19	6.57	6.19	6.78	6.51
Line 16	6.53	6.94	7.49	6.99	5.96	6.04	6.88	6.29	6.25	6.49	7.19	6.64
Mean	7.01	6.56	6.97	6.84	6.02	5.63	7.03	6.23	6.51	6.09	7.00	6.53
L.S.D. 0.05	NS	0.52	NS		0.53	NS	NS	NS				

D1: 20th November as optimum sowing date, D2: 10th December as moderate late sowing date and D3: 30th December as late sowing date "heat stress"

NS: Not significant

4.3 Performance of Wheat Genotypes in Response …

Table 4.12 Mean performance of twelve bread wheat genotypes for canopy temperature depression at two locations under three sowing dates of 2011/2012 season (Omnya Elmoselhy 2015)

Genotypes	Canopy temperature depression 2011/2012											
	Kafr Al-Hamam				Sids				Overall			
	D1	D2	D3	Mean	D1	D2	D3	Mean	D1	D2	D3	Mean
Misr 1	6.87	6.57	6.95	6.80	5.63	5.39	5.90	5.64	6.25	5.98	6.42	6.22
Misr 2	7.36	7.40	7.84	7.53	5.13	4.86	6.89	5.63	6.25	6.13	7.36	6.58
Sakha 93	6.63	6.55	6.59	6.59	4.40	5.02	6.17	5.20	5.52	5.79	6.37	5.89
Giza 168	6.96	7.05	7.21	7.07	4.88	4.37	6.68	5.31	5.92	5.71	6.94	6.19
Gemmeiza 11	6.81	6.71	6.99	6.83	4.97	5.04	6.58	5.53	5.89	5.88	6.78	6.18
Shandweel 1	7.21	7.26	7.78	7.42	5.58	5.46	6.41	5.81	6.40	6.36	7.10	6.62
Sids 1	6.80	6.57	6.96	6.77	4.49	5.52	6.22	5.41	5.64	6.04	6.59	6.09
Sids 12	7.13	6.85	7.13	7.03	5.43	5.42	6.23	5.70	6.28	6.14	6.68	6.37
Sids 13	6.66	6.80	7.09	6.85	4.40	4.48	6.14	5.01	5.53	5.64	6.62	5.93
Line 27	7.28	7.17	7.27	7.24	5.04	5.50	5.30	5.28	6.16	6.34	6.28	6.26
Line 14	6.88	6.89	7.11	6.96	5.18	5.40	5.52	5.37	6.03	6.15	6.32	6.26
Line 16	7.03	7.96	7.75	7.25	5.09	6.10	6.20	5.80	6.06	6.53	6.98	6.52
Mean	6.97	6.90	7.22	7.03	5.02	5.21	6.19	5.47	5.99	6.06	6.70	6.25
L.S.D. 0.05	NS	NS	NS		NS	NS	0.83					

D1: 20th November as optimum sowing date, D2: 10th December as moderate late sowing date and D3: 30th December as late sowing date "heat stress", NS: Not significant

Table 4.13 Mean performance of parental genotypes and their F₁ crosses for days to heading, yield and its attributes under normal irrigation and drought stress conditions and drought sensitivity index (DSI) in growing season 2015/2016 (Naglaa Qabil 2017)

Genotypes	Flag leaf area(cm^2)		Chlorophyll content (%)		Grain yield/plant (g)		
	I	D	I	D	I	D	DSI
Misr 1 (P1)	35.57	20.44	50.0	45	22.5	13.7	1.03
Gemmeiza 9 (P2)	33.88	22.49	48.4	44.2	24	15.7	0.90
Gemmeiza 11 (P3)	33.79	23.41	44.9	41.2	23.1	12.6	1.20
Line1 (P4)	35.44	27.18	49.7	45.9	26.4	16.9	0.95
Line 2 (P5)	33.18	23.18	53.7	46.8	23.7	15.9	0.87
P1 × P2	31.54	20.8	52.2	46.9	24.6	15	1.03
P1 × P3	31.99	23.82	52.5	46.5	25.8	14.1	1.20
P1 × P4	39.65	22.09	49.6	49.0	26	16.6	0.95
P1 × P5	32.66	21.66	56.4	49.6	26.6	17.8	0.87
P2 × P3	36.95	25.86	51.5	48.1	26	18.3	0.78
P2 × P4	38.80	26.50	56.3	48.0	29.8	19	0.96
P2 × P5	34.87	24.76	51.2	45.7	26.0	14.3	1.19
P3 × P4	34.11	22.87	50.3	46.1	24.8	13.3	1.23
P3 × P5	32.87	24.66	51.0	48.2	26.7	17.8	0.87
P4 × P5	37.68	28.65	52.6	46.3	26.3	17.2	0.91
L.S.D $_{0.05}$	1.22	1.13	2.38	1.73	1.80	1.59	

I = Normal irrigation. D = Drought stress. DSI = Drought sensitivity index

flag leaf area, flag leaf angle and transpiration rate for parental genotypes and their crosses. Estimates of stomatal resistance, chlorophyll a/b ratio and leaf temperature were increased under water stress. The two irrigation regimes behaved differently for all studied traits. Mean squares due to genotypes were highly significant for all traits excluding flag leaf area under water stress situation. Parental genotypes Gemmeiza 12, Misr1, Giza 171 and Giza 168 could be considered as excellent parents in breeding programs aimed to release desirable genotypes more tolerant to drought. The best wheat crosses were; Giza 168 × Gemmeiza 11, Giza171 × Giza 168, Giza 171 × Gemmeiza 11 and Gemmeiza 12 × Giza 168 for almost of traits.

4.3.3 Yield and Its Components

The genotypes that did not show any remarkable reduction in yield, when planted late as heat stress or under water stress conditions seem to be tolerant to stresses. Under rainfed and irrigated situations, Kamal et al. (2003) evaluate mean performance of grain yield characters of eight recent wheat genotypes. The wheat genotypes exhibited significant differences in plant height, 1000-grains weight, grain and straw yields. Genotype Aghrani registered the highest grain yield (2.16 t ha^{-1}) was statistically equal to Kanchan (2.08 t ha^{-1}), Akbar (2.06 t ha^{-1}) and Barkat (2.02 t ha^{-1}) across environments. Grain and straw yields were improved with application of irrigation compared to rainfed conditions. The most promising varieties in grain yield were Barkat, Aghrani, Akbar and Kanchan than the others. However, Menshawy (2007) recorded higher 1000-kernel weight by 7.3 gm. and grain yield by 3.5 ardab/fed. under optimum sowing date (23rd Nov.) than the late sowing one (21st Dec.) as heat stress condition. Mohammadi (2012) found that comparison between less and high-temperature field environments showed 24% difference in wheat grain yield, 9.2% in Kernel number and 23.7% in kernel weight. In this regard, twelve wheat genotypes were tested at two levels of sowing dates i.e. normal sowing date (18th November) and late sowing (11th December) by Sial et al. (2005). There was an overall decline in grain yield of 63.55 kg/ha under late planting. The genotypes produced higher grain yield, when planted on 18th November as compared to the 11th December. Genotypes 7–03, SD-4085/3, ESW-9525, SD-4047 and SD-1200/51 were promising in yield in November sowing, while genotypes RWM-9313 and SD-4047 produced higher grain yield than other entries under late planting. As high as 33% and 31% yield reduction was observed in genotypes SD-1200/11 and SD-1200/51, respectively. The genotype RWM-9313 did not show any significant decrease in yield, when planted late and appears to own tolerance to high temperature to some extent.

Hamam and Khaled (2009) found that 1000-kernel weight and grain yield were significantly influenced by years, locations, sowing dates, nitrogen fertilizer levels and genotypes. The highest 1000-kernal weight was recorded by Johara 19 genotype (52.66 gm) at Assiut location, at the favorable sowing date and 100 kg/fed N fertilizer but the lowest (26.39 gm) was registered by of TRI 4157 genotype at Sohag location and late sowing date with an overall average (40.32 gm). Grain yield of the different genotypes varied from 10.40 to 15.47 ardab/fed. for TRI 2586 and Bocro-4//kauz "S" genotypes, respectively. To study the effect of temperature on development and formation of grain, Riza-Ud-Din et al. (2010) found significant genotypic differences for number of tillers/m^2, grains/spike, 1000-grain weight and grain yield/plot. Maximum reduction of 53.75% was noted for grain yield/plot, while tillers/m^2 was suppressed less by high temperature with a minimum reduction of 15.38%. Variety AS-2002 showed better performance than other varieties/lines under late planting by maintaining the tillering ability, number of grains/spike and grain yield/plot. While, Omnya Elmoselhy (2015) showed that grain yield (ard./fed.) and its components were significantly reduced by delaying sowing date from optimum (20th November, D1), moderate stress (10th December, D2) to sever stress (30th December, D3) with values of (6.52 and 0.88%), (10.83 and 17.56%), (6.05 and 11.49%) and (11.39 and 24.69%) for number of spikes/m^2, number of grains/spike, 1000-grain weight and grain yield/fed. compared to optimum date. Wheat genotypes Misr 1, Misr 2, Sids 1 and Sids 13 gave greater number of spikes/m^2 under various environments. Whereas, Gemmeiza 11, Sids 12, Line 14 and Line 16 produced the greatest number of grains/spike overall environments, otherwise Sakha 93 and Sids 13 were the lowest one. Genotypes Gemmeiza 11 and Line 16 under different twelve environments in addition to Sids 12 at the 2nd season and the combined were the heaviest weight, whereas Sids 13 was the lowest one among the studied wheat geno-types. The most promising wheat genotypes exhibited higher productivity were Misr 2, Misr 1, Gemmeiza 11 and Line 16 with 23.85, 23.62, 23.62 and 23.42 ard./fed., respectively. Otherwise, genotypes Sakha 93, Sids 1 and Giza 168 gave the lowest values of productivity with 20.51, 21.74 and 21.82 ard./fed., respectively. Cultivar Misr 1, Giza 168 and Sids 12 under Loc 2 during 2010/2011 and Gemmeiza 11 and Shandweel 1 under Loc. 1 during 2011/2012 season, all maintained their behavior from D2 to D3, therefore could be classified as more tolerant to heat stress.

Testing adaptability parameters of 43 bread wheat genotypes under 10 diverse environments in Egypt during the two growing seasons 2013/2014 and 2014/2015. These environments distributed along the Egypt map from 215 m above sea level (asl) at Toshka in south Egypt to 22 m asl at El-Nobaria in the north studied by Ibrahim and Hamada (2016). They showed that the diversity of conditions was reflected by large variation of a grain yield/plant of the 43 genotypes, which varied from 1.50 g/plant to 77.34 g/plant. In continuos, Hamada and Ibrahim (2016) evaluated a panel of 40 wheat genotypes for 8 yield and yield-contributing traits under recommended sowing date of the Egyptian Ministry of Agriculture as a control and two other different sowing dates as plants will face heat-stressed conditions at anthesis and grain-filling phases. They detected a continuous phenotypic variation in all measured traits, indicating a polygenic inheritance of traits. Also, highly significant genotype × environment interaction which is expected for quantitative traits. Cluster analysis showed two distinct groups with respect to stress tolerance index with substantial diversity among genotypes either sensitive or tolerant to heat stress. Under El-Kattara and Ghazalla regions of Egypt, Ali and Abdul-Hamid (2017) conducted several field experiments to screen 38 wheat genotypes in respect to grain yield under twelve different environments for water and heat stresses in drip and sprinkler irrigation systems in sandy soils and surface flood irrigation system in old clay soils. Wheat genotypes exhibited higher grain yield at drip irrigation than sprinkler and surface flood irrigation systems. Drought stress and delay sowing date led to reduced grain yield in all wheat genotypes rather than optimum water irrigation and favorable sowing date. Grain yield varied from 15.06 for Line 2 to 20.02 for Line 13. Wheat Lines 9, 18 and 21 displayed the desired drought and heat sensitivity indices under all irrigation systems (SI < 1).

ElBasyoni (2018) showed that heat stress had a significant adverse impact on the yield trait. The analysis of variance showed highly significant effect for different six environments on grain yield. The late sown condition displayed a significant adverse effect on grain yield. Cultivar "Gemmeiza12" produced the highest grain yield (8.8 ton/hectare) under the recommended sowing date. While cultivar "Gemmeiza 9" produced the highest grain yield (4.87 ton/hectare) under the late sown condition. At the same time, under normal and water stress treatment, El-Gammaal (2018) registered decrease in number of spikes/plant, number of grains/spike, 1000–grain weight and grain yield/plant for parents and their hybrids. Irrigation mean squares were significant for yield traits, demonstrating that the two irrigation regimes behaved differently for yield characters. Also, mean squares due to genotypes were highly significant for all traits. The four Egyptian parental genotypes Gemmeiza 12, Misr1, Giza 171 and Giza 168 were classified as excellent parents in breeding programs to release new genotypes more tolerant to drought. The best parental combinations were; Giza 168 × Gemmeiza 11, Giza171 × Giza 168, Giza 171 × Gemmeiza 11 and Gemmeiza 12 × Giza 168 for almost traits studied.

Latently, assessment mean performance of grain yield for eleven genotypes under different environments i.e. Ghazaleh (improved environment) and El-Khattara (stress environment), and three seeding rates (350 seeds/m^2 D1), (400 seeds/m^2 D2) to (450 seeds/m^2 D3) during two seasons was performed by Awaad (2021). Based on general mean, performance of grain yield (ard./fed.) for the studied eleven genotypes (Table 4.14) showed slight increase under Ghazaleh location from D1 26.08 to D2 26.47 (ard./fed.), but decrease in D3 valued 23.76 (ard./fed.) in the first season. Whereas, grain yield exhibited maximum average value 26.86 (D2) in the 2nd season rather than D3 which recorded 24.42 (ard./fed.). Also, under Ghazaleh location, Line 1, Line 8 and Line 7 gave maximum grain yield valued 30.53 and 29.81; 30.77 and 30.71 as well as 28.84 and 28.20 (ard./fed.) in D1 and D2 in the 1st season as well as 29.03 and 28.27; 30.97 and 29.30 as well as 27.14 and 31.53 (ard./fed.), in the same respective order in the 2nd season. Line 3 valued 31.82, and Line 5 valued 29.10 (ard./fed.) in D2 in the 2nd season, respectively. A higher estimate of grain yield under Ghazaleh location was attributed to suitable environments with soil fertility, adequate water and other inputs. Where Ghazaleh region is rich in organic matter, 10.34 g kg^{-1}, macro and micro-nutrients with suitable temperatures are factors that led to increasing grain yield. Under El-Khattara location, general mean was decreased and valued 21.01, 19.82 and 18.78 as well as 20.64, 20.33 and 19.38 from D1, D2 and D3, in both seasons, respectively. Line 6 in D2 and Line 7 in D1 produced higher grain yield with values of 25.77 and 25.40 ard/fed. in the 2nd season, respectively. Furthermore, Line 1, Line 5, Line 6, Line 7 and Sakha 94 in D1; Line 6 in D2 in 1st

Table 4.14 Mean performance for eleven wheat genotypes under different environments for grain yield (Grain yield ard°/fed.⁻) (Awaad 2021)

| Genotype | Ghazaleh | | | | | | El-Khattara | | | | | | Combined |
| | 2014/2015 | | | 2015/2016 | | | 2014/2015 | | | 2015/2016 | | | |
	D1	D2	D3	D1	D2	D3	D1	D2	D3	D1	D2	D3	
Line 1	30.53	29.81	25.43	29.03	28.27	22.83	22.27	21.37	21.13	19.77	22.17	20.23	24.40
Line 2	23.31	20.55	22.07	22.82	24.70	25.27	20.20	19.28	19.77	17.87	17.97	18.57	21.03
Line 3	25.83	27.38	28.13	26.87	31.82	25.88	19.12	19.07	19.17	18.20	19.81	17.87	23.26
Line 4	26.30	27.93	25.70	26.23	26.47	25.87	19.77	21.93	20.54	22.96	20.80	22.67	23.93
Line 5	25.76	23.98	21.17	28.53	29.10	25.00	22.65	18.56	19.10	22.87	18.37	20.47	22.96
Line 6	24.70	25.98	21.07	25.75	23.19	23.73	22.40	22.18	17.40	17.77	25.77	18.03	22.33
Line 7	28.84	28.20	21.08	27.14	31.53	22.43	22.10	18.17	17.24	25.40	20.47	20.27	23.57
Line 8	30.77	30.71	23.38	30.97	29.30	26.60	18.84	18.03	16.93	21.63	19.53	18.63	23.77
Sakha 94	22.64	22.53	18.46	21.18	22.10	22.87	22.47	18.68	18.07	22.15	19.33	17.99	20.71
Giza 168	27.10	28.23	29.90	28.97	25.00	26.30	21.07	21.30	17.51	16.50	19.77	18.30	23.33
Misr 1	21.07	25.90	23.94	27.23	23.97	21.84	20.23	19.57	19.78	21.92	19.63	20.16	22.10
Mean	26.08	26.47	23.76	26.79	26.86	24.42	21.01	19.83	18.78	20.64	20.33	19.38	22.85
L.S.D. 0.05	4.930	6.794	5.602	4.697	3.661	4.138	4.806	4.065	3.073	5.589	2.766	3.894	1.257

D1, D2 and D3 refer to 350 seeds/m², 400 seeds/m² and 450 seeds/m² Ardab = 150 kg Feddan = 4200 m²

season and L4, L5, Sakha 94 in D1 and Line 1 in D2 as well as Line 4 in D3 in the 2nd season, all produced grain yield about 22.00 ard/fed. Wheat genotypes Line 1, Line 4, Line 8 and Giza 168 were early coupled with higher grain yield, with good level of protein content for both L8 and Giza 168 only, among them Line 1 was moderate tolerant and Line 4 was tolerant to stress based on TOL values. Whereas, Line 7 attained a good level of yield averaged 23.57 ard./fed., but it was late 85.0 day in heading and moderate tolerant to stress. Therefore, in regions of the terminal stress, breeder seeks for genotypes of shorter grain filling duration to escape or at least to minimize the determine effect of heat stress on grain yield like Line 1, Line 4, Line 8 and Giza 168.

Evaluating independent effects of drought, heat and combined heat + drought on wheat yield and its relevant traits were evaluated on collection of wheat germplasm by Qaseem et al. (2019). Among evaluated yield traits, drought stress had greater reduction effects by 45% on grain yield. Harvest index came in the second most severely influenced trait and it was decreased by 37% due to drought. Heat stress affected 53% reduction in grain yield followed by 39% reduction in harvest index. Simultaneously effect of both drought and heat stress caused reduction by 56% in grain yield, and 41% in harvest index. Genotype SAWSN_3059 and SAWSN_3052 give higher yield and performed well under drought stress, whereas under heat stress MILLAT-2011 and HTWSN_4427 were well performing genotypes. Both genotypes ESWYT_116 and EBWYT_529 produced higher values for yield traits under both heat and drought stress.

Differential response of wheat genotypes to salinity for yield characters was assessed by Yassin et al. (2019). Based on combined analysis across normal and salinity conditions, wheat genotype Misr 2 verified the highest estimates during the two seasons for each of biological yield (2.79 and 2.49 kg/m^2), straw yield (3.18 and 2.82 kg/m^2) and grain yield (1.02 and 0.93 kg/m^2). The lowest values were recorded by Sids 12 for grain yield (0.71 and 0.44 kg/m^2) in the two seasons, while Line 7 recorded the lowest values for both of biological and straw yield (1.86 and 2.03 kg/m^2) in the first season and Sids 12 (1.38 and 1.69 kg/m2) in the second season, respectively.

Genetic variation has been detected among wheat genotypes concerning heavy metals accumulation under East Kazakhstan conditions by Alybayeva et al. (2014). They determined heavy metals in soil and plant samples by atomic absorption spectrophotometry. Registered genotypic variances in the accumulation of copper, lead, cadmium and zinc among wheat genotypes. The lowest amount of heavy metals was accumulated in the seeds of winter wheat varieties Mironovskaya-808 and Krasnovodopadskaya-25, with the highest yield for Mironovskaya-808. Hence, can be recommended to be grown under the technologically disadvantaged regions, with soil polluted by heavy metals, as they accumulate not much heavy metal, and have good indicators for yield. In Zhejiang province, China Lu et al. (2020) estimated Cd accumulation in grains of wheat genotypes in Cd-contaminated soils. They identified Bainong207, Aikang58, Huaimai23, and Yannong21 as respectable candidates of low-Cd genotypes. Low-cadmium genotypes have a great biomass and high accumulation of Cd in straw however low-Cd accumulation in grains.

4.3.4 Protein Content

Grain quality was significantly influenced by changing in environmental conditions. Different degrees of varietal responses to environmental changes were identified in several experimental studies under different geographical locations. Mean performance of grain quality of eight recent wheat genotypes grown under rainfed and irrigated situations was evaluated by Kamal et al. (2003). The wheat genotypes exhibited significant differences in protein, amylose, ash and K contents in kernels. Maximum protein content was assessed in Barkat (13.75%), followed by Ananda (12.65%), Akbar (12.55%), Sonalika (12.5%), Aghrani (12.35%) and Kanchan (12.25%) however, Balaka took the lowest content (11.55%). Barkat gave the highest amylose content (24.4%) followed by Balaka (24.25%), Aghrani (23.6%), Akbar (23.2%), Ananda (23.05%) and kanchan (21.75%). Wheat protein, amylose, P, K and S contents in grains were improved with application of irrigation compared to rainfed conditions. The most promising varieties in grain and protein yields were Barkat, Aghrani, Akbar and Kanchan than the others.

Moreover, twelve wheat genotypes were assessed for quality parameters at two levels of sowing dates i.e. normal sowing date (18th November) and late sowing (11th December) by Sial et al. (2005). Protein content was significantly higher in late sown wheat grains as compared to early sown. The local check variety Sarsabz gave significantly higher protein content at normal sowing date, while, at late sowing, Kiran 95 had higher protein content. Sattar et al. (2010) showed that protein content was influenced by heat stress created by late sowing date (10th December). Maximum protein content was recorded in cultivar Inqlab-91 (10.97%) which did not differ significantly from Fisalbad-2008 (10.71%) sown on 10th Dec., whereas, Shafaq-2006 planted on 10th Nov. gave minimum protein content (10.97%). Moreover, Omnya Elmoselhy (2015) recorded significant differences in protein content due to seasons, locations, sowing dates, genotypes and their first, second and third interactions. Maximum grain protein content was recorded in cultivars Misr 1, Shandweel 1 and Line 16 with 13.19%, 12.99% and 12.93%, respectively overall environments. Whereas, minimum grain protein content (11.76%) was registered by Sakha 93. Furthermore, Awaad (2021) found that grain protein content was significantly differed from environment to another and from genotype to another when subjected to 350 seeds/m^2 (D1), 400 seeds/m^2 (D2) and 450 seeds/m^2 (D3) under Ghazaleh and El-Khattara regions during two seasons (Table 4.15). At Ghazaleh location, grain protein content was significantly higher in D2 as compared to D1 and D3. Otherwise under El-Khattara location, grain protein content was higher in D3 as compared to D1 and D2, during the two seasons, as heat shock proteins are synthesized at very high rates under high-temperature stress and are believed to have a protective role to heat stress. Therefore, it is obvious that the combined effect of sandy soil condition, low rainfall and higher temperatures during grain filling period have been the main cause of higher protein content in the stress environments as a result of dilution effect. So, under El-Khattara location, maximum grain protein content (around 12%) was recorded in Line 4, Giza 168 and Misr 1 in D3 in the 1st and 2nd seasons as well as Line 7 and Line 8 in D1, D2

Table 4.15 Mean performance for eleven wheat genotypes under different environments for grain protein content (Awaad 2021)

Genotype	Ghazaleh						El-Khattara						Combined
	2014/2015			2015/2016			2014/2015			2015/2016			
	D1	D2	D3	D1	D2	D3	D1	D2	D3	D1	D2	D3	
Line 1	10.63	10.70	10.43	10.37	11.03	10.63	11.47	11.67	12.10	11.40	11.47	11.87	11.15
Line 2	10.57	10.80	10.27	10.73	10.93	10.20	10.83	10.70	12.17	11.17	11.67	11.73	10.98
Line 3	10.67	10.97	10.73	11.03	10.97	10.13	11.50	11.90	11.97	11.53	12.03	11.80	11.27
Line 4	10.40	10.67	10.83	10.60	10.93	10.80	11.00	11.67	12.80	11.03	11.57	12.73	11.25
Line 5	10.53	10.80	10.50	11.10	10.87	10.40	11.43	11.93	12.10	11.53	11.57	12.43	11.27
Line 6	10.97	11.97	10.57	11.03	11.60	10.37	12.10	11.97	12.40	11.50	11.60	11.50	11.46
Line 7	11.17	11.90	11.88	11.27	11.83	12.10	11.87	11.93	12.80	12.27	12.40	12.87	12.02
Line 8	11.37	11.57	10.87	11.13	12.03	11.83	11.70	11.60	12.70	12.67	12.17	12.83	11.87
Sakha 49	11.07	11.07	10.83	10.33	11.87	10.87	10.93	11.50	12.80	11.27	10.63	11.67	11.24
Giza 168	11.60	11.23	11.97	10.73	11.70	11.50	11.50	11.83	12.73	10.87	11.83	12.53	11.67
Misr 1	12.03	11.63	11.10	11.97	11.63	10.60	11.47	12.00	12.20	12.27	11.27	12.23	11.70
Mean	11.00	11.21	10.91	10.94	11.40	10.86	11.48	11.79	12.43	11.59	11.66	12.19	11.44
L.S.D. 0.05	0.546	0.554	0.570	0.651	0.704	0.478	0.701	0.543	0.517	0.566	0.693	0.913	0.171

D$_1$, D$_2$ and D$_3$ refer to 350 seeds/m^2, 400 seeds/m^2 and 450 seeds/m^2

and D3 in the 2nd one. As a result of the dilution effect due to environmental stress. Whereas, under Ghazaleh location, maximum grain protein content was recorded in Line 6 (11.97(% in D2 in the 1st season, Line 7 (12.10%) in D3 in the 2nd season, Misr 1 (12.03 and 11.97%) in D1in both seasons, respectively.

On the light of the previous results, which show the differences in the wheat genotypes for different economic characteristics, are of great importance in the perspective of climate change tolerance. And thus the possibility of selecting the promising genotypes to be introduced in the breeding programs to release new cultivars more adapted and stability.

4.4 Adaptability and Yield Stability in Relation to Environmental Changes

Stability is a key word for plant breeders who analyze genotype × environment interaction because it improves the progress of selection in any single environment (Yau 1995). Stable genotype is defined as genotype that performs relatively the same over a wide range of environments. Lin et al. (1986) mentioned that the concept of stability is defined in several ways according to the way the scientist wants to study the problem. Information about phenotypic stability is beneficial for the selection of wheat genotypes and for planning appropriate breeding schemes. Wheat grain yield and environmental stress tolerance are a complex and quantitative traits that has strong G × E interactions. Thus, measuring the adaptability and stability of giving cultivars became one of the plant breeding strategies before releasing the cultivar. The advantage of selecting superior genotypes using stability analysis instead of average performance is that stable genotypes are reliable across environments which reduce G × E interaction. Studies have shown that stability analyses according to various measures can result in better identification of stable genotypes (Abbas Mosavi et al. 2013).

In general, various statistical procedures have been described to determine the adaptability and stability of new cultivars. The following stability measurements were performed to find out yield adaptability and stability of grain yield and other economic characters under the diverse environments i.e., coefficient of variability (CVi) (Pinthus 1973), regression coefficient (bi) and (S^2d) (Eberhart and Russe 1966), Wricke's ecovalance (Wi) (Mulusew et al. 2014), superiority measure (Pi) (Lin and Binn 1988), regression stability parameters (β and dj) (Perkins and Jinks 1968), and average absolute rank difference of genotype on the environment (Si) (Akcura and Kaya 2008). Moreover, the additive main effects and multiplicative interaction model (AMMI) (Romagosa and Fox 1993) was applied. Then the genotype main effect plus G × E interaction (GGE biplot) (Akcura and Kaya 2008) was used to visualize the G × E interaction.

Applied studies in this area using fifteen local and introduced bread wheat geno-types were evaluated under eighteen diverse environments which were the combi-nations between two seasons × three sowing dates × three locations by Awaad and Aly (2002). Phenotypic stability parameters (Table 4.16) indicated that wheat culti-vars, Gemmeiza 5 and Gemmeiza 9 were classified as highly adapted to favorable environments for grain yield/ fed., number of grains/spike and flag leaf area as well as ACSAD 941 for grain yield/fed. and flag leaf area. Whereas, Sakha 8, Tsi-Vee "S" and ACSAD 949 could be grown under Khattara or East Bitter Lakes (Sinai) as less favorable environmente for gain yield/fed. and flag leaf area. The most desired and stable genotypes were Gemmeiza 7, ACSAD 903 and Sakha 69 for grain yield/fed.; Gemmeiza 7 and Sakha 69 for both grain yield/fed. and 1000-grain weight as well as Sids 6 and Gemmeiza 7 for number of grains/spike, days to 50% heading and flag leaf area. Genotypic stability estimates (Table 4.17) revealed that, the most average stable genotypes were, Sakha 69 and ACSAD 903 for grain yield/fed; Tsi-Vee 'S' and ACSAD 935 for number of spikes/plant, number of grains/ spike and flag leaf area; ACSAD 925 and ACSAD 939 for 1000-grain weight as well as Sahel 1 for days to 50% heading. However, Gemmeiza 7 had above average degree of stability for flag leaf area. And complemented with the previous study, Aly and Awaad (2002) showed that the most adapted genotypes to be grown under favorable conditions were the Egyptian cultivars Sakha 69, Gemmeiza 9, as well as the exotic one ACSAD 925, ACSAD 935 and ACSAD 949 for grain protein content. Whereas, Sakha 8 and Tsi/Vee 'S' performed well under less favorable environments for The quality char-acter. Hereby, it could be useful for growing under stress environments. Based on all stability parameters, the most desirable and stable genotypes were Giza 168 and Gemmeiza 5 for grain protein content.

Forty bread wheat genotypes from divers origin that were chosen for their toler-ance to heat stress grown under eight environments (the combinations of 2 years × 2 sowing dates × 2 locations) by Tawfelis (2006) to evaluate their stability. The joint regression analysis of variance indicated highly significant differences among geno-types for number of kernels/spike, 1000-grain weight and grain yield (ton/ha). The regression coefficient "b" was positively correlated with the mean performance, indi-cating that high yielding genotypes had generally, positive "b" values and revealed a good response to the improving environments regarding number of kernels/spike. The both genotypes No. 18 and No. 19 had significant "b" values that they posi-tively responded to more favorable conditions. The "b" values for genotypes No. 10, 11, 29 and 35 gave significantly negative values meaning that they were adapted to stress condition resulted from late sowing. While, for grain yield, regression coeffi-cient values ranged from 0.286 to 1.410, reinforcing that these genotypes responded differently to the studied environments. Also, genotype × environment interaction for grain yield, number of tillers/meter row, number of grains/spike and 1000-grain weight was assessed by planting ten wheat cultivars in six diverse sowing dates over two consecutive years under irrigated conditions (Anwar et al. 2007). Stability anal-ysis of variance showed highly significant differences for environments, genotypes and environment "linear" for all the studied traits except grain yield in which, non-significant result was obtained for genotypes. Different genotypes remained stable

Table 4.16 Phenotypic and genotypic stability parameters for grain yield, days to 50% heading and flag leaf area of fifteen bread wheat genotypes under eighteen environments (Awaad and Aly 2002).

Character parameter genotype	Grain yield (ard./fed.)					Days to 50% heading					Flag leaf area (cm²)				
	$\overline{X}i$	bi	S^2di	α	λ	$\overline{X}i$	bi	S^2di	α	λ	$\overline{X}i$	bi	S^2di	α	λ
1- Sakha 8	12.249	0.792*	2.359**	-0.014	19.168*	83.018	1.212*	10.623**	0.059	20.567*	35.093	0.809*	11.389**	-0.015	6.083*
2- Sakha 69	13.596	0.951	1.224	-0.0136	2.48	84.555	0.954	10.750**	-0.017	20.322*	36.377	0.877	12.449**	-0.021	6.212*
3- Giza 168	13.445	0.955	2.019*	-0.0123	10.035*	86.251	0.596**	26.411**	-0.109*	21.203*	39.392	0.979	40.618**	-0.024	15.519*
4- Gemmeiza 5	14.937	1.297*	3.339***	-0.0007	18.808*	83.566	0.847	6.173	-0.039	11.662*	41.77	1.256*	21.938***	0.069*	14.336*
5- Gemmeiza7	14.332	1.108	1.751	0.029	13.942*	85.011	0.942	4.485	-0.02	8.655*	39.891	1.152	7.274	-0.161*	2.506
6- Gemmeiza9	14.104	1.379*	1.635	0.103*	18.062*	91.628	0.902	8.232	-0.022	15.921*	42.407	1.290**	8.621	0.051	18.587*
7- Sahel 1	13.458	0.914	4.359***	-0.023	26.273*	81.889	1.201	43.124**	0.052	2.365	38.422	0.915	22.128***	-0.033	7.640*

(continued)

Table 4.16 (continued)

Character parameter genotype	Grain yield (ard./fed.)					Days to 50% heading					Flag leaf area (cm²)				
	$\bar{X}_i$	b_i	S^2d_i	α	λ	$\bar{X}_i$	b_i	S^2d_i	α	λ	$\bar{X}_i$	b_i	S^2d_i	α	λ
8- Sids 6	12.503	1.147	3.814**	0.043	27.319*	82.006	1.149	7.213	0.044	14.437*	39.037	1.072	7.465	-0.009	7.006*
9- Tsi/Vee 'S'	11.637	0.258**	5.036**	-0.202*	8.502*	84.996	1.17	6.226	0.047	13.187*	35.946	0.802*	12.582**	0.003	2.151
10- ACSAD 903	13.648	0.944	0.581	-0.014	2.367	83.364	1.015	12.371**	0.00067	21.849*	34.946	1.038	26.162**	-0.034	11.223*
11- ACSAD 925	12.683	1.141	2.307*	0.037	18.085*	85.969	1.047	15.237**	0.014	21.386*	38.222	1.001	48.732**	0.053	21.885*
12- ACSAD 935	13.026	1.089	3.167**	0.024	23.210*	89.988	0.943	20.594**	-0.05	20.791*	33.699	0.816*	32.832**	-0.047	2.412
13- ACSAD 939	12.753	0.865	2.060*	-0.036	16.089*	92.368	1.066	14.752**	0.012	19.920*	36.552	0.883	16.123**	0.009	12.794*
14- ACSAD 941	14.279	1.414**	5.897**	0.111*	27.685*	90.883	0.866	16.481**	0.05	18.840*	34.483	1.223*	37.739**	-0.025	9.502*
15- ACSAD 949	10.486	0.741*	3.378**	-0.072	23.454*	88.202	1.086	9.840**	0.021	15.590*	34.022	0.805*	31.497**	-0.078*	9.234*
Grand mean	13.142					86.246					37.35				
L.S.D.0.05	1.155					2.561					3.389				

$\bar{X}_i$: Mean overall environments, b_i: Regression coefficient, S^2d_i: Deviation from regression coefficient, α_i: Measures the linear response of the environmental effect and λ_i measures the deviation from the linear response in terms of the magnitude of the error variance

Table 4.17 Genotypes average performance over twelve environments and stability parameters of the twelve bread wheat genotypes for grain yield and its components (Omnya Elmoselhy et al. 2015)

Genotypes	Number of spikes/m²			Number of grains/spike			1000-grain weight			Grain yield (ard.°/fed.')		
	X̄	Bi	S²di	X̄	bi	S²di	X̄	bi	S²di	X̄	bi	S²di
Misr 1	446.61	1.02	119.13**	59.19	1.18	9.87**	46.20	1.79**	1.41	23.62	0.94	0.62
Misr 2	453.50	1.41*	233.64**	61.83	0.93	11.04**	45.30	1.43*	1.15	23.85	1.37**	0.78
Sakha 93	416.81	0.84	810.56**	58.89	1.14	8.71**	45.44	1.18	4.49**	20.51	1.09	2.13
Giza 168	428.53	0.78	403.45**	62.50	0.97	4.44**	44.17	0.72	6.06**	21.83	0.83	1.12
Gemmeiza 11	393.61	0.70	337.25**	65.92	1.13	9.80**	51.51	0.53	6.55**	23.62	0.85	0.27
Shandweel 1	444.50	1.22	592.92**	61.47	0.81	3.31	44.73	1.23	3.73**	22.54	1.19*	0.02
Sids 1	459.58	1.80*	706.94**	59.92	0.68	7.82**	43.97	0.89	4.79**	21.75	1.00	3.33*
Sids 12	424.36	0.99	222.06**	64.19	1.23	11.05**	46.43	0.62	3.30**	22.93	1.17	1.51
Sids 13	468.33	0.51*	386.66**	59.75	1.02	8.86**	42.51	1.15	4.50**	22.88	0.75*	0.14
Line 27	442.22	1.36	600.03**	61.00	0.99	6.48**	45.07	0.57	3.30**	22.06	0.94	0.12
Line 14	435.39	0.79	455.05**	63.94	1.16	3.90	44.91	0.94	3.24**	23.30	1.09	1.20
Line 16	387.47	0.57	378.69**	69.42	0.76	28.45**	50.55	0.96	3.66**	23.42	0.78	1.64
Mean	433.41			62.34			45.90			22.69		
L.S.D 0.05	8.53			1.29			0.887			0.639		

Xi: Mean overall environments, bi: Regression coefficient, S²di: Deviation from regression coefficient, o Ardab = 150 kg ■ Feddan = 4200 m²

and adaptive for different traits. For grain yield, three varieties Inqilab-91, AS-202 and v-97112 with "b" value near to unity, non-significant deviation from regression and above average yield were proved to be stable genotypes. As regards number of grains/spike, Inqilab-91 and SH-202 were classified as stable ones. The varieties U906-2000 and Iqbal-2000 for 1000-grain weight and Chenab-2000 and MH-97 for number of tillers showed stability. Whereas, Hamam and Khaled (2009) indicated that joint regression analysis of variance revealed highly significant differences among genotypes for yield traits. Wheat genotypes Sids 1, local Kah-91, Johara-19, Sahel 1, BocRo-4/Kauz "s", TRI 5646 and TRI 258 were stable for 1000-kernel weight as well as Sids 1, Sakha 93, Giza 168 and TRI 4157 for grain yield. Whereas, genotypes local Kah-91, Johara-19, Sahel 1, TRI 18,686, TRI 3399 and TRI 5649 gave significant S^2d and instable. The "b" values of local kah-91, TRI 4157 and TRI 258 genotypes were less than one and these genotypes were adapted for stress environments.

Yield stability of twelve wheat genotypes under four environments (two sowing dates and two years) to assess the high temperature tolerance were investigated by Al-Otayk (2010). Stability analysis revealed that the genotypes YR-19 and YR-20 gave high mean yield and regression coefficient "b" not significantly different from the unit. Hereby these genotypes performed well under wide range of conditions. On the other hand, YR-2 and YR-3 showed above average stability (b = 1.582 and 1.594), indicating that these genotypes performed well under favorable conditions, whereas, its grain yield reduce markedly under stress conditions.

During two years under five diverse locations of North West of Pakistan, Parveen et al. (2010) estimated stability parameters for tiller/m^2, grain weight and yield of thirteen spring wheat cultivars. Pooled analyses of variance revealed significant differences among locations, years and location × year interaction for previous traits. Wide range of stability statistics were obtained among cultivars for stability parameters. The two cultivars Tramb-73 and Nowshera-96 had regression coefficient above 1.0 for tillers/m^2, 1000-kernel weight and grain yield, revealing that these genotypes adapted to favorable environments. Whereas, the cultivar Saleem 2000 had regression coefficient below 1.0 for these traits, hence it was adapted to unfavorable environments. Arain et al. (2011) evaluated twelve bread wheat genotypes at different locations having different agro-climatic conditions. Stability analysis showed a wide variation among genotypes; some genotypes exhibited wide adaptation while other showed specific adaptation either to favorable or unfavorable environments. Genotypes MSH-03 and MSH-05 produced grain yield over range of environments showed below regression coefficient (b = 0.78 and 0.69, respectively) and higher deviation from regression (S^2d = 1.076 and 1.29 respectively), indicated specific adaptability of these genotypes to stress environments. Genotype NIA-47 with above average regression coefficient (b = 1.23), could produce higher yield at favorable environments.

To assess heat tolerance under four environments, El-Ameen (2012) estimated stability analysis of 17 wheat genotypes. Genotypes No. 7, 3 and 1 produce high grain yield and regression coefficient not significant from the unity. Conversely, genotypes No. 15, 14, 12, 11 and Giza 168 displayed "b" values more than unity (1.16, 2.24, 1.78, 1.82 and 1.14), showing that these genotypes performed well under favorable environments, however their grain yield are reduce clearly under stress environments. Concerning 1000-grain weight, late sowing which represented the heat stress resulted in a significant reduction by 22.22%. Three genotypes displayed minor S^2d estimates and were insignificant, whereas genotypes No. 9 and 14 showed large estimates and were significant. Genotypes No. 7, 3 and 6 had great number of kernels/spike and regression coefficient "b" not significant from the unity. In contrast, genotypes No. 12, 14 and 11 had "b" (1.98, 1.88 and 1.52), demonstrating that they performed well under favorable environments. Furthermore, ten bread wheat genotypes were verified across 8 environments (the combinations of 2 years × 2 irrigation regimes × 2 sowing dates) by Abd El-Shafi et al. (2014). They found a wide range of regression coefficients between genotypes, showed a differential response to environmental fluctuations. The regression coefficient "b" fluctuated from 0.70 to 1.33 for grain yield. The genotypes Bouhouth 8, Cham 8 and Sahel 1 were categorized as highly stable through environments because the regression of these genotypes did not significantly different from 1.0. Additionally, the S^2d estimates of LR-40 were not significantly different from zero, thus could be considered as good stability. Bouhoth 4 had larger "b" estimates, displaying greater sensitivity to environmental fluctuations and was relatively suitable under favorable environments. All genotypes, except LR-40 were unstable based on S^2d estimates, demonstrating high sensitivity to climate changes. Moreover, Omnya Elmoselhy et al. (2015) showed that wheat genotype Misr 2 was highly adapted to favorable conditions for number of spikes/m^2, 1000-grain weight and grain yield (ard./fed.) as well as Sids 1 for number of spikes/m^2, Misr 1 for 1000-grain weight and Shandweel 1 for grain yield (ard./fed.). However, Sids 13 could be grown under stress conditions for number of spikes/m^2 and grain yield (ard./fed.). The most desired and phenotypically stable genotypes were Line 14 for number of grains/spike and Misr 1, Gemmeiza 11, Sids 12, Line 14 and Line 16 for grain yield (ard./fed). It is importance to note that, wheat genotype Misr 1 which appeared to be more preferred and stable in grain yield, was also more stable for 1000-grain weight. Moreover, Line 14 was the most preferred and stable for both number of grains/spike and grain yield (Table 4.17 and Figs. 4.3, 4.4, 4.5 and 4.6).

Figures 4.3, 4.4, 4.5, 4.6 Graphical of stability parameters (bi) and mean performance (-Xg) of individual genotypes for number of spikes/m^2, number of grains/spike, 1000-grain weight and grain yield (Omnya Elmoselhy et al. 2015).

From the perspective that grain yield is a quantitative and complex trait that was found to be affected by genotype by environment interaction (G × E). Further-more, grain yield is the most critical factor that determines a cultivar's accep-tance by farmers. ElBasyoni (2018) showed that under twelve environments *i.e.* two sowing dates, two locations during three years, Gemmeiza 9 displayed high stability according to coefficient of variation (C.V%), superiority measure (P_i) and Wrike's ecovalence (W_i). Gemmeiza10 was the most stable one for stability parameters; regression coefficient (b_i), and Perkins and Jinks (D_i). Stability parameters Wrike's ecovalence (W_i) and average absolute rank difference of genotype on environment (S_i) recognized Gemmeiza12 to be the most stable. Besides, regression coefficient (b_i) and Perkins and Jinks (D_i) identified Gemmeiza10 to be high stable one. In

No.	Genotypes
1	Misr1
2	Misr 2
3	Sakha 93
4	Giza 168
5	Gemmeiza 11
6	Shandweel 1
7	Sids 1
8	Sids 12
9	Sids 13
10	Line 27
11	Line 14
12	Line 16

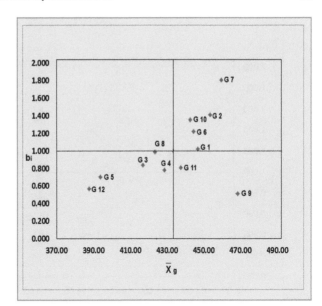

Fig. 4.3 Number of spikes/m^2

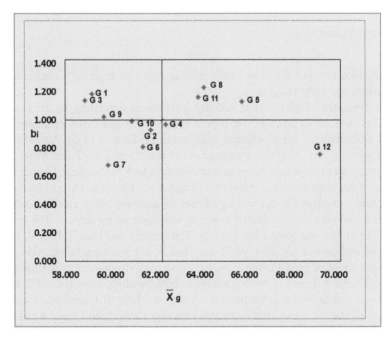

Fig. 4.4 Number of grains/spike

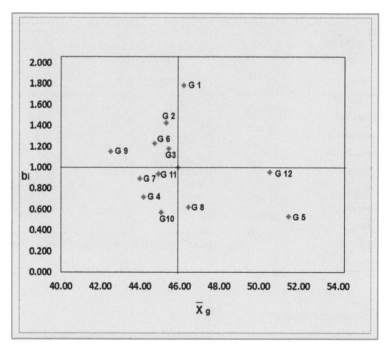

Fig. 4.5. 1000-grain weight

addition, Giza168 was the most stable cultivar based on regression coefficient (b_i) and Perkins and Jinks (D_i).

The estimates of phenotypic stability parameters according to Eberhart and Russell (1966) were computed for eleven bread wheat genotypes evaluated under twelve environments for mentioned traits given in Table 4.18 by Awaad (2021). Phenotypic stability parameters showed that Line 5 and Line 7 and Misr 1 were highly adapted to improved environments for days to 50% heading; Line 2, Line 4 and Line 5 for grain protein content as well as Line 3, Line 8 and Giza 168 for grain yield as they displayed $b_i < 1$ and significant. In contrast, wheat cultivars Sakha 94 and Giza 168 were highly adapted to stress environments for days to 50% heading; Line 7, Giza 168 and Misr 1 for grain protein content and Line 2, Line 4, Line 6, Sakha 94 and Misr 1 for grain yield, they had $b_i > 1$ and significant. Also, wheat genotypes could be grown under a wide-ranging of environments were Line 1, Line 2, Line 3, Line 4, Line 6, Line 8 for days to 50% heading; Line 1, Line 3, Line 6, Line 8 and Sakha 94 for grain protein content and Line 1, Line 5 and Line 7 for grain yield. The most preferred and stable genotypes were Line 1, Line 4 and Line 8 for earliness; Line 8 for grain protein content as well as Line 1 and Line 5 for grain yield. These preceding genotypes presented mean values above grand mean, regression coefficients (b_i) did not differ significantly from unity with lowest deviation mean squares (S^2_{di}).

The above-mentioned results, regarding differences in wheat genotypes in adaptation and stability, are of great importance in the perspective of tolerance to different

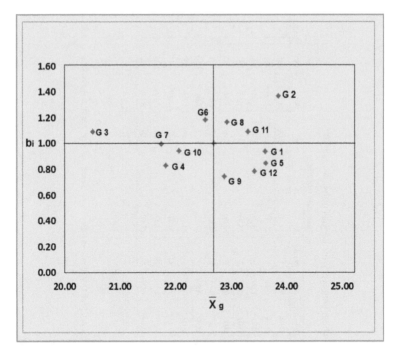

Fig. 4.6 Grain yield

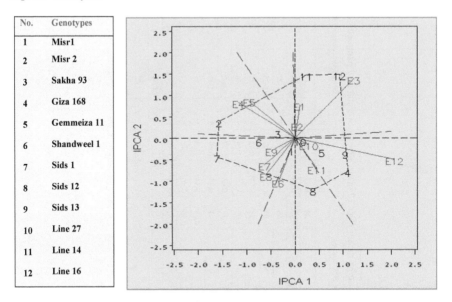

Fig. 4.7 Graphic display of the GE biplot for twelve bread wheat genotypes across twelve environments for grain yield (ard./fed.) assessed (G1-G12) and the twelve environments considered (E1-E12) in the AMMI model (Omnya El Moselhy et al. 2015)

Table 4.18 Genotype means over 12 environments and stability parameters of the 11 wheat genotypes for various traits (Awaad 2021)

Genotypes	Days to 50% heading					Grain protein content %					Grain yield (ardᵒ/fed.)				
	X̄	b_i	S^2di	ASV	Rank	X̄	b_i	S^2di	ASV	Rank	X̄	b_i	S^2di	ASV	Rank
L 1	79.24	1.00	0.48	1.07	7	11.15	1.14	0.04	0.34	1	24.40	1.06	3.17	0.66	2
L 2	81.29	0.92	0.16	0.13	1	10.98	1.20*	0.04	0.71	4	21.03	0.66**	2.66	0.94	3
L 3	82.94	1.16	0.18	0.39	2	11.27	1.03	0.12*	1.09	9	23.26	1.42**	5.31*	2.02	7
L 4	80.62	0.96	0.21	0.43	4	11.25	1.34**	0.07	0.67	2	23.93	0.82*	1.56	0.32	1
L 5	82.53	1.21*	0.64*	0.42	3	11.27	1.23*	0.09	0.87	6	22.96	1.01	3.94	1.15	4
L 6	81.00	1.03	1.10**	1.44	9	11.46	0.98	0.17**	0.76	5	22.33	0.67**	6.31**	1.58	6
L 7	85.00	1.27*	0.63*	0.97	6	12.02	0.75*	0.10*	0.89	7	23.57	1.11	8.06**	2.60	10
L 8	81.88	0.91	0.37	0.69	5	11.87	0.97	0.09	0.68	3	23.77	1.67**	1.84	2.09	8
Sakha 94	80.43	0.65*	0.54*	1.59	11	11.24	1.07	0.17**	1.15	11	20.71	0.63*	6.00*	2.41	9
Giza 168	79.23	0.56*	0.40	1.47	10	11.67	0.69**	0.19**	1.14	10	23.33	1.31*	7.19**	2.73	11
Misr 1	80.89	1.32*	0.73**	1.36	8	11.70	0.62**	0.19**	1.08	8	22.10	0.64*	3.14	1.26	5
Mean	81.26					11.44					22.85				
L.S.D	0.42					0.17					1.26				

*, **Significant at 0.05 and 0.01 levels of probability, respectively. X̄: Mean overall environments, b_i: Regression coefficient, S^2di: Deviation from regression coefficient and ASV: AMMI Stability Value, o Ardab = 150 kg ∎ Feddan = 4200 m²

Table 4.19 Scores of 12 bread wheat genotypes across 12 environments for grain yield (ard./fed.) assessed (G1-G12) for two first axes of the biplot representation (IPCA 1 and IPCA 2) and AMMI stability value (ASV) (Omnya El Moselhy et al. 2015)

Genotypes	Grain yield (ard°/fed.˙)		
	IPCA 1	IPCA 2	ASV
Misr 1	−0.07	−0.27	0.30
Misr 2	−1.57	0.39	2.75
Sakha 93	−0.36	0.12	0.64
Giza 168	1.10	0.77	2.06
Gemmeiza 11	0.56	−0.31	1.02
Shandawell i	−0.75	−0.07	1.30
Sids 1	−1.61	−0.44	2.82
Sids 12	0.37	1.21	1.37
Sids 13	1.05	−0.35	1.85
Line 27	0.11	−0.05	0.20
Line14	0.24	1.47	2.19
Line 16	0.93	1.49	2.01
Kafer Al-Hamam (D1) 2010/2011	0.10	0.78	0.79
Kafer Al-Hamam (D2) 2010/2011	0.05	0.30	0.31
Kafer Al-Hamam (D3) 2010/2011	1.23	1.38	2.53
Sids (D1) 2010/2011	−1.17	0.83	2.18
Sids (D2) 2010/2011	−0.95	0.87	1.85
Sids (D3) 2010/2011	−0.36	−1.02	1.19
Kafer Al-Hamam (D1) 2011/2012	−0.63	−0.62	1.26
Kafer Al-Hamam (D2) 2011/2012	−0.60	−0.86	1.35
Kafer Al-Hamam (D3) 2011/2012	−0.49	−0.29	0.89
Sids (D1) 2011/2012	0.28	−0.15	0.51
Sids (D2) 2011/2012	0.46	−0.72	1.07
Sids (D3) 2011/2012	2.07	−0.50	3.61

D1: 20th November as optimum sowing date, D2: 10th December as moderate late sowing date and D3: 30th December as late sowing date "heat stress" ○ Ardab = 150 kg ■ Feddan = 4200 m^2

environmental stresses i.e. high temperature, drought, salinity and others. And thus the possibility of recommending the cultivation of more promising genotypes to overcome fluctuations in climate conditions or introduce them in hybrid breeding programs to produce more adapted and stable cultivars.

4.5 Additive Main Effects and Multiplicative Interaction

Due to the results of stability parameters might be inconsistency in describing the stability and adaptability of genotypes. The additive main effects and multiplicative interaction (AMMI) model combines the analysis of variance for the genotype and environment main effects with the principle components analysis of the genotypes-environments interaction. The model use the standard analysis of variance (ANOVA) method, where after the AMMI model separates the additive variance from the multiplicative variance (interaction), and then applies PCA to the interaction (residual) portion from the ANOVA to extract a new set of coordinate axes which account more efficiently for the interaction patterns (Shafii et al. 1992). A genotype is considered as stable if its first and second correspondence-analyses scores are near zero (Lopez 1990 and Kang 2002). The AMMI is an informative method of stability and adaptation analysis to be employed in applied crop breeding and subsequent cultivar recommendations. Many investigators studied additive main effects and multiplicative interactions analyses (AMMI) to recognize the most stable genotypes in yield performance under different environments of them, Kaya et al. (2002). They exposed twenty bread wheat genotypes to six environments and found that 90.76% of the total sum of squares was due to environmental effects, only 2.5% to genotypic effects and 7.12% to GEI effects. A large sum of squares of environments revealed that the environments were divers, with great differences between environmental means causing most of the variation in grain yield. Results from AMMI analysis also indicated that the first principle component axis (PCA 1) of the interaction seized 50.78% of the interaction sum of squares in 24.21% of the interaction degrees of freedom. Likeness, the second principle component axis (PCA 2) showed a further 27.86% of the GEI sum of squares superior than that of genotypes. The mean squares for the PCA 1 and PCA 2 were significant and cumulatively contributed to 78.64% of the total GE.

Stability of 13 winter bread wheat genotypes across 4 locations and 2 years with regard to grain yield were verified in Lithuania by Tarakanovas and Ruzgas (2006). The analysis of variance of the 13 genotypes in 8 environments displays that genotype (G), location (L), crop-year (Y) and their interactions were significant for grain yield. Highly significant G × L effects revealed the necessity for evaluating wheat genotypes under several locations. Additive main effects and multiplicative interaction (AMMI) model was effective for evaluating winter wheat genotype-environment interaction (GEI). The first bilinear AMMI model terms accounted for 76.8%. The biplot displays that the genotypes Zentos, Compliment, LIA3948, Elfas and Marshal are best appropriate for cultivation in a wide range of environments. While, genotypes Cubus, Aristos, Marshal and LP.790.1.98 are best suitable for cultivation in favorable environments. Genotype Meunier is well-suitable for cultivation in poor environments. Genotype Elfas was the best at combining yield stability and productivity. Whereas, genotypes Aristos, LP 790.1.98 and Marshal were more stable but lower yielding than Elfas.

Ten bread wheat genotypes were used to assess the stability under arid conditions of Northern Sudan by Mohamed (2009). Mean squares from AMMI analysis of variance indicated significant variation among the genotypes, environments and their interaction for grain yield. The GEI is highly significant and accounting for 16.6% of the sum of squares implying the need for investigating the nature of differential response of the genotypes to environments. The genotype × environment interaction was partitioned into four interaction principle components analysis axis. The IPCA 1 score is highly significant explaining 65.32% of the variability relating to GEI. The IPCA 2 was significant accounting for 16.9% of the variability. The cheek cultivar El-Nielain exhibited high yield associated with good stability and would be expressed to perform well in a wide range of environments. The cheek Wadi Elniel ranked first in grain yield but trend to be unstable. Selaim environment was the best for testing wheat genotypes. Moreover, fourteen bread wheat genotypes grown in six different environments (3 years × two different conditions i.e. irrigated and rainfed) were used to determine the AMMI stability value and yield stability by Farshadfar et al. (2011). The results indicated that the first four principal component axes (IPCA 1, IPCA 2, IPCA 3 and IPCA 4) were highly significant. The partitioning of total sum of squares exhibited that the environment effect was a predominant source of variation followed by GE interaction and genotype effect. The GE interaction was higher than that of the genotype effect, suggesting the possible presence of different environment groups. AMMI stability values discriminated genotypes No. 10 and No. 6 as the stable accessions, respectively.

Also, Mohamed et al. (2013) utilized additive main effects and multiplicative interaction (AMMI) and GEE- biplot analysis of genotype × environment interactions for grain yield of ten bread wheat genotypes at 12 environments (2 years × 2 sowing dates × 3 drought treatments). The results showed that bread wheat grain yields was significantly affected by environments E, which explained 75.01% of the total treatment (G + E + GE) variation, whereas, the G and GEI were significant and accounted for 9.48 and 15.5%, respectively. Three IPCAs were significant, IPCA 1, IPCA 2 and IPCA 3 accounted for 65.49%, 17.10% and 10.11% of the G × E interaction, respectively. The three IPCAs represent a total 92.70% of the interaction variation. Based on the two analyses AMMI and GGE-biplot models, Giza 168, Sakha 8 and Sids 1 categorized by high yield and stability. So, Giza 168 closes to ideal genotype, and is adapted for a wide range for drought and heat stress conditions. At the same time, ten bread wheat genotypes were evaluated at five wheat growing locations by Hagos and Abay (2013). The AMMI analysis of variance for grain yield perceived significant effects for genotype, location and genotype by location interaction. Location effect was responsible for the greatest part of the variation, followed by genotype and genotype by location interaction effects. Based on AMMI stability value, G4, G10, G8 and G9 were the most stable genotypes, while G1, G2 and G3 were the most responsive genotypes. Omnya El Moselhy et al. (2015) computed mean squares from AMMI analysis for grain yield of twelve bread wheat genotypes across twelve environments. Environments (E), genotypes (G) and the G × E interaction mean squares were highly significant for yield. Also, the AMMI analysis of variance showed that (70.79%) of the total sum of squares were due to environmental effects, (6.70%) to genotypic effects and (12.83%) to GEI effects for grain yield/fed., respectively. A high sum of squares for environments indicated that the environments

were different, with large variances among environmental means producing most of
the differences in the evaluated characters. The amount of the GEI sum of squares
was higher than that for genotypes in respect to grain yield/fed., demonstrating the
existence of considerable differences in genotypic response across environments.
The genotype × environment interaction (GEI) was divided into five interaction
principle components analysis axis (IPCAs) for grain yield. The five IPCAs were
highly significant, and IPCA 1, IPCA 2, IPCA 3, IPCA 4 and IPCA 5 accounted
for 36.00, 20.04, 14.87, 14.15 and 6.67%, respectively. IPCA scores of genotypes
and environments exhibited positive and negative estimates. A genotype with great
positive IPCA score in some environments must have great negative interaction in
some other environments. Consequently, these scores presented a disproportionate
genotype response, which was the main source of variation for any quantitative interaction. They added that the AMMI stability value "ASV" showed that genotypes Line
27, Misr 1 and Sakha 93 were the most desirable and stable, from them, Misr 1 have
high yield potentiality, while Line 27 was moderate yield potentiality and Sakha 93
was the lowest one. Moreover, cultivars Gemmeiza 11 and Shandweel 1 seemed to
be moderate stability, however, genotypes Misr 2, Sids 1, Line 14, Line 16, Giza
168 and Sids 12 were unstable and more responsive, from them, Misr 2, Sids 12,
Line 14 and Line 16 have high yield potentiality, whereas Sids 1 was moderate yield
potentiality and Giza 168 was the lowest one as showing in Table (19) and Fig. (7).
The best genotypes with respect to E2 and E10 were Misr 1, Sakha 93, Gemmeiza 11
and Line 27; E9 for Shandweel 1; E1 for Line 14; Line 16 for E3; Giza 168 and Sids
13 for E12; Sids 1 for E6, E7 and E8 and Misr 2 for E4 and E5. The environments
E3 and E12 were the discriminative ones.

Under El-Kattara and Ghazalla districts of Egypt, Ali and Abdul-Hamid (2017)
conducted some field experiments to evaluate 38 bread and durum wheat genotypes
for grain yield adaptability and stability under twelve environments for drought and
heat stresses. The AMMI analysis of variance showed that environments explained
77.21% of total variation and it was greater than genotypes (5.30%) and genotype
× environment (GEI) (12.54%). IPCA 1 score explained 25.08% and IPCA 2 had
17.81% of the total GEI for AMMI model. Whereas, IPCA 1 score explained 36.02%
and IPCA 2 had 17.56% of the total GGEI for SREG model. According to the ASV
ranking, the bread wheat genotypes, Line 10, Giza 168, Line 15, Line 8 and Sakha
93 and durum line 31 were more stable. GGE biplot displayed Line 3 as ideal wheat
genotype for grain yield.

To complement the results of stability measures, genotype by environment (G ×
E), ElBasyoni (2018) using the additive main effect and multiplicative interaction
(AMMI) analyses and genotype main effect plus genotype × environment interaction
(GGE). He found that the two models were fitted in the AMMI and GGE biplot;
the first was by considering sowing dates as portion of the environments, i.e., 12
environments, whereas the second was by running the AMMI analysis across years
and locations within each sowing date, i.e., six environments. In the first model (12
environments), the analysis of variance for AMMI model revealed significant effect
of the environments, genotypes, and genotype × environment interaction. However
the variance of the environment was 63.2%, but the variance due to genotypes was

14.6% and those for genotype × environment interaction were 22.2%. In the second model (6 environments within each sowing date), the variance of the AMMI model for the recommended sowing date was 29.52, 46, and 24.48% for the environments, genotypes and genotypes × environment interaction, respectively. Moreover, the variance of the AMMI model for the late sown condition valued 12.99, 48.57, and 38.44% for the environments, genotypes and genotypes × environment interaction, in the same respective order. In both models, the genotype × environment interaction was highly significant, advising variance response of genotypes to environments. Sizable variance for the environments in the first model compared to the second model was recognized, shows an intensification effect of sowing dates on the environmental effect. Finally, Awaad (2021) showed that analyses of variance for environments (E), wheat genotypes (G) and the G × E interaction mean squares were highly significant for grain yield (ard./fed.). Variance components of the sum of squares differed from genotypes, environments and GEI viz. 6.52%, 43.06% and 23.33%, respectively (Table 4.20). IPCA 1 score had 33.31% and IPCA 2 had 26.56% of the total GEI for AMMI models. For the SREG model, IPCA 1 score displayed 39.63%, and IPCA 2 had 22.80% of the total GGEI. As given in Table 20 and Fig. 4.8 revealed that Line 4, Line 1, Line 2, Line 5 and Misr 1 were the most desired and stable genotypes exhibited ASV 0.32, 0.66, 0.94, 1.15 and 1.26, respectively. GE biplot graph for

Table 4.20 AMMI analysis of variance over twelve environments for grain yield (ard./fed.) (Awaad 2021)

Source of variation	df	Grain yield (ard./fed.*)		
		Sum of square	Mean of square	Percent
Environment (E)	11	3458.42	314.40**	43.41
Reps/Env	24	399.99	16.67**	
Genotype (G)	10	523.78	52.38**	6.52
G × E	110	1874.16	17.04**	23.33
IPCA1	20	624.26	31.21**	33.31
IPCA2	18	497.83	27.66**	26.56
IPCA3	16	247.59	15.47**	13.21
IPCA4	14	155.14	11.08	8.28
IPCA5	12	110.65	9.22	5.9
IPCA6	10	91.49	9.15	4.88
IPCA7	8	77.70	9.71	4.15
IPCA8	6	54.42	9.07	2.9
IPCA9	4	13.39	3.35	0.71
IPCA10	2	1.67	0.83	0.09
Pooled error	240	1775.58	7.40	
Total	395	8031.93		

*Ardab = 150 kg, Feddan = 4200 m^2, **Significant at 0.01 level of probability

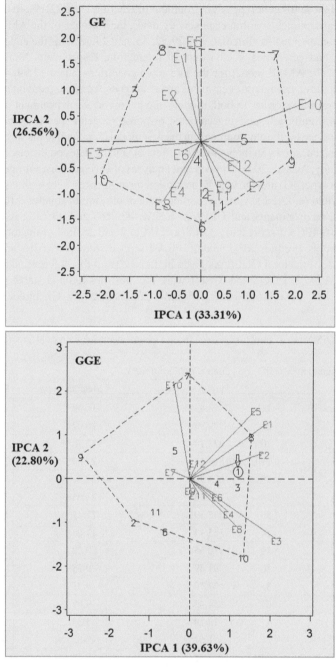

Fig. 4.8 Graphics display of the GE and GGE biplots for eleven wheat genotypes (assessed G1 - G11) and 12 environments (assessed E1- E12) in the AMMI and SREG models, respectively grain yield (ardab/fed.) (Awaad 2021)

the AMMI model illustrated that environments E_5, E_{10}, E_7 and E_3 were the most differentiating ones. The vertex Line 7, Line 9, Line 6, Line 10, Line 3 and Line 8 were located far away from the origin. Line 4, Line 1, L2, L5 and Line 11 were the desirable and stable. GGE biplot graph for the SREG model showed that, Line 1 was ideal genotype, it had the highest vector length of the high yielding genotypes with zero GE, as represented by the dot with an arrow pointing to it in Fig. 4.8. Line 3 and Line 4 were desirable genotypes. The environments E_7 with E_{10}, E_{12} with (E_5 and E_1) and E_6 with E_8 were positively correlated. Whereas, the environments E_7 and E_{10} had negatively correlated with E_6, E_4, E_8, E_3 and E_{11}. The ideal test environment was E_2 and the favorable environments were E_1 and E5, but the unfavorable ones were E_7, E_8 and E_9.

4.6 Gene Action, Genetic Behavior and Heritability for Wheat Traits Related to Environmental Stress

Environmental stress tolerance is a complicated parameter in which performance of crop varieties can be affected by several features (Ingram and Bartels 1996). So, it is of interest to mention that traits with high heritability are easier to improve than those with lower heritability (Saba et al. 2001). Assessment of heritability of a character and its relationship with grain yield is beneficial in formulating appropriate breeding strategy and reliable indices for genetic improvement especially under stress conditions. Hence, indices with high heritability might be used as indirect selection criteria to improve yield in stress environments (Jatoi et al. 2011). In this regard, six diverse wheat genotypes have been employed and crossed in a half diallel fashion by Awaad (2001). He assessed some phonological and morpho-physiological characters related to stress conditions. Results indicated that both additive and dominance genetic components were involved in the inheritance of days to heading and grain filling rate, flag leaf chlorophyll content and grain yield/plant. The additive gene effect was the main type controlling the inheritance of flag leaf area, whereas the dominance gene effect was the prevailed type controlling the inheritance of grain filling period. The dominant increasing alleles were more frequent than recessive ones for flag leaf area, flag leaf chlorophyll content, grain filling period, grain filling rate and grain yield/plant. Low narrow sense heritability was recorded for flag leaf chlorophyll content, grain filling period and grain filling rate, moderate for both flag leaf area and grain yield/plant, while, it was high for days to heading. Moreover, at Ismailia Agricultural Research Station, Egypt, as stress environment, Salem et al. (2003) determine the adequacy of genetic model and gene action controlling relative water content, transpiration rate, osmotic pressure, proline content, leaf chlorophyll content, flag leaf area, days to heading and grain yield/plant. They employed six population (P_1, P_2, F_1, F_2, B_1 and B_2) of five wheat crosses, namely (1) Skha 69 × Sahel 1, (2) Sakha 69 × Shi# 4414/GOW "s" // Seri 82, (3) Sahel 1 × Bocro-4, (4) Gemmeiza 5 × Giza 168 and (5) Shi# 4414/Gow "s"// Seri 82 × Bocro − 4

(Tables 4.21 and 4.22). The results indicated the importance of additive genetic variance (D) in the genetic control of days to heading in all crosses, relative water content in 1st cross; transpiration rate in 1st and 3rd crosses; proline content in 1st, 2nd and 4th crosses; leaf chlorophyll content in 1st, 4th and 5th crosses; flag leaf area 2nd, 3rd, 4th and 5th crosses as well as osmotic pressure and grain yield/plant in 3rd and 5th ones. Whereas, the dominance genetic variance (H) was found to be the prevalent type controlling the remaining crosses. They obtained high (> 50%) narrow sense heritability for relative water content, transpiration rate, osmotic pressure, proline content, leaf chlorophyll content, flag leaf area and days to heading in most cases, while it ranged from low (25.25%) to moderate (46.95%) for grain yield/plant.

Six Egyptian bread wheat cultivars and lines were involved by Al-Naggar et al. (2007) under full and stress irrigation. Both additive and dominance genetic variance were importance in the inheritance of previous characters under both situations. The magnitude of additive was greater than that of dominance one for the studied traits, and played the major role in the inheritance of earliness and grain filling traits under the two irrigation regimes. Average degree of dominance $(H_1/D)^{0.5}$ was less than unity for days to heading, days to maturity and grain filling period, indicating that partial dominance played the most important role in the inheritance of that traits. Narrow sense heritability estimates were very high for days to heading, grain filling period, high for days to maturity and moderate for grain filling rate. Moreover, in order to assess efficiency of drought tolerance indices, El-Rawy and Hassan (2014) evaluated 50 bread wheat genotypes under three environments: normal (clay fertile soil, E_1), 100% (E_2), and 50% (E_3) field water capacity in sandy calcareous soil. A moderate to high broad-sense heritability was found for 1000-kernel weight (0.47), spike length (0.38), plant height (0.54), flowering time (0.73), stomata frequency (0.59) and stomata length (0.54). Meanwhile, low broad-sense heritability estimates were found for grain yield/plant (0.15); number of tillers (0.09) and stomata width (0.03).

Genetic components of variance and their derived parameters under normal irrigation and drought stress conditions has been estimated by Naglaa Qabil (2017). Results indicated that (Table 4.23) additive component (D) was significant for days to heading under two environments, revealing the importance of phenotypic selection in improving earliness. On the other hand, dominance genetic components (H1 and H2) were significant for flag leaf area under normal irrigation condition as well as grain yield/plant under drought stress environment. Meantime, both additive (D) and dominance (H1 and H2) genetic components were significant for chlorophyll content under both normal irrigation and drought stress environments. The dominance components were higher in magnitude than the corresponding additive one for flag leaf area, chlorophyll content and grain yield/plant under both two environments. The average degree of dominance was more than unity, confirming the importance of over-dominance, so the non-fixable gene type could be exploited efficiently through hybrid breeding method. F values coupled with KD/KR showing more frequent of the increasing dominance alleles than the recessive ones in the parental populations for days to heading under normal irrigation condition, as well as chlorophyll content and grain yield/plant under two environments. The environmental variance (E) was

Table 4.21 Additive (D) dominance (H) and environmental (E) variances and heritability (T_b and T_n) for relative water content %, transpiration rate mg H_2O/ g. FW./h., osmotic pressure and proline content μmoles/g. F.W. in five wheat crosses (Salem et al. 2003)

Character	Parameter					
	D	H	E	$\sqrt{H/D}$	Tb%	Tn%
Relative water content						
Cross						
1	155.063	158.235	15.086	1.01	88.586	58.658
2	9.674	98.123	6.836	3.185	81.118	13.36
3	90.667	151.217	65.229	1.29	56.035	30.555
4	254.72	176.566	25.474	0.833	87.067	64.658
5	147.789	342.151	38.096	1.522	80.714	37.41
Transpiration rate						
Cross						
1	10984.32	10425.255	436.303	0.974	94.888	64.435
2	9953.36	34386.452	3737.723	1.859	78.408	28.749
3	8188.57	6215.269	265.642	0.871	95.508	69.233
4	423.927	2641.341	1142.278	2.496	43.299	10.521
5	3172.755	4283.65	289.985	1.162	90.161	53.825
Osmotic pressure						
Cross						
1	11.091	19.19	0.372	1.315	96.528	51.754
2	0.523	2.559	0.368	2.212	71.006	20.603
3	7.761	1.945	0.31	0.501	93.371	82.974
4	0.2003	4.354	0.245	4.662	82.911	6.985
5	6.998	1.329	0.345	0.436	91.739	83.783
Proline content						
Cross						
1	68.151	52.851	0.403	0.881	99.155	71.145
2	51.948	0.597	0.5	0.107	98.122	97.561
3	8.776	20.809	3.612	1.539	72.641	33.237
4	78.88	31.699	0.441	0.634	99.078	82.501
5	6.538	22.079	1.445	1.838	85.88	31.943

significant for days to heading under drought stress environment. The proportion of genes with positive and negative effects in the parents as indicated by (H2/4H1) were less than its maximum value (0.25) for days to heading, flag leaf area, under two environments, chlorophyll content under drought stress environment, suggesting asymmetrical distribution of positive and negative alleles among the parental population. On the other hand, it was near to its maximum value (0.25) for chlorophyll

Table 4.22 Additive (D) dominance (H) and environmental (E) variances and heritability (T_b and T_n) for chlorophyll content (SPAD value), flag leaf area cm^2, days to heading and grain yield plant (g) in five wheat crosses (Salem et al. 2003)

Character	Parameter					
	D	H	E	$\sqrt{H/D}$	Tb%	Tn%
Leaf chlorophyll content						
Cross						
1	28.454	9.51	2.167	0.578	88.456	75.79
2	180.1	197.696	8.861	1.048	94.026	60.707
3	16.852	22.036	13.494	1.144	50.804	30.719
4	39.029	13.695	2.787	0.592	89.166	75.857
5	59.214	24.87	8.564	0.648	80.707	66.699
Flag leaf area						
Cross						
1	3.485	62.825	1.335	4.246	92.893	9.277
2	105.989	52.788	2.724	0.706	96.043	76.894
3	50.042	1.848	0.616	0.192	97.64	95.87
4	46.273	15.959	0.947	0.587	96.627	82.415
5	58.46	15.778	0.688	0.52	97.968	86.32
Days to heading						
Cross						
1	8.098	1.371	1.283	0.412	77.399	71.357
2	2.607	2.263	1.631	0.932	53.413	37.245
3	38.559	33.524	1.646	0.932	94.383	65.932
4	10.419	1.327	1.549	0.357	77.599	72.954
5	2.952	1.143	1.476	0.622	54.412	45.588
Grain yield/plant						
Cross						
1	0.142	0.156	0.0727	1.049	60.241	38.871
2	0.024	0.111	0.008	2.149	83.582	25.255
3	0.399	0.251	0.227	0.793	53.602	40.777
4	0.036	0.076	0.002	1.439	95.583	46.957
5	0.114	0.096	0.08	0.916	50.272	35.431

content, under normal irrigation environment, as well as grain yield/ plant under two environments. This result suggesting equally distribution of positive and negative alleles among the parental population for these characters. Narrow sense heritability was high for days to heading under two environments, flag leaf area under drought stress environment as well as moderate for flag leaf area under normal irrigation condition. Thus, phenotypic selection might be effective for improving both characters. Heritability estimates in narrow sense were low for chlorophyll content and grain yield/plant under both environments, so selection was ineffective.

Under normally irrigated "four times at different stages", as well as "one irrigated only once at tillering stage" El-Gammaal (2018) estimated gene action for flag leaf area, flag leaf angle, transpiration rate, stomatal resistance, chlorophyll a/b ratio, leaf temperature and grain yield. Mean squares of combining ability were highly significant for most characters in the two environments and the combined, representative the existence of both additive and non-additive types of gene effects. The four parents Gemmeiza 12, Misr1, Giza 171 and Giza 168 might be deliberated as excellent general combiner parents in breeding programs to improve drought tolerance. The best parental combinations were; Giza 168 × Gemmeiza 11, Giza171 × Giza 168, Giza 171 × Gemmeiza 11 and Gemmeiza 12 × Giza 168 for almost of the studied traits. Hence, selection will be used in the segregation generation to determine the most parents and crosses with high grain yield and drought tolerant.

Under three natural photo-thermal environments created by three different sowing dates (10th Oct., 29th Oct. and 26th Nov.), using six bread wheat genotypes and their F_1's by Menshawy (2005). Hayman's model indicated that additive and dominance effects had significant role in the inheritance of earliness component characters i.e., days to heading, days to anthesis, days to maturity, grain filling period and grain filling rate. However additive gene effects were the major part of the genetic variability. Therefore, selection in early segregating generations would be effective in developing promising early lines. The F value was positive and significant reflecting that dominant genes were more frequent than recessive ones among the parental genotypes. Broad and narrow sense heritability values were high for all earliness characters at the three sowing dates. Furthermore, seven parents of bread wheat and their F_1's were tested under two sowing dates, 8th Nov. and 8th Dec. and two nitrogen levels (40 and 80 kg N/fed.) by Salama and Manal Salem (2006). They showed that the magnitude of additive was greater than that of dominance for flag leaf area. The average degree of dominance $(H_1/D)^{0.5}$ was less than unity, suggesting the presence of partial dominance in the F_1 hybrid under both sowing dates. Narrow sense heritability estimates were decreased from 0.84 to 0.31% under both conditions, respectively.

To identify gene action controlling grain yield/plant and its components under two sowing dates, 26th of November (normal sowing date) and 26th of December (late sowing date), a half diallel set of crosses was established among seven wheat genotypes by Ahmed and Mohamed (2009). Additive and non-additive gene effects were involved in controlling all traits under both sowing dates. Most of the variation was attributed to the non-additive gene effects. Heritability in narrow sense was low for grain yield/plant. Under three environments (early, normal and late sown conditions) through an 8 × 8 diallel cross by Ahmed et al. (2011) stated that both additive and dominance gene action were significant for days to heading under early, normal and late planting as well as the dominance under normal and late planting. Significant additive and dominance genetic variation were observed for grain yield/plant under various situations. Unequal values of H_1 and H_2 indicated presence of positive

Table 4.23 Additive (D), dominance (H) genetic variances and their derived parameters for days to heading, yield and its attributes under normal irrigation and drought stress conditions in growing season 2015/2016 (Naglaa Qabil 2017)

Grain yield/plant (g)	Days to heading (day)		Flag leaf area (cm²)		Chlorophyll content (%)		Grain yield/plant (g)	
	I	D	I	D	I	D	I	D
D	18.03**	16.21**	0.95	5.82*	9.50*	4.26*	1.84*	2.83
H1	3.34	2.21	23.88*	12.56*	24.99**	14.24**	9.42**	14.47*
H2	2.64	2.03	18.88*	11.11*	23.73*	12.97**	9.40**	14.33*
F	2.18	-2.27	-0.77	-0.18	6.37	4.33	0.42	1.25
h²	2.05	1.77	1.27	1.63	22.23**	19.84**	13.61**	4.63
E	0.27	0.48*	0.19	0.16	0.63	0.34	0.39	0.29
(H1/D)$^{1/2}$	0.43	0.37	5.02	1.47	1.62	1.83	2.26	2.26
H2/4H1	0.20	0.23	0.20	0.22	0.24	0.23	0.25	0.25
KD/KR	1.33	0.68	0.85	0.98	1.52	1.77	1.11	1.22
h2/H2	0.78	0.87	0.07	0.15	0.94	1.53	1.45	0.32
h(n.s)	89.86	90.42	40.60	55.83	25.06	14.38	20.81	18.18

*and**, Significant at 0.05 and 0.01 levels of probability, respectively. I = Normal irrigation. D = Drought stress

and negative alleles in unequal frequencies. Narrow sense heritability estimates for days to heading valued 71.09 and 50.89% under early and normal planting, respectively, whereas, it was low (25.54%) under late planting, while for grain yield valued 10.13% under normal condition.

Also, seven wheat (*Triticum aestivum* L.) cultivars obtained after screening against heat stress classified as tolerant, moderately and sensitive to heat stress, were mated in a complete diallel system by Farooq et al. (2011). The magnitude of additive was greater than that of dominance for days to heading and days to maturity and played the major role in the inheritance of earliness. The average degree of dominance $(H_1/D)^{0.5}$ was less than unity. The estimates of narrow sense heritability were decreased from 67 to 62 and 77 to 64% for days to heading and days to maturity under normal (10th Nov.) and heat stress (25th Dec.) environments, respectively. The magnitude of additive was greater than that of dominance for flag leaf area, with average degree of dominance $(H_1/D)^{0.5}$ was less than unity, suggesting the presence of partial dominance in the F_1 hybrid under both environments.

In F_1 progenies of 7×7 diallel cross fashion comprising four high temperatures tolerant and three sensitive spring wheat parental genotypes were evaluated under normal and heat stress conditions (10th Nov. and 20th Dec.) by Irshad et al. (2012). Genetic analysis under both conditions showed additive gene action with partial dominance for days to heading, spike index at anthesis, plant height, spikes/plant, spikelets/spike and grain yield/plant. Advising that these characters might be beneficial for the improvement of terminal heat tolerant cultivars by modified pedigree selection. Conversely over dominance type of gene action was registered for spikelets/spike suggesting that further improvement in this trait may be achieved by biparental mating joined with few cycles of recurrent selection. Environmental variance was significant for grain yield/plant under both conditions. High narrow sense heritability estimates were also recorded under both normal and heat stress conditions for these traits, considerably large additive proportion in the heritable genetic variation. Meanwhile, mode of gene action and heritability for earliness, yield and some its attributes of bread wheat genotypes grown under different sowing dates were assessed by Eman Abdallah et al. (2015). Narrow-sense heritability was found to be 67.01, 72.56 and 65.85% for days to heading; 64.86, 67.65 and 61.49% for days to maturity on the early, normal and late sowing dates, respectively. Narrow-sense heritability valued 53.14, 76.36 and 45.31% for plant height; 18.54, 19.66 and 18.38% for number of grains/spike; 15.0, 12.14 and 17.5% for 1000-grain weight and 17.0, 20.0 and 8.05% for gain weight/plant on the early, normal and late sowing dates, respectively. Graphical analysis revealed that additive gene action with partial dominance was governed the inheritance of days to heading and days to maturity under the three sowing dates. Negative intercepts of regression lines indicated nonadditive gene action with partial dominance for plant height on the early and the late sowing dates, number of grains/spike, 1000-grain weight and grain weight/plant under the three sowing dates.

Moreover, EL-Gammaal (2019) carried out three experiments under the three sowing dates under three sowing dates i.e., of 25 October (D_1 sowing date), 15th November (D_2 sowing date) and 5th December (D_3 sowing date) at El-Gemmeiza

Agricultural Research Station, El-Gharbiah Governorate, Egypt during the two successive growing seasons 2014/15 and 2015/16. A half diallel fashion among eight common wheat lines and cultivars was built for estimation combining ability for grain yield and its components. Mean squares due to general combining ability GCA, specific combining ability SCA, GCA × sowing dates and SCA × sowing dates were highly significant and/or significant for grain yield and some of its components in the three sowing dates and combined analyses. The ratio of GCA/SCA were largely exceeded the unity, so the largest portion of the total genetic variability related with these traits was owing to additive and additive × additive gene action. Five crosses (Gemmeiza 12 × Line 1), (Sakha 94 × Sids 1), (Gemmeiza 12 × Shandaweel 1), (Sakha 94 × Misr 1) and (Sakha 61 × Sakha 94) gave significant positive specific combining ability effects ($\hat{S}_{ij}$) estimates for grain and straw yields and some of its components in the three sowing dates and the combined analyses. Thus breeding program through pedigree method should be effective to isolate early lines with high stable grain and straw yields to a wide range of environments. Furthermore, under both normal and salinity stress conditions, Yassin et al. (2019) recorded high values of broad sense heritability were coupled with high estimates of genetic advance as a percent of mean at 10% selection intensity for grain filling rate, spikes/m^2 and biological yield. Whereas, broad sense heritability and genetic advance tended to decrease from 49 and 16.37% under normal to 32 and 12.52% under salinity stress conditions for 1000-grain weight as a result of salinity effect.

4.7 How Can Measure Sensitivity of Wheat to Environmental Stress?

4.7.1 Stress Sensitivity Measurements

It is important to measure the sensitivity of the wheat genotypes to environmental stresses. This is useful in identifying the most tolerant genetic makeup that can be recommended for growing in severe environments to provide satisfactory yield levels. There are several measures used in this regard. Stress tolerance could be assessed as follows using different measurements:

1. Drought sensitivity index (DSI) (Fischer and Maurer 1978)
   DSI = 1 − (Ys / Yp) / SI, while SI = 1 − (Ys / Yp).
2. Stress tolerance index (STI) (Fernandez 1992; Kristin et al. 1997)
   STI = (Yp × Ys) / (Yp)^2.
3. Yield stability Index (YSI) (Bouslama and Schapaugh 1984)
   YSI = Ys / Yp.
4. Harmonic mean (HM) (Kristin et al. 1997)
   HM = 2(Yp × Ys)/(Yp + Ys).

Where, Ys is the grain yield of a genotype under stress; Yp is the grain yield under normal condition; SI is stress intensity; Ys and Yp are the means of all genotypes under stress and normal conditions, respectively.

In order to assess efficiency of drought tolerance indices, El-Rawy and Hassan (2014) evaluated 50 bread wheat genotypes under three environments in Egypt: normal (clay fertile soil, E_1), 100% (E_2), and 50% (E_3) field water capacity in sandy calcareous soil. Grain yield/plant was strongly positively correlated with grain yield/spike, No. of tillers, plant height, flowering time, stomata length, stress tolerance index, yield stability index, and harmonic mean, while negatively correlated with stomata frequency and drought susceptibility index in E_2 and E_3, respectively. Thus, highly heritable traits strongly correlated with grain yield under stress conditions especially stomata frequency and length can be used as reliable indices for selecting high-yielding genotypes tolerant to drought stress.

A wide range of response to drought tolerance in wheat genotypes was registered based on drought sensitivity index (DSI). Naglaa Qabil (2017) showed that the parental wheat cultivar Gemmeiza 11 and F1 crosses (Line 1 × Gemmeiza 11), (Misr 1 × Line 2) and (Gemmeiza 9 × Gemmeiza 11) displayed DSI values less than unity. Thus, these genotypes are considered as more tolerant to drought stress as regards to their grain yield/plant. Moreover, parental wheat cultivars Misr 1 and Gemmeiza 9 and F1 crosses (Line 1 × Gemmeiza 9) and (Misr 1 × Gemmeiza 9) had DSI values near one. So these genotypes are considered as moderate tolerant to drought stress. Otherwise, Line 1 and Line 2 and F1 crosses (Line 1 × Misr 1), (Line 1 × Line 2), (Misr 1 × Gemmeiza 11) and (Line 2 × Gemmeiza 9) give DSI values more than 1.0, hence classified as sensitive to drought stress as previously shown in Table (4.13).

Heat sensitivity index was measured by Menshawy (2007) and varied in genotypes from 0.30 and 0.42 in the early genotype Line 6 and Line 4, respectively to 1.67 and 1.76 in Sakha 93 and Line 12 which were relatively late maturing genotypes. In general, all early maturing genotypes except Sids 4 had low H.S.I compared with the late genotypes and expressed as tolerant to heat stress conditions. Al-Otayk (2010) revealed that heat sensitivity index values ranged from 0.49 to 1.38. The genotypes YR-20 and YR-19 were relatively heat resistant (HSI < 1), while local cultivar "Sama" and YR-2 were relatively heat sensitivity (HSI > 1). Sikder and Paul (2010) identified cultivars Gourab, Kanchan, and Shatabdi as heat tolerant (H.S.I < 1), whereas, cultivars Sonora and Kalyansona were regarded as heat sensitive (H.S.I > 1). El-Ameen (2012) evaluated seventeen wheat genotypes to heat tolerance under four environments. Two cultivars Giza 168 and Sids 12 and genotypes No. 13, 14 and 12 displayed H.S.I values > 1 indicating relative sensitivity to heat stress, whereas, the other genotypes displayed H.S.I values < 1 with relative resistance to heat stress.

Eman, Abdallah et al. (2015) calculated heat sensitivity index (HSI) for determining the stress tolerance of wheat genotypes based on minimization of yield losses due to late sowing compared to optimum sowing date. The parental wheat cultivar Sids 12 (0.78) and F_1 crosses Gemmeiza 9 × Misr 1 (0.78), Gemmeiza 9 × Line 1 (0.68) and Line 1 × Misr 2 (0.71) exhibited HSI values less than unity. Hence these genotypes were considered as more tolerant to heat stress. Furthermore, the

Table 4.24 Tolerance index (TOL) of eleven wheat genotypes for grain yield (Awaad 2021)

Genotypes	Ghazaleh	El-Khattara	Tolerance index (TOL)
Line 1	27.65	21.16	6.49
Line 2	23.12	18.79	4.33
Line 3	27.56	18.87	8.69
Line 4	26.37	21.45	4.92
Line 5	25.59	20.33	5.26
Line 6	24.07	20.59	3.47
Line 7	26.54	20.60	5.94
Line 8	28.62	18.93	9.69
Sakha 94	21.63	19.78	1.85
Giza 168	27.58	19.00	8.58
Misr 1	23.99	20.21	3.78

genotypes showing HSI values near 1.0 are moderate to late sowing, in this respect, parental wheat genotype Line 1 (0.98) and F_1 crosses Sids 12 × Gemmeiza 9 (0.95), Line 2 × Line 1 (0.94) and Line 2 × Misr 2 (0.93) had HSI values near one. On the contrary, Gemmeiza 9 (1.14) and F1 crosses Sids 12 × Line 2 (1.14), Sids 12 × Line 1 (1.23), Gemmeiza 9 × Line 2 (1.20) and Misr 1 × Line 1 (1.16) had HSI values more than 1.0, they were classified as sensitive to heat stress.

Tolerance index (TOL) provides a measure of yield stability based on yield loss under stress as compared to non-stressed condition (Rosielle and Hamblin 1981). In this respect, Awaad (2021) mention that wheat genotypes could be classified into three categories based on TOL values (Table 4.24), the first category was tolerant to stress condition and include Sakha 94, Line 6, Misr 1, Line 2 and Line 4, these genotypes exhibited lower TOL values. The second category was moderate tolerant and include Line 5, Line 7 and Line 1, whereas, the third category was sensitive and comprise Line 8, Line 3 and Giza 168.

Based on the both indices of GMP and STI for salt tolerance, Yassin et al. (2019) showed that Line 11, Line 17, Misr 1, Line 4 and Misr 2 were classified as salt tolerant genotypes. These genotypes took greater estimates of GMP and STI, whereas Line 7, Line 9, Line 20 and Sids 12 were recognized as sensitive ones, as they exhibited low values for GMP and STI. In the same situation, the two indices i.e. TOL and SSPI showed that genotypes Line 10, Line 11 and Line 13 were more tolerant to salinity, whereas Line 9, Line 15 and genotype Sids 12 were more sensitive in comparison with the others. As well as, similar ranking pattern of tolerant /sensitive genotypes were attained by the three indices of YSI, SSI and CV.

4.8 Role of Recent Approaches

4.8.1 Biotechnology

Biotechnology tools are important in understanding the biosynthesis pathways of genes that control the response to drought and heat stress and other environmental stresses. Biotechnology has employed practical methods to engineer crop plants that tolerant to environmental stress conditions (Umezawa et al. 2006 and Awaad 2009). Environmental stress tolerance is a complex process governed by numerous minor effect genes or QTLs and is often confused by variances in plant morphology and physiology at diverse situations (Fu et al. 2016).

To distinguish genomic regions that contribute to grain protein content GPC accumulation and identify accessions with high and low grain protein content under well-watered and water-deficit growth conditions, ElBasyoni et al. (2018) evaluated 2111 accessions of spring wheat using near-infrared spectroscopy (NIR). Results indicated significant influences of moisture, genotype, and genotype × environment interaction on the GPC accumulation. Furthermore, genotypes exhibited a wide range of variation in GPC, indicating the presence of high levels of genetic variability among the studied accessions. Around 366 (166 with high GPC and 200 with low GPC) wheat genotypes performed relatively the same across environments, which implies that GPC accumulation in these genotypes was less responsive to water deficit. Seven single nucleotide polymorphism (SNPs) were linked with GPC under well-watered growth conditions, while another six SNPs were linked with GPC under water-deficit conditions only. Moreover, 10 SNPs were linked with GPC under both well-watered and water-deficit conditions.

Molecular genetics proved that overexpression of TaFBA1, an F-box protein gene extracted from wheat, improved thermo tolerance of transgenic tobacco plants (Li et al. 2018). Overexpression of the TaGASR1 gene in Arabidopsis and wheat plants improved the heat stress tolerance of transgenic plants (Zhang et al. 2017). Additionally, ectopic overexpression of TaWRKY1 and TaWRKY33 from wheat contributed to enhanced tolerance to high temperatures in Arabidopsis (He et al. 2016).

Analyzed the gene expression a profile of wheat under heat stress was implemented by Su et al. (2019). They identified a total of 1705 and 17 commonly difference expressed genes (DEGs) in wheat grain and flag leaf, respectively using transcriptome analysis. Also applied Gene Ontology (GO) and pathway enrichment to clarify the functions and metabolic pathways of DEGs controlling thermo tolerance in both wheat grain and flag leaf. They revealed that transcriptional regulation of zeatin, brassinosteroid and flavonoid biosynthesis pathways may play a significant role in wheat's heat tolerance. Real-time PCR technique has been reported in recognition the expression of three potential genes responsible for the flavonoid biosynthesis process.

Cytosine methylation is a well- known epigenetic mark. Wang et al. (2014) showed that methylation status of a salinity-tolerant wheat cultivar SR3 was explored worldwide and within a set of 24 genes responsive to salinity stress. Comparison was made between DNA isolated from plants grown under control and salinity stress. They detected difference between cv. SR3 and JN177 genomes in their global methylation level, which reduced by salinity stress. Genetic stress induced methylation pattern of 13 loci in non-stressed SR3; the same 13 loci were detected to undergo methylation for salinity-stressed JN177. Also, the salinity-responsive genes SR3 and JN177 displayed diverse methylation changes.

From another technology point of view, Lehnert et al. (2018) showed that Arbuscular mycorrhizal symbiosis is identified to increase water stress tolerance in wheat. A different set containing of 94 bread wheat genotypes was phenotyped at water stress and adequate situations with and without mycorrhizae. Also, wheat accessions were genotyped using 90 k iSelect chip, resultant in a set of 15,511 polymorphic and mapped SNP markers, utilized for genome-wide association studies (GWAS). They detected several QTL regions on diverse chromosomes linked to grain yield and its contributing traits drought stress situations. Also, two genome regions were identified on chromosomes 3D and 7D significantly linked with the response to mycorrhizae in water stress situations.

4.8.2 Nano Technology

Nanotechnology is one of the recent developments that help to manage with wheat crop problems and to meet the challenge of food security (Kashyap et al. 2015). Nanotechnology has a broader application in the area of crop production, sustainability and climate change. The application of Nanotechnology can reduce production costs and increase the crop productivity and the stability of crop production by reducing the losses due to abiotic and biotic stresses (Kashyap et al. 2017). Recent research advises that nanotechnology in wheat has positive prospects for effective nutrient use through Nanoformulations of fertilizers, control and management of pests and diseases, development of new-generation pesticides, and etc. (Jasrotia et al. 2018). Mahmoodzadeh and Aghili (2014) revealed that Nanotitanium dioxide with concentration (1200 ppm) has a stimulating effect on root and shoot of the wheat, where fresh and dry weights of the root are positive influenced by nTiO2. Jaberzadeh et al. (2013) reported that TiO2NPs increased growth and yield components of wheat under water stress condition. This due to role of TiO2 NPs in regulation enzymes activity controlling nitrogen metabolism i.e., nitrate reductase, glutamate dehydrogenase, glutamine synthase and glutamic-pyruvic transaminase that helps in the formation of protein and chlorophyll, which can increase the fresh weight and dry weight of the plant. Yasmeen et al. (2017) showed that Cu NPs increase wheat genotypes to water stress tolerance. Cu NPs enhanced the yield and drought stress tolerance of wheat through starch degradation, glycolysis, and tricarboxylic acid cycle. Furthermore, Mohamed et al. (2017) showed that seed priming with Ag NPs alleviate the salt stress in wheat through reducing the oxidative stress. The lower concentration (2-5 mM NaCl(of Ag NPs might be an effective approach to mitigate the negative influence of salt stress on wheat. Abdel-Aziz et al. (2016) investigated the delivery of chitosan Nanoparticles loaded with nitrogen, phosphorus and potassium for wheat genotypes by foliar uptake. They found that wheat plants grown on sandy soil with Nano chitosan-NPK fertilizer showed significant increases in harvest index, crop index and mobilization index of wheat yield parameters. Under El Wadi El Gadeed governorate, Egypt as a hot climate area using two cultivars, Sids1 classified as heat tolerant and Gimmeiza 7 as heat sensitive. Hassan et al. (2018) exploited the Nanoparticles of Zn NPs (80 nm) and Fe NPs (50 nm) with several concentrations (0, 0.25, 0.50, 0.75, 1.0 and 10 ppm). Some concentrations from NPs enhancing several biochemical markers and increasing antioxidant enzymes activities such as Glutathione S transferase, superoxide dismutase, peroxidase and catalase, the appearance of new bands in some isozymes and decreasing of lipid peroxidation product malondialdehyde. These effects led to the greatest survival of wheat plants under heat stress conditions. The highest mean value of yield production was achieved at 10 ppm Zn NPs and 0.25 ppm Fe NPs with Gimmeiza 7 cultivar. Whereas, Astaneh et al. (2018) showed that replacing urea by Nano chelated nitrogen led to an increase wheat grain yield even under drought stress and decline quantity of required fertilizer. Application of 14, 27 and 41 kg. /ha Nano fertilizer of urea caused 31, 44 and 98% improve in wheat grain yield rather than control, respectively. Interaction between stress, urea and Nano fertilizer on grain yield, showed that under normal and stress condition, 110 kg. /ha urea and 41 kg. /ha Nano fertilizer gave the maximum grain yield with values of 7591 and 4091 kg. /ha, respectively.

Saadah et al. (2000) showed that the foliar spraying of Egyptian wheat cultivar Gemmeiza 12 twice after 35 and 50 days from sowing with a solution of Nano-zinc at a rate of 400 mg / liter per spray in addition to mineral fertilization at a rate of 80 kg N/fed resulted in obtaining the maximum growth and quality. The study recommended that to maintain the growth and high quality of grains, reduce production costs and reduce environmental pollution, foliar spraying with a solution of Nano-zinc at a rate of 400 mg / liter per spray is carried out in addition to mineral fertilization at a rate of 64 kg N/fed under the environmental conditions of Talkha Center, Dakahlia Governorate, Egypt.

4.9 Agricultural Practices to Mitigate Environmental Stress on Wheat

Statistical estimates indicate that each degree of temperature rise will lead to a 6% reduction in global wheat production (Asseng et al. 2015). Therefore, it is important to study agricultural procedures and processes that increase tolerance and develop varieties of wheat more resistant to unfavorable environmental conditions to promote sustainable agriculture.

1. Current research recommends that inoculation of wheat plants with the beneficial soil fungi like arbuscular mycorrhizae affects modulation of proteins related to drought stress response and consequently decreases osmotic pressure

(Bernardo et al. 2017). Inoculation of wheat with mycorrhizal fungi improves drought stress tolerance in wheat (Lehnert et al. 2018).

2. The date of sowing, fertilization and irrigation plays an important role in improving the tolerance of wheat genotypes to environmental stresses. In this respect, Menshawy (2007) revealed that the optimum sowing date (23rd Nov.) produced higher 1000-kernel weight by 7.3 gm, and grain yield by 3.5 ardab/fed. compared to the late sowing (21st Dec.). He added that, early genotypes are more tolerant to heat stress. Under El-Khattara region represents sandy soil condition, Ali (2017) tried three nitrogen levels (50, 80 and 120 kg/fed) splitted in six equal doses with sprinkler irrigation system, also added phosphate and potassium fertilizer at amounts of 150 kg/fed (15.5% P2O5) and 50 kg/fed (48% K2O), respectively. They found that the highest responsive genotypes to the tested environments were Line 1, Giza 168, Gemmeiza 9 and Misr 1 with higher grain and biological yields.

Under the conditions of the Desouk region, Kafr El-Sheikh Governorate, Egypt as affected by the salt stress, El-Seidy et al. (2020) showed significant effects of nitrogen fertilization levels (60, 75, 90 and 105 kg/fed) on the characteristics of wheat grain yield and its contributions. Level of 105 kg N/fed recorded the highest grain yield, followed by 90 kg N/fed and the lowest at 60 kg N/fed. The two varieties, Giza 171 and Misr 2, outperformed Misr 1 and Gemmeiza 11 in grain yield in the trial seasons. The seed rate at 450 grains/m^2 produce the highest grain yield per unit area.

Farooq et al. (2020) showed that nitrogen fertilizer rates affected dry weight per plant, leaf area index, net assimilation rate and crop growth rate of wheat cultivar Giza 171 in both seasons. The highest amount of nitrogen fertilizer treatment (120 kg N/ fed) gave the highest growth and physiological characters in both seasons. They added that spraying wheat plants with methanol and potassium bicarbonate significantly increased growth and physiological characters in both seasons as compare with the control. They concluded that the higher growth and physiological characters for drip-irrigated wheat could be obtained from the amount of 120 kg N/ fed and spraying with potassium bicarbonate at the rate of 0.07 g/l.

Genedy et al. (2020) concluded that cut off irrigation at 90% of irrigation slide length rather than 85 and 80% of the two wheat cultivars Misr 3 and Giza 171 creates the highest productivity and simultaneously saving irrigation water under the environmental conditions of North Delta compared to Shandaweel 1 and Gemmeiza 9 cultivar.

3. Application of mixtures of selenium and humic acid under saline soil led to an increase tolerance of wheat plants to salinity conditions (Mona Nossier et al. 2017)

4. Under Egyptian conditions, crop breeders succeeded to produce a range of bread wheat cultivars tolerant to environmental stresses such as Giza 160, Giza 168, Sids 1, Sids 6, Sids 12, Sids 13, Sids 14, Sakha 93, Sakha 94, Sakha 95, Misr 3 and durum wheat varieties i.e. Sohag 3, Beni Suef 1 and Beni Suef 3 (Anonymous 2020).

4.10 Conclusions

Environmental stress affects wheat productivity in many regions of the world, and caused significant yield reduction in wheat production. In the light of the increase in Egyptian population and extension of agriculture to marginal lands where wheat plants face the effects of heat stress and water shortage, it is of important to reduce genotype × environment interaction. There are several options to deal with environmental stress include release new wheat cultivars and determine their adaptability and stability to cope with environmental changes. Also, approaches should be advanced to manage with the climate change for alleviating the harmful effects of environmental stress on wheat through exploit biotechnology techniques, adjusted agricultural practices with meteorological data, and follows appropriate fertigation programs to avoid harmful effects of high heat waves and water stresses on wheat production.

4.11 Recommendations

Utilized Joint regression analysis and additive main effects and multiplicative interaction AMMI and GEE-biplot analyses are considered more important in determining the nature and significant of genotype × environment effects. Due to the differences existed in wheat genotypes in adaptation and stability and tolerance to various environmental stresses of high temperature, drought, salinity and others, it is necessity testing wheat varieties at multiple locations. Therefore, it is recommended to the cultivation of more promising genotypes adaptable and stable to overcome fluctuations in climate conditions. Also, integration of recent improvement procedures as DNA-marker assisted selection and gene manipulation with traditional crop breeding aids in improving adaption in wheat genotypes.

References

Abbas MA, Babaiean JN, Kazemitabar K (2013) Environmental responses and stability analysis for grain yield of some rice genotypes. World Appl Sci J 21:105–108

Abd El-Shafi MA, Gheith EM, Abd El-Mohsen AA, Suleiman HS (2014) Stability analysis and correlations among different stability parameters for grain yield in bread what. Sci Agric 2(3):135–140

Abdel-Aziz HM, Hasaneen MN, Omer AM (2016) Nano chitosan-NPK fertilizer enhances the growth and productivity of wheat plants grown in sandy soil. Span J Agric Res 14(1):0902

Ahmed F, Khan S, Ahmed SQ, Khan H, Khan A, Muhmmad F (2011) Genetic analysis of some quantitative traits in bread wheat across environments. Afr J Agric Res 6(3):686–692

Ahmed MSH, Mohamed SMS (2009) Genetic analysis of yield and its components in diallel crosses of bread wheat (*Triticum aestivum* L.) under two sowing dates. In: 6th international plant breeding conference, Ismallia. Egypt. Abstract, p 8

Akcura M, Kaya Y (2008) Nonparametric stability methods for interpreting genotype by environment interaction of bread wheat genotypes (*Triticum aestivum* L.). Genet Mol Biol 31:906–913

Akter N, Islam MR (2017) Heat stress effects and management in wheat a review. Agron Sustain Dev 37(37):1–17

Ali MMA (2017) Stability analysis of bread wheat genotypes under different nitrogen levels. J Plant Prod, Mansoura Univ 8(2):261–275

Ali MMA, Abdul-Hamid MIE (2017) Yield stability of wheat under some drought and sowing dates environments in different irrigation systems. Zagazig J Agric Res 44(3):865–886

Allard RW, Bradshaw AD (1964) Implication of genotype × environmental interaction in applied plant breeding. Crop Sci 503–508

Allard RW (1999) Principles of plant breeding. John Willey, New York, p 485p

Al-Naggar AM, Moustafa MA, Atta MMM, Shehab Eldeen MT (2007) Gene action of earliness and grain filling in bread wheat under two irrigation regimes. Egypt J Plant Bread 11(3):279–297

Al-Naggar AMM, Sabry SRS, Atta MM, El-Aleem A (2015) Effect of salinity on performance, heritability selection gain and correlation in wheat (*Triticum aestivum* L.) doubled hapliods. Scientia 10(2):70–82

Al-Otayk SM (2010) Performance of yield and stability of wheat genotypes under high stress environments of the central region of Saudi Arabia. Meteorol Environ Arid Land Agric Sci 21:81–92

Aly AA, Awaad HA (2002) Partioning of genotype × environment interaction and stability for grain yield and protein content in bread wheat (*Triticum aestivum* L). Zagazig J Agric Res 29(3):999–1015

Alybayeva RA, Kenzhebayeva SS, Atabayeva SD (2014) Resistance of winter wheat genotypes to heavy metals. IERI Procedia 8:41–45

Anonymous (2020) Recommendations techniques in wheat cultivation. Agricultural Research Center, Giza, Egypt

Anonymous (2021) Freshwater crisis. National geographic https://www.nationalgeographic.com/environment/freshwater/freshwater-crisis/

Anwar J, Bozakhan S, Rasul I, Zulkhiffa M, Hussain M (2007) Effect of sowing dates on yield and yield components in wheat using stability analysis. Int J Agri Biol 9(1):129–132

Arain MA, Sial MA, Rajput MA, Mirbahar AA (2011) Yield stability in bread wheat genotypes. Pak J Bot 43(4):2071–2074

Asseng S, Ewert F, Martre P, Rötter RP, Lobell DB, Cammarano D, Kimball BA, Ottman MJ, Wall GW, White JW et al (2015) Rising temperatures reduce global wheat production. Nat Clim Chang 5:143–147

Astaneh N, Bazrafshan F, Zare M, Amiri B, Bahrini A (2018) Effect of nano chelated nitrogen and urea fertilizers on wheat plant under drought stress condition. Nativa, Sinop 6(6):587–593

Awaad HA (2001) The relative importance and inheritance of grain filling rate and period and some related characters to grain yield of bread wheat *(Triticum aestivum* L.) proceed. In: The second plant breeding conference, Oct 2, (Assiut University), pp 181–198

Awaad HA (2009). Genetics and breeding crops for environmental stress tolerance, I: drought, heat stress and environmental pollutants. Egyptian Library, Egypt

Awaad HA (2021). Performance, adaptability and stability of promising bread wheat lines across different environments. In: Awaad HA, Abu-hashim M, Negm A (eds) Handbook of mitigating environmental stresses for agricultural sustainability in Egypt, Springer Nature Switzerland AG, pp 178–213

Awaad HA, Aly AA (2002) Phenotypic and genotypic stability parameters for grain yield and its contributing characters in bread wheat (*Triticum aestivum* L.). Zagazig J Agric Res 29(3):983–997

Bernardo L, Morcia C, Carletti P, Ghizzoni R, Badeck FW, Rizza F et al (2017) Proteomic insight into the mitigation of wheat root drought stress by arbuscular mycorrhizae. J Proteomics 169:21–32

Bouslama M, Schapaugh WT (1984) Stress tolerance in soybean. Part. 1: evaluation of three screening techniques for heat and drought tolerance. Crop Sci 24:933–937

Campbell BT, Jones MA (2005) Assessment of genotype × environment interactions for yield and fiber quality in cotton performance trials. Euphytica 144:69–78

Cooper M, De Lacy IH (1994) Relações entre os métodos de análise utilizados para estudar a variação genotípica e pelo genótipo-ambiente de interação no melhoramento de plantas multi-experimentos ambiente. Theor Appl Genet 88:561–572

Cruz CD, Carneiro PCS (2006) Modelos biométricos aplicados ao melhoramento genético, vol 2, UFV, Viçosa, 585 p

Cruz CD, Castoldi FL (1991) Decomposição da interação genótipos × ambientes em partes simples e complexa. Revista Ceres 38:422–430

Eberhart SA, Russell WA (1996) Stability parameters for comparing varieties. Crop Sci 6:36–40

El-Ameen T (2012) Stability analysis of selected wheat genotypes under different environments in Upper Egypt. Afr J Agric Res 7(34):4838–4844

ElBasyoni IS (2018) Performance and stability of commercial wheat cultivars under terminal heat stress. Agronomy 8(4):5–29

ElBasyoni IS, Morsy SM, Ramamurthy RK, Nassar AM (2018) Identification of genomic regions contributing to protein accumulation in wheat under well-watered and water deficit growth conditions. Plants 7(56):1–15

El-Gammaal AA (2018) Combining ability analysis of drought tolerance screening techniques among wheat genotypes (Triticum aestivum L). J Plant Prod, Mansoura Univ 9(11):875–885

EL-Gammaal AA (2019) Genetic analysis of yield and its attributes in wheat (Triticum aestivum L.) under three sowing dates conditions. Egypt J Plant Breed 23(1):93–118

El-khamissi H (2018) Morphological and biochemical evaluation of some Egyptian wheat genotypes under salinity stress conditions. https://www.researchgate.net/publication/328346410_Jou rnal_MORPHOLOGICAL_AND_BIOCHEMICAL_EVALUA

El-Rawy MA, Hassan MI (2014) Effectiveness of drought tolerance indices to identify tolerant genotypes in bread wheat (Triticum aestivum L.). J Crop Sci Biotechnol 17(4):255–266

El-Seidy ELS, Mokhtar M, Ghoneem MMA (2020) Effect of nitrogen fertilizer and seeding rates on yield and yield components on some wheat cultivars. In: 16th international conference for crop science, Agronomy Department, Faculty of Agriculture, Al-Azhar University, Egypt, Oct 2020. Abstract p 103

Abdallah E, Ali MMA, Yasin MAT, Salem AH (2015) Combining ability and mode of gene action for earliness, yield and some yield attributes of bread wheat (Triticum aestivum L.) genotypes grown under different sowing dates. Zagazig J Agric Res 42(2):215–235

Fadia Chairi N, Aparicio MD, Serret JL, Araus, (2020) Breeding effects on the genotype × environment interaction for yield of durum wheat grown after the Green Revolution: the case of Spain. Crop J 8(4):623–634

FAOSTAT (2020) Food and agriculture organization statistical databases. http://faostatfao.org/

Farias FJC, Ramalho MAP, Carvalho LP, Moreira JAN (1996) Parâmetros de estabilidade em cultivares de algodoeiro herbáceo avaliadas na Região Nordeste do Brasil. Pesq Agrop Brasileira 31:877–883

Farooq AS, Hassanein AM, Hagras AM, Azb AM (2020) effect of effect of nitrogen fertilizer, methanol and potassium bicarbonate on some physiological characters under newly reclaimed soil. In: 16th international conference for crop science, Agronomy Deppartment, Faculty of Agriculture, Al-Azhar University, Egypt, Oct 2020, Abstract, p 37

Farooq M, Bramley H, Palta JA, Siddique KHM (2011) Heat stress in wheat during reproductive and grain-filling phases. Crit Rev Plant Sci 30:491–507

Farshadfar E, Mahmodi N, Yaghotipoor A (2011) AMMI stability value and simultaneous estimation of stability and yield stability in bread wheat (Triticum aestivum L.). AJCS 5(13):1837–1844

Fatma FM (2019) Breeding parameters for grain yield and some morpho-pysiological characters related to water stress tolerance in bread wheat. Thesis, Agronomy Department, Faculty of Agriculture Zagazig University, Egypt, M. Sc

Fernandez GCJ (1992) Effective selection criteria for assessing plant stress tolerance. In: Adaptation of food crops to temperature and water stress. In: Kuo CG (ed) AVRDC publication, Shanhua, Taiwan, ISBN: 92-9058-081-X, pp 257–270

Fischer RA, Maurer R (1978) Drought resistance in spring wheat cultivars. I. Grain yield responses. Aust J Agric Res 29:897–912

Fu G, Feng B, Zhang C, Yang Y, Yang X, Chen T, Zhao X, Zhang X, Jin Q, Tao L (2016) Heat stress is more damaging to superior spikelets than inferiors of rice (*Oryza sativa* L.) due to their different organ temperatures. Front Plant Sci 7, 1637

Genedy MS, Sharshar AM and El-Ghannam MK (2020) Impact of cut off irrigation on water saving and productivity of some bread wheat cultivars in North Delta Egypt. In: 16th international conference for crop science, Agronomy Department, Faculty of Agriculture, Al-Azhar University, Egypt, Oct 2020 Abstract, p 35

Hagos HG, Abay F (2013) AMMI and GGE biplot analysis of bread wheat genotypes in the Northern part of Ethiopia. J Plant Breed Genet 01:12–18

Hamada A, Ibrahim Kh (2016) Heat stress impact and genetic diversity among some bread wheat genotype. Egyp J Agron 38(3):389–412

Hamam KA (2013) Response of bread wheat genotypes to heat stress. Jordan J Agric Sci 9(4):486–506

Hamam KA, Khaled AGA (2009) Stability of wheat genotypes under different environments and their evaluation under sowing dates and nitrogen fertilizer levels. Aust J Basic Appl Sci 3(1):206–2017

Hassan SN, Salah El Din TA, Hendawey MH, Borai IH, Mahdi AA (2018) Magnetite and zinc oxide nanoparticles alleviated heat stress in wheat plants. Curr Nanomater 3(1):32–43

He GH, Xu JY, Wang YX, Liu JM, Li PS, Ming C, Ma YZ, Xu ZS (2016) Drought-responsive WRKY transcription factor genes TaWRKY1 and TaWRKY33 from wheat confer drought and/or heat resistance in Arab. BMC Plant Biol 16:116

Hereher ME (2016) Time series trends of land surface temperatures in Egypt: a signal for global warming. Environ Earth Sci 75:1218

Hopkins AA, Vogel KP, Moore KJ, Johnson KD, Carlson IT (1995) Genotype effects and genotype by environment interactions for traits of elite switchgrass populations. Crop Sci 35:125–132

Ingram J, Bartels D (1996) The molecular basis of dehydration tolerance in plants. Annu Rev Plant Physiol Plant Mol Biol 47(1):377–403

Iqbal M, Raja NI, Yasmeen F, Hussain M, Ejaz M, Shah MA (2017) Impacts of heat stress on wheat: a critical review. Adv Crop Sci Technol 5:251

Irfaq M, Mohammad T, Amin M, Jabar A (2005) Performance of yield and other agronomic characters of four wheat (*Triticum aestivum* L.) genotypes under natural heat stress. Int J Bot 1(2):124–127

Irshad M, Khaliq I, Khan AS, Ali A (2012) Genetic studies for some agronomic traits in spring wheat under heat stress. Pak J Sci 49(1):11–20

Jaberzadeh A, Moaveni P, Moghadam HRT, Zahedi H (2013) Influence of bulk and nanoparticles titanium foliar application on some agronomic traits seed gluten and starch contents of wheat subjected to water deficit stress. Notulae Botanicae Horti Agrobotanici Cluj-Napoca 41(1):201

Jasrotia P, Kashyap PL, Bhardwaj AK, Kumar S, Singh GP (2018) Scope and applications of nanotechnology for wheat production: a review of recent advances. Wheat Barley Res 10(1):1–14

Jatoi WA, Baloch MJ, Kumbhar MB, khan NU, Kerio MI. (2011) Effect of water stress on physiological and yield parameters at anthesis stage in elite spring wheat cultivars. Sarhad J Agric 27:332–339

Kamal AMA, Islam MR, Chowdhury BLD, Maleque Talukder MA (2003) Yield performance and grain quality of wheat varieties grown under rainfed and irrigated conditions. Asian J Plant Sci 2:358–360

Kashyap PL, Kumar S, Srivastava AK (2017) Nanodiagnostics for plant pathogens. Environ Chem Lett 15:7–13

Kashyap PL, Xiang X, Heiden P (2015) Chitosan nanoparticle based delivery systems for sustainable agriculture. Int J Biol Macromol 77:36–51

Kaya Y, Palta C, Taner S (2002) Additive main effects and multiplicative interactions analysis of yield performances in bread wheat genotypes across environments. Turk J Agric for 25:275–279

Khan MA, Ahmad N, Akbar M, Rehman A, Iqbal MM (2007) Combining ability analysis in wheat. Pak J Agri Sci 44:1–5

Kristin AS, Senra RR, Perez FI, Enriquez BC, Gallegos JAA, Vallego PR, Wassimi N, Kelley JD (1997) Improving common bean performance under drought stress. Crop Sci 37:43–50

Lehnert H, Serfling A, Friedt W, Ordon F (2018) Genome-wide association studies reveal genomic regions associated with the response of wheat (*Triticum aestivum* L.) to mycorrhizae under drought stress conditions. Front Plant Sci 04 Dec 2018. https://doi.org/10.3389/fpls.2018.01728

Li Q, Wang W, Wang W, Zhang G, Liu Y, Wang Y, Wang W (2018) Wheat F-box protein gene TaFBA1Is involved in plant tolerance to heat stress. Front Plant Sci 9:521

Lin CS, Binns MRA (1988) superiority measure of cultivar performance for cultivar × location data. Can J Plant Sci 68:193–198

Lin CS, Binns MR, Lefkovitch LP (1986) Stability analysis: where do we stand? Crop Sci 26:894–900

Lu M, Cao X, Pan J, Li T, Khan MB, Gurajala HK, He Z, Yang X (2020) Identification of wheat (*Triticum aestivum* L.) genotypes for food safety on two different cadmium contaminated soils. Environ Sci Pollut Res 27:7943–7956

Mahmoodzadeh H, Aghili R (2014) Effect on germination and early growth characteristics in wheat plants (*Triticum aestivum* L.) seeds exposed to TiO_2 nanoparticles. J Chem Health Risks 4(1):29–36

Menshawy AMM (2005) Genetic analysis for earliness components in some wheat genotypes of different photothermal response. In: Proceedings fourth plant breeding conference, (Ismailia). Egypt. J Plant Breed 9(1):31–47. 5 Mar 2005

Menshawy AMM (2007) Evaluation of some early bread wheat genotypes under different sowing dates: 2 agronomic characters. Egypt J Plant Breed 11(1):41–55

Mohamed AKS, Qayyum MF, Abdel-Hadi AM, Rehman RA, Ali S, Rizwan M (2017) Interactive effect of salinity and silver nanoparticles on photosynthetic and biochemical parameters of wheat. Arch Agron Soil Sci 63(12):1736–1747

Mohamed MI (2009) Genotype × environment interaction in bread wheat in Northern Sudan using AMMI analysis. Am-Euras J Agric Environ Sci 6(4):427–433

Mohamed NEM, Said AA, Amein KA (2013) Additive main effects and multiplicative interaction (AMMI) and GGE biplot analysis of genotype × environment interaction for grain yield in bread wheat (*Triticum aestivum* L.). Afr J Agric Res 8(42):5197–5203

Mohammadi M (2012) Effect of kernel weight and source-limitation on heat grain yield under heat stress. Afr J Biotech 11(12):2931–2937

Mohsen A (1999). Sustainable development in the light of climate change. In: IIIEE network, the first global environmental experience sharing conference, 24–26 Sept 1999. http://www.iiiee.lu.se/networkconference/2.4ahmed.htm. [Accessed 1 Feb 2004]

Mona Nossier I, Gawish ShM, Taha TA, Mubarak M (2017) Response of Wheat plants to application of selenium and humic acid under salt stress conditions. Egypt J Soil Sci 57(2):175–187

Mondal S, Singh RP, Crossa J, Huerta-Espino J, Sharma I, Chatrath R, Singh GP, Sohu VS, Mavi GS, Sukaru VSP, Kalappanavarge IK, Mishra VK, Hussain M, Gautam NR, Uddin J, Barma NCD, Hakim A, Joshi AK (2013) Earliness in wheat: a key to adaptation under terminal and continual high temperature stress in South Asia. Field Crop Res 151:19–26

Mulusew F, Bing DJ, Tadele T, Amsalu A (2014) Comparison of biometrical methods to describe yield stability in field pea (*Pisum sativum* L.) under south eastern Ethiopian conditions. Afr J Agric Res 9:2574–2583

Qabil N (2017) Genetic analysis of yield and its attributes in wheat (*Triticum aestivum* L.) under normal irrigation and drought stress conditions. Egypt J Agron 39(3):337–356

Elmoselhy O (2015) Gene action and stability for some characters related to heat stress in bread wheat. Ph.D. thesis, Agronomy Department, Faculty of Agriculture Zagazog University, Egypt

Elmoselhy O, Ali AA-G, Awaad HA, Swelam AA (2015) Phenotypic and genotypic stability for grain yield in bread wheat across different environments. Zagazig J Agric Res 42(5):913–926

Parveen L, Khalil IH, Khalil SK (2010) Stability parameters for tillers, grain weight and yield of wheat cultivars in North-West of Pakistan. Pak J Bot 42(3):1613–1617

Perkins JM, Jinks JL (1968) Environmental and genotype-environmental components of variability. 3. Multiple lines and crosses. Heredity (Edinb) 23(3):339–56

Pinthus MJ (1973) Estimate of genotypic value: a proposed method. Euphytica 1973(22):121–123

Qaseem MF, Qureshi R, Shaheen H (2019) Effects of pre-anthesis drought, heat and their combination on the growth, yield and physiology of diverse wheat (*Triticum aestivum* L.) genotypes varying in sensitivity to heat and drought stress. Sci Rep 9:1–13

Rajaram S (2001) Prospects and promise of wheat breeding in 21st century. Euphytica 119:3–15

Ramalho MAP, Santos JB, Zimmermann MJ (1993) Genética quantitativa em plantas autógamas. UFG, Goiânia, 272 p

Riza-Ud-Din R, Subhani GHM, Ahmed N, Hussain M, Rehman AU (2010) Effect of temperature on development and grain formation in spring wheat. Pak J Bot 42(2):899–906

Romagosa I, Fox PN (1993) Genotype × environment interaction and adaptation. In: Plant Breeding; Springer: Dordrecht, The Netherlands, pp 373–390, ISBN 9401046654

Rosielle AA, Hamblin J (1981) Theoretical aspects of selection for yield in stress and non-stress environment. Crop Sci 21:943–946

Saadah SE, El-Khateeb AY, Salama AMA, Mohamed ASMA (2000) Effect of chelated and nano zinc foliar application and nitrogen fertilizer levels on growth and grains quality of wheat. In: 16th international conference for crop science, Agronomy Department, Faculty of Agriculture, Al-Azhar University, Egypt, Oct 2020 Abstract, p 39

Saba J, Moghaddam M, Ghassemi K, Nishabouri MR (2001) Genetic properties of drought resistance indices. J Agric Sci Technol 3:43–49

Salama SM, Salem MM (2006) Gene action and combining ability over sowing dates in bread wheat (*Triticum aestivum* L.). Egypt, J Appl Sci 21(118):526–641

Salem AH, Eissa MM, Bassyoni AH, Awaad HA, Morsy AM (2003) The genetic system controlling some physiological characters and grain yield in bread wheat (*Triticum aestivum* L.). Zagazig J Agric Res 30(1):51–70

Sattar A, Cheema MA, Farooq N, Wahid MA, Wahid A, Babar BH (2010) Evaluating the performance of wheat cultivars under late sown conditions. Int. J Agric Biol 12:361–365

Shafii B, Mahler KA, Price WJ, Auld DL (1992) Genotype × Environment interaction effects on winter rapeseed yield and oil content. Crop Sci 32:922–927

Sial MA, Azal Arain M., Khanzada S, Naqvi MH (2005) Yield and quality parameters of wheat genotypes as affected by sowing dates and high temperature stress. Pak J Bot 37(3):575–584

Sikder S, Paul NK (2010) Evaluation of heat tolerance of wheat cultivars through physiological approaches. Thai J Agric Sci 43(4):251–258

Su P, Jiang C, Qin H, Rui Hu, Feng J, Chang J, Yang G, He G (2019) Identification of potential genes responsible for thermotolerance in wheat under high temperature stress. Genes 10(174):1–15

Tarakanovas P, Ruzgas V (2006) Additive main effect and multiplicative interaction analysis of grain yield of wheat varieties in Lithuania. Agron Res 4(1):91–98

Tawfelis MB (2006) Stability parameters of some bread wheat genotypes (*Triticum aestivum*) in new and old lands under uper Egypt conditions. Egypt J Plant Breed 10(1):223–246

Umezawa T, Fujita M, Fujita Y, Yamaguchi-Shinozaki K, Shinozaki K (2006) Engineering drought tolerance in plants: discovering and tailoring genes to unlock the future. Curr Opin Biotechnol 17(2):113–122

Vencovsky R, Barriga P (1992) Genética biométrica no fitomelhoramento. Revista Brasileira de Genética, Ribeirão Preto, 486 p

Vencovsky R, Ramalho MAP, Toledo FHRB (2012) Contribution and perspectives of quantitative genetics to plant breeding in Brazil. Crop Breed Appl Biotechnol 12:7–14

Wang M, Qin L, Xie C, Li W, Yuan J, Kong L, Yu W, Xia G, Liu S (2014) Induced and constitutive DNA methylation in a salinity-tolerant wheat introgression line. Plant Cell Physiol 55:1354–1365

Xie C, Mosjidis JA (1996) Selection of stable cultivars using phenotypic variances. Crop Sci 36:572–576

Yasmeen F, Raja NI, Razzaq A, Komatsu S (2017) Proteomic and physiological analyses of wheat seeds exposed to copper and iron nanoparticles. Biochem Biophys Acta 1865:28–42

Yassin M, Fara SA, Hossain A, Saneoka H, El Sabagh A (2019) Assessment of salinity tolerance bread wheat genotypes using stress tolerance indices. Fresenius Envi Bull 28(5):4199–4217

Yau SK (1995) Regression and AMMI analyses of genotype × environment interactions: an empirical comparison. Agron J 87:121–126

Zhang L, Geng X, Zhang H, Zhou C, Zhao A, Wang F, Zhao Y, Tian X, Hu Z, Xin M et al (2017) Isolation and characterization of heat-responsive gene TaGASR1 from wheat (*Triticum aestivum* L.). J Plant Biol 60:57–65

Chapter 5
Approaches in Rice to Mitigate Impact of Climate Change

Abstract Egypt suffers from limited water resources, and rice is one of the main water-consuming crops. Since, our share of the Nile water is not enough for irrigation purposes, this has prompted approach to find ways to economize water use without any loss in grain yield. This is exemplified by the cultivation of short-dutation cultivars, which can save about 20–30% of irrigation water. So, under climat change, the phenotypic performance of the genotype is influenced by environmental factors; some genotypes may perform well in one environment but fail in several others. Hereby, studies of adaptability and stability provide information about the behaviour of each genotype under different environmental conditions. This chapter highlights on approaches to mitigate effect of climate change on rice production, new genotypes stable across environments or suited to target regions, biotechnology approache and appropriate technical recommandations to reduce the impact of climate change on sustainable rice production.

Keywords Rice · Climate change · Performance · Adaptability · Stability · Biotechnology · Sensitivity measurements · Mitigation · Sustainability

5.1 Introduction

Rice crop is one of the most important strategic food crops at the local and the global levels, as it provides food for the vast majority of the world's population. It is also an important food for animals and birds through the production of fodder produced by rice straw pressing. Rice is grown in an area of 162.055.938 million hectares worldwide gave total production 755.473.800 million metric tons as shown in Fig. 5.1. Asia is the largest producer of rice in the world, followed by America, Africa, Europe and Oceania at a proportion of 90.6, 5.2, 3.5, 0.6 and 0.1%, respectively as shown in Fig. 5.2 (FAO 2021). In the meantime, in Egypt, the area of rice cultivation in the 2019 season reached about 0.545 million hectare, producing 5.27 million tons, with an average of 9.595 tons per hectare, Fig. 5.3 shows the development of rice production in Egypt (Anonymous 2020).

Climate change is one of the global environmental issues of the twenty-first century, especially with the increase in greenhouse gas emissions resulting from

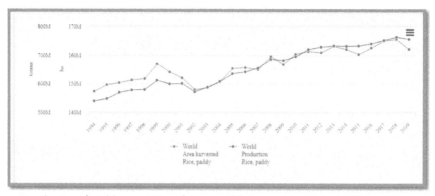

Production / yield quantities of rice, paddy in world + (Total) 1994 – 2021

Fig. 5.1 World area harvested and production of rice, paddy in world + (Total) 1994–2019 (FAO 2021)

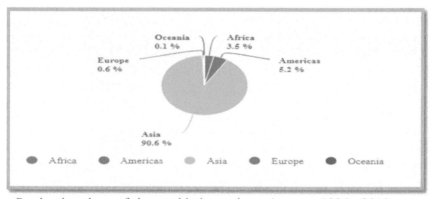

Production share of rice, paddy by region - Average 1994 - 2019

Fig. 5.2 Grain of rice, paddy by region average 1994–2019 (FAO 2021)

human activities. The mitigation of greenhouse gases in the atmosphere of carbon dioxide, methane and nitrous oxide that contribute significantly to global warming is a major task in recent agriculture. Methane emissions from submerged rice cultivation are influenced by many properties, for instance soil texture, soil management practices, previous yield and selected varieties. Moreover, up to 90% of the methane gas generated in the cultivation of submerged rice is released into the atmosphere. The remaining 10% of CH_4 in soil is often re-oxidized to CO_2 and differs broadly between rice genotypes such as hybrids, lines and conventional varieties. Therefore, the selection of the varieties and the treatments that will reduce the anthropogenic GHG emissions is considered an important goal in this regard (Wu et al. 2018). From another point of view, climate change is contributing to various abiotic pressures, thus limiting rice productivity globally. Among them is drought, which is one of the most

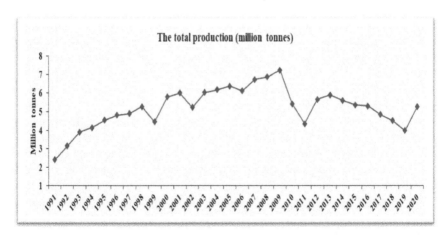

Fig. 5.3 The development of total rice production in Egypt (Anonymous 2020)

serious abiotic stresses that constantly threaten food security in the world (Passioura 2007). Therefore, it is imperative to discover the genotypes that can grow under water-limited conditions to expand rice-growing areas. This is useful in addressing the challenge of the ever-increasing global demand for food (Anis et al. 2019).

Climate change, impact the regularity and level of environmental fluctuations, is a major hazard to agriculture sustainability particularly in developing nations, and causes different abiotic stresses for crop plants. The main constraint of rice cultivation in Egypt is the limited source of irrigation water from Nile River and the shortage of available water, especially in terminal canals in North Delta, whereas rice cultivation is concentrated. In addition, rice consumes large quantities of water during its growing season, which could effect on the reclamation of new lands and cultivation of more crops. Because of adopting of the new short-duration varieties (110–125 days) and irrigation water stress tolerant rice varieties, about 30% of the irrigation water consumption could be saved every year instead of the old, long-duration varieties, which have a 165-day a lifetime (Anonymous 2020). Therefore, one of the main objectives of the Rice Research Department in Egypt is to release new highly productive varieties characterized by tolerant to unfavorable environmental conditions as well as with high grain quality characteristics to suit local consumption and the global market.

Production and cultivation areas are concentrated in the coastal governorates. Egypt is one of the African countries suffering from water poverty, particularly after the building of the Ethiopian renaissance dam. Therefore, the development of new rice varieties that are adapted and tolerate to stress conditions has become an urgent research trend. This will only succeeded if these genotypes prove to be highly associated and have very high genetic stability under target environments (Gaballah et al. 2021).

It is of interest to mention that, rice yield stability requires an attenuation of the reduction of yield losses caused by environmental stresses. Although, rice in

Egypt has the highest productivity (about 10 tons/ha) among the world, despite this it faces many pressures such as biotic stresses for instance diseases (sheath blight, blast, straight head, kernel smut, false smut, bacterial panicle blight, brown spot, sheath rot, stack burn etc.), and insects such as (rusty grain beetle, red flour beetle, confused flour beetle, saw-toothed grain beetle, granary weevil, rice weevil, yellow mealworm, lesser grain borer, etc.), and the abiotic stresses especially problems of water, salinity and alkalinity of the soil. Efforts of crop breeding in the past, generally concentrated on improving high yielding genotypes. Now day, adapted, stable and sustainable yields under different environments have consistency gained importance. The release of genotypes that are adapted to a wide range of environmental conditions is one of the aims of plant breeders in improvement programs (Awaad 2021).

Hence, this chapter focuses on the interaction between genotype × environment for economic traits, the performance, adaptability and stability of rice genotypes in response to different environments. In addition, different methods to measure tolerance in rice genotypes, genetic behavior of tolerance traits and the role of biotechnologies. Also, different methods to measure tolerance in rice genotypes, genetic behavior of tolerance traits and the role of biotechnologies. Also, part on agricultural managements to mitigate environmental stress on sustainable rice production.

5.2   Genotype × Environment Interaction and Its Relation to Climate Change

The Genotype × environment GE interaction is the response of each genotype to variations in the environment and it has been one of the principal subjects of study in plant breeding, allowing the generation of different methodologies for genetic improvement and recommendation of stable genotypes. Selection of early maturing genotypes in breeding programs and making recommendations for high-yielding varieties to farmers is difficult due to Genotype-Environment Interaction (G × E), given the scarcity of obtaining an adaptive variety in all ecological locations (Bernardo 2002). For this reason, identifying the interaction of the genotype × environment and separating it into its components is necessary to distinguish the contribution of the factors under study i.e. locations, seasons, and diverse genotypes, and determining the performance of genetic materials, tested in different locations through different years (Awaad 2021).

Grain yield was influenced by genotype, environment and, genotype × environment interaction, so the maximum potential of rice will seem when cultivated in a suitable environment. Raman et al. (2012) recorded that the variance analysis for grain yield indicated a highly significant genotype × degree of stress severity interaction.

In Egypt, Abdallah et al. (2014) displayed highly significant differences among environments, genotypes, and environments × genotype interaction for root and shoot traits in both normal and drought environments. Under 18 environments El-Degwy and Kmara et al. (2015) tested 13 rice genotypes. They found that mean squares attributed to genotypes, sowing dates, nitrogen levels, first and the second order interactions were significant or highly significant for days to heading, plant height, panicle length, number of panicles/plant, number of spikelets/panicle, sterility percentage, number of grains/panicle, 1000- grain weight and grain yield (ton/fed).

Also, the variation owing to interaction between year and genotype was insignificant for morphological and physiological traits of large-panicle rice genotypes with high filled-grain percentage (Meng et al. 2016). Moreover, seven promising rice genotypes, plus two check varieties Shiroudi and 843, were analyzed in three consecutive years from 2012 to 2014 under three locations that differed in altitude along the temperature gradient in northern Iran by Sharifi et al. (2017). The combined analyses of variance indicated significant effects of environment, genotype and genotype × environment (GE) interactions on grain yield. The significant effect of GE interaction reflected on the differential response of genotypes in various environments and demonstrated that GE interaction had remarkable effect on genotypic performance in different environments.

Behaviour of 23 paddy (*Oryza sativa* L.) genotypes were tested under five diverse environments in India during 2015 and 2016 seasons (Table 5.1) by Das et al. (2018a). They registered highly significant mean squares due to environments, genotypes and interaction between genotypes × environments for grain yield. Combined analyses of variance revealed that genotypes (G), environments (E), and GE interaction (GEI) contributed 6.0, 74.2 and 18.76%, respectively, to the total sum of squares. Results indicated that environmental factor (E) followed by GEI contributed to the maximum variation in the yield performance of genotypes which may be due to the diverse nature of five environments representing different agroecological regimes and differential sensitivities of different genotypes to the different test environments.

Under Rice Research and Training Center Farm, Sakha Research Station, Agricultural Research Center, Egypt, Gaballah et al. (2021) conducted field experiment in successive two rice-growing seasons on 16 rice genotypes. They studied growth characteristics under normal and drought diverse environments i.e., root length, root

Table 5.1 Combined analyses of variance for grain yield response (Das et al. 2018a)

Source of variation	d.f	Sum of squares	Mean of squares	Explained % of TSS
Environment (E)	4	576,191,374.9	144,047,843.7**	74.20
Replication (Environment)	10	189,482.8	18,948.2 NS	0.02
Genotypes (G)	22	46,591,422	2,117,791.9**	6.00
G × E	88	145,672,298	1,655,367.0**	18.76
Error	220	7,780,424.1	35,365.5	
Mean (kg/ha)	4827.3			

**Significant at P* 0.001, NS: Non-significant, d.f: Degrees of freedom, TSS: Total Sum of Squares

volume, root thickness, number of roots/plant, root: shoot ratio, days to heading, leaf rolling, grain yield and related traits i.e. flag leaf area, plant height, relative water content %, number of panicles/plant, 100-grain weight, sterility percentage and water use efficiency. Mean squares attributed to years, environments and genotypes were significant or highly significant for growth characteristics as well as grain yield and related traits, except, leaf rolling, flag leaf area and plant height for years. Thus, would show overall vast variances between the years, environments and genotypes. Mean squares due to genotype × environment interactions were highly significant for the traits except days to heading, relative water content and 100-grain weight. Therefore, the verified genotypes differed in ranks from the environment to environment. Mean squares attributed to genotype × year interactions were significant for all grain yield characters and days to heading. Mean squares due to year × environment interactions were significant for the evaluated traits, except root volume, number of panicles/plants, sterility percentage and grain yield/plant. Significant differences among rice genotypes indicated genetic variability among the material and provided an excellent yield improvement opportunity.

5.3 Performance of Rice Genotypes in Response to Environmental Changes

The performance of any character is a combined result of the genetic makeup (G) of the variety, the environment (E) and the interaction between genotype and environment (GE). Tested genotypes varied from the environment-to-environment and ranked differently from the normal condition to the stress one. Some genotypes surpassed the others once, representative that each genotype's performance in one environment will be changed from one year to another or from environment to another.

5.3.1 Earliness and Photoperiod Characteristics

Heading time is an important ecological character in rice, and is associated to regional and seasonal adaptabilities. Earliness is governed by genetic factors beside the environmental ones, mainly day length and temperature. Several genes and/or alleles control photoperiod sensitivity and length of basic vegetative growth in rice (Rana et al. 2019).

With increasing the stress of salinity levels from control, 4000, 6000 and 8000 ppm., Hassan et al. (2013) found tendency toward earliness which increased by increasing the salinity level in all tested genotypes. The most affected genotype was Sakha101 recording 24 and 23 days earliness under 8000 ppm during the two seasons, respectively. Meanwhile, 10 and 11 days changes were detected for Giza 182 and Sakha104 under the same conditions.

Earliness characteristics represent an important role in adaptation of varieties and avoiding environmental stress conditions. Under different photoperiods and heat stress environments induced through three sowing dates i.e. 15th April, 1st May and 15th May under the effect of three levels of nitrogen fertilization i.e. 40, 60 and 80 kg N/fed. Among 13 rice genotypes, the Egyptian cultivars Sakha 105, Sakha 106 and Giza 177 were the earliest valued 88.11, 91.2 and 91.9 day, respectively. Whereas, both imported genotypes WAB 878 and Moroberekan were the latest in heading with values of 122.2 and 120.4 day, respectively, the other rice genotypes were in between them (El-Degwy and Kmara et al. 2015).

Moreover, sixteen rice genotypes were evaluated for agronomic characters under normal and drought stress conditions by Gaballah et al. (2021). The genotypes Sakha 102, Sakha 103, Giza177 and Sakha 105 registered the least number of days to heading and have earliness values of 94.42, 94.33, 94.17, and 93.92 days, respectively. On the contrary, days to heading took significantly longer time for E. Yasmine, Giza 181, GZ1368-S-5-4 and IET1444 with the values of 118.25, 114.08, 112.67 and 112.50 days, respectively. These variances between rice genotypes may be due to their genetic background. The contrary approach was observed in other cultivars, which took a significant delay in maturity with drought. Heading delay is a typical drought response noticed in rice (Gaballah and Abd Allah 2015), which is expected to confer a benefit in those environments where stress is temporary, if development and flowering resume after the stress are relieved.

5.3.2 Yield and Its Components

The substantial differences among rice genotypes revealed genetic differences give the potential to improve yield. Genotype characteristics that confer an advantage in some water stress environments may prove useless or may even be a liability in other environments. In this respect, several studies has been investigated to assess the constancy of rice grain yield and improve genotypes that react optimally and steadily through years and geographic areas. In Egypt under the two stresses, salinity and water pressure to recognize promising genotypes can be grown under both conditions and to be exploited as a donor to establish an Egyptian super rice varieties program, Hassan et al. (2013) tested comprise 10 local and exotic rice cultivars. Results obviously exhibited significant and highly significant variances between genotypes concerning all deliberated yield characters as affected by the different salinity levels and water stress environments during the two seasons. Under both conditions, all characters were reduced significantly by increasing salinity levels excluding sterility % increased by increasing the intensity of salinity. Rice genotype GZ 1368-S-5-4 followed by Giza 179 and Giza 178 were the highest tolerant once to salinity levels alongside to its significant degree of drought tolerance. Thus, could be grown either under salt effected soil or under water stress environments and used as a donor for those types of stresses in rice program towards breeding for Egyptian super rice varieties. Whereas, although, Sakha 102, Sakha 101 and Giza 182 which produced high yield under normal conditions, however, they were very sensitive under conditions of salinity and water stress. El-Degwy and Kmara et al. (2015) conducted field experiment on 13 rice genotypes under 18 environments. Results showed that across environments, Egyptian Hybrid rice 2, Giza 178 and IET 1444 recorded the maximum estimates in grain yield. Hybrid rice 2 registered the most desirable values of number of panicles/plant, number of grains/panicle, either under first sowing date 15th April or the second one 1st May rather than 15th May associated with the highest nitrogen level 80 kg N/fed.

Grain yield stability of 20 genotypes of rice was assessed under dry season of 2014 in three locations of Indonesia i.e. Cilacap, Kebumen and Sukabumi by Sitaresmi et al (2020). The average rice grain yield of the three locations fluctuated from 5.60 to 7.89 ton/ha. The combined analyses indicated that, the interaction between genotype × environment was significant. Six lines exhibited grain yield were not significantly different from Ciherang, which were BP5168F-KN-16–3, BP4114-7f-Kn-22–2-KLT-2*B-SKI- 1*B, BP5168f-kn-16–3-KLT-2*B-SKI-1*B, BP11282f-Kn-4–3-KLT-2*B, BP5478-1f-Kn-19–1-2-KLT-2*B-SKI-1*B, and HHZ12-SAL2-Y3-Y1. Only line OBS 8412 produce grain yield superior than Ciherang. Moreover, Gaballah et al. (2021) appraised sixteen rice genotypes for grain and its agronomic characters under normal and drought stress conditions. Rice genotypes Giza179, IET1444, Hybrid 1, and Hybrid 2 displayed the greatest number of panicles/plant. The percentage of required sterility was registered by the genotypes IET1444, Giza178, Hybrid 2, and Giza179. However, rice genotypes Giza179, IET1444, Hybrid 1 and Hybrid 2 achieved the greatest values for grain yield/plant with maximum values of water use efficiency. The genotypes IET1444, Giza 178 and Giza179 were appropriate resources for developing drought breeding. In another direction, Karwa et al. (2020) screening collection of rice genotypes and physiologically characterize the component traits using contrasting set of genotypes. Genotypes have been evaluated for tolerant or sensitive to heat stress. Genotype IET 22,218 recorded high spikelet fertility (>85%), pollen viability (>95%) at high temperature (39–44 °C) with relative humidity (>60–80%). This genotype verified higher photosynthesis, canopy temperature depression, and accumulation of endogenous level of polyamines under both optimum and heat stress environments. Moreover, IET 22,218 registered lower H_2O_2 accumulation, cell membrane damage with higher activity of antioxidant enzymes. Heat stress tolerance in IET 22,218 was at par with heat tolerant checks, i.e. Nagina 22 and Nerica L-44. Stimulatingly, IET 22,218 likewise maintained lower chalkiness (<34%) and higher head rice yield (>85%) under heat stress. So, IET 22,218 was designated as the unique donor for heat tolerance.

Furthermore, under Egyptian conditions, many progresses has been achieved by rice breeder in developing water-saving early maturity cultivars characterized by tolerant to abiotic stress conditions especially salinity, drought and heat and biotic stress resistance, high yielding ability with good milling quality as shown in Table 5.2 (Anonymous 2020).

Table 5.2 The Egyptian rice genotypes, characteristics and features (Anonymous 2020)

Genotype	Characteristics and features
Giza 177	Japonica type—short grains—early duration125 days—plant height 95–100 cm—resistant to blast—sensitive to drought and salinity—high yield potential (4–4.3 tons fed^{-1})—high milling outturn (73%)—amylose content 18% and excellent cooking and eating qualities
Giza 178	Indica/Japonica type—medium grain—medium maturing (120–135 days)—semi-dwarf (95 cm plant height) with erect flag leaves—resistant to blast—tolerant to drought and salinity—high yield potential (4.5–5.0 tons fed^{-1})—milling outturn about 68%—amylose content is 17% and has good cooking and eating quality
Giza 179	Indica/Japonica type—short grains—short growth duration (110–120 days)—plant height 95 cm—resistant to blast—tolerant to salinity and drought—high yield potential (4.5–5 tons fed^{-1})—milling outturn about 68%—amylose content is 18%
Giza 181	Indica type—long-grain—late duration (145 days from planting to harvest)—short stature (90–95 cm)—resistant to blast—moderate tolerant to salinity—high yield (4–5 tons fed^{-1})—milling outturn about 69%—grains are transparent and cooking qualities are excellent
Giza 182	Indica type—long grains—early duration (125 days)—short stature (95 cm)—resistant to blast—sensitive to drought—high yielding ability (4.5–5.5 tons fed^{-1})—70% milling outturn—18% amylose content—its grains are transparent and cooking qualities are excellent
Sakha 101	Japonica type—short grains—140 day growth duration—short stature (90 cm plant height)—sensitive to blast - moderate tolerant to salinit—high yield potential (4.5–5.0 tons fed^{-1})—72% milling outturn—19% amylose content and excellent cooking quality
Sakha 102	Japonica type—short grains—growth duration 125 days—long stature (110 cm)—resistant to blast—sensitive to drought—yield potential (4–4.25 tons fed^{-1})—72% milling outturn—19% amylose content and excellent cooking quality
Sakha 103	Japonica type—short grains—early maturing (120 days)—short stature (90–95 cm)—sensitive to drought—resistant to blast—high yield potential (4–4.5 tons fed^{-1})—high milling outturn (73%)—amylose content is 18%—excellent cooking quality

(continued)

Table 5.2 (continued)

Genotype	Characteristics and features
Sakha 104	Japonica type—short grains—moderate duration (125–140 day)—plant height is 105 cm)—sensitive to blast and brown spot—resistant to stem borers—sensitive to drought—moderate tolerant to salinity—high yield potential (4–4.5 tons fed^{-1})—high milling outturn (72%) - translucent milled grains and excellent cooking and eating qualities
Sakha 105	Japonica type—short grains—early duration (125 days)—plant height is 100 cm—resistant to blast—sensitive to drought—high yield potential (4–4.5 tons fed^{-1})—72% milling outturn—17% amylose content and excellent cooking quality
Sakha 106	Japonica type—short grains—early maturity 110–128 days—plant height is 105 cm—resistant to blast—moderately tolerant to salinity—sensitive to drought—high yield potential (4.5 tons fed^{-1})—72% milling outturn—17% amylose content and excellent cooking quality
Sakha 107	Japonica type—short grains—early duration (110–125 days)—plant height 90 cm—resistant to blast—tolerant to drought—moderate tolerant to salinity—high yield potential (4–4.5 tons fed^{-1}) under normal irrigation conditions but under water deficit (3–3.5 tons fed^{-1})—72% milling outturn—17% amylose content and excellent cooking quality
Sakha 108	Japonica type—short grains—duration (120–135 days) according to the method of cultivation—short stature (90–95 cm)—resistant to blast—moderate sensitive to salinity—high yield potential (4.5–5 tons fed^{-1})—high milling outturn (72%)—excellent cooking quality
Sakha 109	Japonica type—short grains—early duration (124 days)—plant height (94–98 cm)—resistant to blast—sensitive to brown spot—high yield potential (4.0–4.5 tons fed^{-1})—72% milling outturn—excellent cooking and eating qualities
Sakha super 300	Japonica type - short grains—growth duration from 114 to 147 day, and plant height from 105 to 115 cm depend on planting time—tolerant to drought and salinity—high yield potential (4–5.5 tons fed^{-1})—73% milling outturn and excellent grain quality
Black rice	Japonica type—short grains—early duration (120 days)—short stature—resistant to blast-grain yield (3–3.5 tons fed^{-1})—72% milling outturn—excellent cooking quality—high nutritional value grains contain zinc, iron and potassium—wide adapted
Hybrid Rice	Indica type—short grains—growth duration is 135 days—plant height is 100 cm—resistant to blast—tolerant to drought and salinity—very high yield (5.5–6 tons fed^{-1})—high tillering ability—70% milling outturn—19% amylose content and acceptable cooking quality

(continued)

Table 5.2 (continued)

Genotype	Characteristics and features
Egyptian Yasmine (Aromatic rice)	Indica type—long grains—late duration (150 days)—short stature (100 cm) with completely erect flag leaves—resistant to blast—moderate tolerant to salinity—yield potential is lower (about 3.125 tons fed^{-1}) however, it is characterized by its high price—grains with aromatic scent—amylose content is 19%—Egyptian Yasmin has excellent cooking quality

Source Rice Research and Training Center Farm, Sakha Research Station, Agricultural Research Center, Egypt

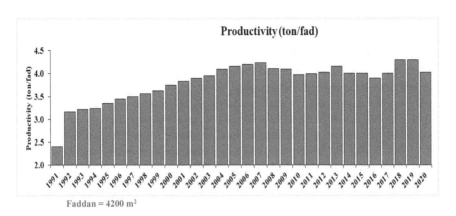

Faddan = 4200 m²

Fig. 5.4 The development of rice productivity in Egypt (Anonymous 2020)

Under Egyptian conditions, in the recent years, progress occurred in a great rice production per faddan as a result of the development of high-yielding cultivars, as illustrated in Fig. 5.4.

5.4 Adaptability and Yield Stability in Relation to Environmental Changes

In order to act more successfully on this new approach, the authors target the rational assessment and determination of the adaptability and stability of the different genotypes of rice. The identification of genotypes with high yield, stability and wide adaptability to various environments is one of the main targets of rice breeding programs. Thus, Gaballah et al. (2016) computed yield adaptability and stability of twelve rice genotypes at Sakha, Gemmeiza and Zarzoura locations of Egypt during 2014 and 2015 seasons. Adaptability and stability were determined by Eberhart and Russell 1966 and revealed that, genotypes Giza179, GZ9399-4–1-1–2-1–2, IR74512-128–2-1–2, IR77734-93–2-3–2, IR78555-3–2-2–2 and IR84233-30–2-3 gave great assessment of grain yield but differed in their stability. Rice genotypes varieties Giza179, Milyang 97 and Suweon 351 displayed a high grain yield over the population mean, with regression coefficient and deviation from regression was minimum, so that those genotypes were stable than others.

Whereas in Brasil, stability parameters of rice grain yield trait in 24 rice genotypes were evaluated under various environments by Júnior et al. (2019). Among the tested genotypes of irrigated rice, Jasmine, SCS 119 Ruby, BRS 358 and BRS AG were genetically higher in terms of grain yield to IAS 12–9 Formosa cultivar, joining high yield, adaptability and genetic stability. Whereas in West Java, Lia et al. (2019) on the stability and adaptability of nine genotypes of hybrid rice in two locations. The results of the experiments showed that the nine genotypes of hybrid rice are SW-907, SW-804, SW-902, US-915, SW-82, SW-923, Intani-2, SL 8 SHS, and SHS 04 WM measured the height of the plants, number of productive tillers, length of the panicles, number of grains per panicle, number of filled grains per panicle, weight of 1000-grains, and the grain yield per plot, and were unstable and unadaptive in Bandung and Tasikmalaya. Whereas, Shrestha et al. (2020) evaluated promising rice genotypes for yield stability at different mid-hill environments of Nepal. The rice genotype NR10676-B-5-3 produced the maximum grain yield (6.72 t/ha) amongst genotypes. Phenotypic stability showed that NR 10,676-B-5-3 and NR 11,011-B-B-B-B-29 were adapted under wide range of environments and more stable. Both genotypes 08FAN10 and Khumal-4 could be grown under favorable environments. Whereas, 08FAN10 was unstable as it exhibited higher and significant ($S^2di = 2.8016*$).

Based on the coefficient of regression and coefficient of variability, Sitaresmi et al. (2020) found that rice genotypes, HHZ12-SAL2-Y3-Y1 and OBS 8412 were stable and have general adaptability in Cilacap, Sukabumi, and Kebumen. Whereas, BP12244-20-3-1–3 and HHZ12-SAL8-Y1-Y2 were specifically adapted in the low yielding environment. However, BP5480-3f-Kn-4–1-KLT-2*B, Ciherang, and INPARI 32 HDB were specifically adapted in high yielding environment.

5.5 Additive Main Effects and Multiplicative Interaction

The additive main effects and multiplicative interaction AMMI model is a hybrid analysis that integrates both additive and multiplicative components of the two-way data structure. AMMI is the only model that discriminates clearly between the main and interaction effects and this is usually desirable to make reliable yield assessments (Gauch 1992). AMMI biplot analysis is considered an operative tool to diagnose GE interaction pattern graphically. The AMMI model designates the GE interaction in more than one dimension and it offers better opportunities for studying and understanding GE interaction than analysis of variance (ANOVA) and regression of the mean. In this connection, Hasan et al. (2014) evaluated seventeen hybrid rice genotypes over 12 environments in Bangladesh. They stated that GE interaction patterns revealed by AMMI biplot analysis designated that the hybrid rice genotypes are widely adapted. Genotypes BRRI53A/BRRI26R, Jin23A/507R, Jin23A/BR7881-25–2–3–12 and IR79156A/F2277R were finest for the environments Gazipur and Rangpur during second and third year. Genotypes Jin23A/PR344R, BRRI11A/AGR and IR79156A/BRRI20R displaying great yield performance and broadly adapted to all environments. Moreover, Sharifi et al. (2017) analyzed seven promising rice genotypes, plus two check varieties in three consecutive years from 2012 to 2014 under three locations in Iran. The application of AMMI model for partitioning the GE interaction effects revealed that only the first two terms of AMMI were significant based on Gollob's F-test. The lowest AMMI-1 was recorded for G7, G2 and G6, G7 and G6 had greater grain yield. According to the first eigenvalue, G1, G6, G2 and G9 were the most stable genotypes. The values of the sum of first two interaction principal component scores could be useful in identifying genotype stability, and G6, G5 and G2 were the most dynamic stable genotypes. AMMI stability value presented G6 as the most stable one. In relation to AMMI biplot, G6 was high yielding and highly stable genotype. Das et al. (2018a) assessed stability and adaptability behavior of 23 paddy genotypes under five diverse environments of Odisha state in India during kharif season of 2015 and 2016 for grain yield. Based on AMMI and GGEBiplot tools, results revealed that G17 (OR 2573–12), G2 (OR 2487–13), G14 (OR 2546–9) and G16 (OR 2573–11) were the most stable genotypes with above-average grain yield among the tested genotypes as they exhibited lower value of AMMI stability parameters (ASV, Di) and lesser degree of deviation due to GE component and greater genotypic contribution (G) in the mean and stability-based ranking of genotypes in GGEBiplot.

5.6 Gene Action, Genetic Behavior and Heritability for Rice Traits Related to Environmental Stress

Conservation of yield in rice under environmental stresses i.e. drought, heat, salinity and pollution is a complicated phenomenon governed by the cumulative effects of several traits. Earliness as a way to escape from environmental stress in rice is controlled by genetic factors with environmental ones. Two earliness genes, *Ef1* and *Efx* controlling basic vegetative phase (BVG), and *m-Ef1*, the enhancer to the former gene; and two lateness genes, *Se1-pat*(t) and *se-pat* controlling photosensitivity and BVG, respectively (Rana et al. 2019). Furthermore, after exposure to water stress in rice, many diverse of genes are expressed with around 5000 genes up regulated and 6000 down regulated (Bin Rahman and Zhang 2016 and Joshi et al. 2016). Most of the genes regulated under water stress are related to the regulation of abscisic acid production (Gupta et al. 2020). Several genes are also associated with osmoregulation under water shortage in rice (Dash et al. 2018; Upadhyaya and Panda 2019). For instance, the gene *DRO1* encourages root elongation and deeper rooting in rice (Uga et al. 2013). Both genes *OsPYL/RCAR5* and *EcNAC67* delay leaf rolling and induce higher root and shoot mass of rice under water stress (Kim et al.

2014; Rahman et al. 2016). Moreover, OsbZIP71and a *bZIP* factor confers salinity and drought tolerance in rice (Liu et al. 2014). Whereas, Sohn-Jaekeun et al. (2002) showed that resistance to ozone damage in rice showed high heritability (>80%), meanwhile, tolerance to cadmium stress governed by additive and overdominance gene action, with moderate to high heritability (Wu-Shutu et al. 2000). Under both salinity and water stress conditions, Hassan et al. (2013) recorded high heritability coupled with high genetic advance for number of filled grains per panicle, number of panicles/ plant and grain yield/ plant indicated the major role of additive gene action in the inheritance of these characters and these characters could be improved by selection in early segregating generations, and may serve as an selection criteria for rice improvement.

5.7   Role of Recent Approaches?

5.7.1   Biotechnology

The development of new biotechnological techniques provides increased support to evaluate genetic variation at DNA level could be used for designing effective breeding programs. Molecular marker technology is a powerful tool for determining genetic variation in rice varieties. SSR markers can distinguish a high level of allelic diversity, and identify genetic variation among rice subspecies (Chukwu et al. 2020). Numerous QTLs in rice with stable effects on grain yield under water-stress conditions were described (Suh et al. 2014). Among them, DTY1.1 was identified in rice genotype N22 and transferred to sensitive lines IR64 and MTU1010 (Vikram et al. 2011), DTY3.1 and DTY2.1 were also recognized and explains about 30 and 15% of the phenotypic variance, respectively (Venuprasad et al. 2009) as well as another two QTLs, DTY2.2 ad DTY12.1 for reproductive stage drought tolerance in rice (Shamshudin et al. 2016).

Furthermore, Gaballah et al. (2021) exploited specific DNA markers related to drought tolerance by Simple Sequence Repeats (SSR) markers and grouping cultivars. The effective number of alleles per locus ranged from 1.20 alleles to 3.0 alleles with an average of 1.28 alleles, and values for all SSR markers varied from 0.94 to 1.00 with an average of 0.98. The polymorphic information content (PIC) values for the SSR were varied from 0.83 to 0.99, with an average of 0.95 along. The highest similarity coefficient between Giza181 and Giza182 was observed and are sensitive to drought stress.

Transgenic approaches in rice help in improving grain yield under stress by introduction certain genes into the receptor genotypes. Higher water use efficacy, osmolytes accumulation, antioxidant enzyme activity and improved photosynthesis are detected in transgenic rice by interrogation of some genes i.e. *OsTPS1* (Li et al. 2011) and *OsMIOX* (Duan et al. 2012), *EDT1/HDG11* (Yu et al. 2013), *AtDREB1A* (Ravikumar et al. 2014), *OsCPK9* improves drought tolerance by regulate stomatal closure and improved osmoregulation in transgenics (Wei et al. 2014). Overexpression of *OsDREB2A* increases transgenic plants survival under severe drought and saline environments (Cui et al. 2011). The WRKY genes play important roles in plant development by responding to drought stress (Sahebi et al. 2018). Das et al. (2018b) found that gene expression pattern of both OsMYB48 and OsHAC1;1 in root and reproductive organs has a potential significance in restricting toxic arsenates in the cellular vacuoles of root tissues at vegetative and reproductive periods and so limiting the translocation of toxic into rice grains. Using transgenic approaches, several genes have been tested for imparting drought tolerance in rice under laboratory or glass house conditions. However, these genes should be tested under field conditions before use in molecular breeding programs. In the field, Stokstad (2020) showed that the transgenic rice yielded up to 20% more grain and resist to heat waves.

5.7.2 Nano Technology

Nanotechnology approach is essential in agricultural area particularly with climate change, population demand and limited availability of macro- and micronutrients of crops. In this regard, potassium and phosphorus Nano fertilizers had stimulation effect on rice growth (Zhang et al. 2010). Nano-silica fertilizers had an effect on the number of stomata, chlorophyll content and growth of black rice (Putri et al. 2017). Also, Wu (2013) and Meena and Lal (2018) worked on Nano-carbon and slow-released fertilizers on rice crop. They revealed that it is possible to use Nano-carbon as coating agent for different combinations of fertilizers and helpful for reducing the water pollution especially Jingzhengda slow-released fertilizer and Nano-carbon (JSCU + C).

Urea coated with HA nanoparticles which slow down the release of nitrogen because of chemical bonding properties of HANPs and enhance urea uptake into rice crop. Kottegoda et al. (2017) observed that the sizes of NPs with a length of 10–200 nm and diameters of 15–20 showed no penetration effect in the rice crops.

At Sakha in Egypt, Sorour et al. (2020) used four Egyptian rice varieties Sakha106, Sakha107, Giza177, and Giza179. They applied phosphorus fertilizer treatments comprised superphosphate (15% P2O5) as a soil application, Nano-phosphorus as a foliar application, superphosphate as a foliar application and control (distilled water). Revealed that application of phosphorus as superphosphate or Nano-phosphorus improved grain quality characteristics as well as nitrogen and phosphorus content in milled grains. Superphosphate soil application and a foliar application of Nano phosphorus to Giza 179 surpassed the other varieties in grain and straw yields. Application of phosphorus as superphosphate or Nano-phosphorus enhanced grain quality characteristics as well as nitrogen and phosphorus content in milled grains. Furthermore, the use of aluminum nanoparticles had a fungicide effect on *Ustilaginoidea virens* causing false smut disease of rice (Priya et al. 2018).

5.8 How Can Measure Sensitivity of Rice to Environmental Stress

5.8.1 Stress Sensitivity Measurements

Numerous stress tolerance indices used in measuring sensitivity of rice genotypes to environmental stress including stress sensitivity index (SSI), tolerance index (TOL), mean productivity index (MP), geometric mean productivity (GMP), stress tolerance index (STI), yield index (YI), yield stability index (YSI), drought resistance index (DI), yield reduction ratio (YR), abiotic tolerance index (ATI), stress sensitivity percentage index (SSPI), harmonic mean (HARM) and golden mean (GOL) were calculated based on grain yield under normal (Yp) and stress (Ys) conditions by El-Hashash and EL-Agoury (2019). Using mean performances, drought tolerance indices and multivariate analysis, the genotypes G3, G16, G7 and G2 (Group A) in normal and stress conditions as well as the genotypes G14, G12 and G13 in stress conditions were the most drought tolerant genotypes. Based on ranking method, the genotypes G16, G13, G12 and G2 (Group A) appeared as the most drought tolerant, and could be used as parents in hybridization programs for improvement drought tolerance. Meanwhile, Gars and Bhattacharya (2017) found that stress tolerance index (STI) and yield index (YI) were superior in rice genotypes RAU-1421–12-1–7-4–3, RAU-1397–25-8–1-2–5-4, RAU-1428–6-7–3-6 and RAU-1451–35-7–6-9–5-1 representative that they could be used as alternative to select drought tolerant genotypes with great yield performance. The stress sensitivity index (SSI), tolerance index (TOL) and yield stability index (YSI) were superior in the genotypes Rasi, Vandana, RAU-1428–31-5–4-3–2-2–2, closely followed by RAU-1421–15-3–2-5–7-3 and RAU-1428–31-5–4. This indicated that SSI, TOL and YSI could be utilized to select drought tolerant and suitable genotypes under two reproductive stage drought condition. Whereas, under Egyptian conditions, Hassan et al. (2013) comprise 10 local and exotic rice cultivars to be tested under salinity and water stress based on stress sensitivity index. The results revealed that genotype GZ 1368-S-5–4 followed by Giza 179 and Giza 178 were found to be the highest tolerant to salinity levels beside to its significant degree of drought tolerance. These genotypes could be chosen to be grown whichever under salinity or under water stress conditions. On the contrary, while, Sakha 102, Sakha 101 and Giza 182 which exhibited high yielding under normal conditions, they were found to be highly sensitive under both salinity and water stress conditions, the other genotypes gave different behavior.

Heat sensitivity index and cumulative stress response index were exploited in screening set of 36 rice genotypes by Karwa et al. (2020). A greater heat stress tolerance is shown by smaller values of HSI. Based on HSI for grain yield, spikelet fertility, pollen viability, genotypes N22, NL-44, MTU1010, IET 23,334, IET 22,218, IR64 were classified as more tolerant, while the most sensitive ones were IET 22,894, US-312, IET 23,296, PHB-71 and PR-113. PR-113 was more sensitive based on HSI for 1000-grain weight. According to cumulative stress response index (CSRI), six genotypes (N22, NL-44, IET 22,218, IET 23,324, IR64 and MTU1010) ranged between −37 and –53, while for five genotypes (IET 22,894, US-312, IET 23,296, PHB-71 and PR-113) CSRI value fluctuated from −133 to –210, the rest genotypes were recognized in between them as moderately sensitive to heat stress.

Under normal and saline stress in India, Krishnamurthy et al. (2016) evaluated a set of 131 rice accessions. The low values of stress sensitivity index (SSI) indicate those genotypes perform well under stress and have sufficient plasticity to respond to the potential environment. Genotypes ICs 545,004, 545,486 and 545,215 were among the bottom 10 candidates for SSI root length, shoot length and Na^+ content. Greater estimates of STI show the superiority of genotypes due to both high yield and stress tolerance. Rice accession 545,004 seemed amongst the top five contenders for STI root length, shoot length and vigor. Likewise, accession 545,486 had the lowest SSI for shoot length and Na + content and recorded maximum STI for root length among the accessions. Thus, accession 545,486 followed by accession 545,004 and 545,215 performed well under saline stress and were superior over the others.

5.9 Agricultural Practices to Mitigate Environmental Stress on Rice

For achieving sustainable cultivation, the Rice Research and Training Center (RRTC) in Sakha, Kafr El Sheikh, Egypt have contributed to the pronounced achievements and release of the newly developed Egyptian rice cultivars. The national average rose to about 4.0 tons per feddan, which is considered one of the highest rates in the world. However, there is an urgent need to increase rice productivity, especially with the threats of the Ethiopian Renaissance Dam, through procedures for recommending rice cultivation.

- For, **nursery**, the suitable period for planting rice is from April 20 to May 15, and any delay after this time leads to significant yield reduction.
- The area of the nursery is 1/10 of the permanent field, fertilized with 4 kg super-phosphate calcium (15.5 P_2O_5) before plowing, then the nursery is well plowed and left for aeration. Hence, nitrogen fertilizer is added at a rate of 6 kg of ammonia sulfate per carat, then immersion and muddling. Then, zinc sulfate fertilizer is added to the nursery at a rate of one kilogram per carat.
- Seeds are soaked for 48 h and incubated for 24 h before sowing, and planted in the water level is only 2–3 cm above the soil surface for a period of 5 days, then the nursery is drained, hence irrigation after two days, with the water drained again after 4–5 days and left without irrigation for two days to aerate the soil, which helps the growth of the roots. Repeated the irrigation of the nursery every 4–6 days and drain the water one day before it's irrigated.
- Control weeds in the nursery with Saturn pesticide 50%.
- **In permanent field**, apply superphosphate fertilizer before plowing at a rate of 100 kg/fed (15.5 P_2O_5) and it must be taken into account not to add superphosphate in the presence of water, because this encourages the growth and reproduction of

the rem. Then nitrogen fertilizer is added at a rate of 200 kg ammonium sulfate 20% in the case of the cultivars Egyptian Yasmin, Sakha 104, Sakha 106 and Sakha Super 300, while in the case of the varieties Sakha 101, Sakha 107, Giza 177, Giza 178, Giza 179, Giza 182, Sakha 108, Sakha 109, Egyptian Hybrid 1 and Black rice, it increased to 300 kg ammonia sulfate 20%. The nitrogen fertilizer is added in two doses, 2/3 of nitrogen fertilizer before flooding and apply the remaining 1/3 of N fertilizer at panicle initiation.

– Best age for rice seedlings for most of the Egyptian varieties is 25–30 days, by 3–4 seedlings/ hill at 20 × 20 cm apart of high tillering varieties, while seedlings of low tillering varieties, such as Sakha 107 and Giza 177 transplanted at 15 × 15 cm spacing.

– Attention to controlling various pests by using the recommended pesticides with optimal doses at the proper time.

In this respect, under saline soil conditions, El-Serw in the northern region of eastern Delta, Egypt, Omar (2002) found that application of 60 kg N/fad in split on rice cultivar Giza 178 gave the highest values of flag leaf area, chlorophyll content, grain yield (ton/fad), crude protein content and amylose content followed by applying 40 kg N + 0.5% K_2SO_4 + 2% Zn SO_4 rather than the control and the other treatments. Applying 60 kg N/fad increased grain yield by 9 , 10.6, 21.7 and 63% over 40 kg N/fad (split), 40 Kg N/fad., 20 KG N/fad and the control, respectively (Table 5.3).

Metwally et al. (2012) showed that Egyptian Yasmin, IR77510-88-1-3-3, IR78530-45-3-1-3, IR 74052-177-3-3, IR 71137-51-2 and IR65610-38-2-4-2-6-3 could be recommended to be sown in April 24th and May 10th. Sultan et al. (2013) found that the highest values of leaf area index, dry matter weight, plant height, number of panicles/m^2, panicle weight, 1000-grain weight, grain yield (ton/fed) and harvest index were recorded in plots fertilized with full recommended dose of nitrogen (69 kg), phosphorus (15.5 kg) and potassium (24 kg) followed by the fertilized with full dose of NPK without any compost or 75 % recommended dose of NPK + compost.

Khattab (2019) evaluated three varieties of Rice (Sakha 102, Giza 178 and Giza 182), at two sowing methods (direct seeding and seedling transplanting) under treatment of NPK concentration (0.0, 2 and 4%) as form 20:20:20 NPK. The results indicated that increase in NPK concentration with Giza 178 under direct seeding recorded the highest results for vegetative and physiology parameters, yield and yield components, and contents of P, K, Zn, Mn and Fe in leaves and grains, while all parameters recorded lowest with Giza 182 under control NPK and transplanting.

4. Under Egyptian Conditions, Rice Breeder Succeeded to Produce a Range of Rice Cultivars Tolerant to Environmental Stresses as Previously Registered in Table 5.2 (Anonymous 2020).

Table 5.3 Flag leaf area, leaf chlorophyll content, grain yield (ton/fad), crude protein and amylose contents as combined (Modified after Omar 2002)

Treatments	Flag leaf area	Leaf chlorophyll content	Grain yield (Ton/fad)	Crude protein content (%)	Amylose content (%)
Check	24.41g	36.92e	1.680e	6.99g	16.42
20 KG N/fad	25.75f	39.13cd	2.253d	7.21f	15.97
40 KG N/fad	27.05d	39.63cd	2.477bc	7.35de	16
40 KG N/fad (split)	27.42bcd	41.12ab	2.508b	7.59bc	15.75
60 KG N/gad (split)	28.90a	41.39a	2.741a	7.90a	16.48
20 kg N + 0.5% K_2SO_4 + 1% Zn SO_4	26.22ef	39.70cd	2.361cd	7.24ef	16.12
20 kg N + 0.5% K_2SO_4 + 2% Zn SO_4	27.45bcd	39.87cd	2.486bc	7.8bc	16.17
20 kg N + 1% K_2SO_4 + 1% $ZnSO_4$	26.67de	38.88d	2.473bc	7.45cd	16.88
20 kg N + 1% K_2SO_4 + 2% Zn SO_4	26.95de	39.88cd	2.536b	7.48cd	17.3
40 kg N + 0.5% K_2SO_4 + 1% Zn SO_4	27.39bcd	39.36cd	2.497b	7.48cd	16
40 kg N + 0.5% K_2SO_4 + 2% Zn SO_4	28.17ab	40.34abc	2.696a	7.65b	16.13
40 kg N + 1% K_2SO_4 + 1% Zn SO_4	27.29cd	40.12bc	2.553b	7.64b	16.4
40 kg N + 1% K_2SO_4 + 2% Zn SO_4	28.05bc	40.34abc	2.712a	7.64b	16.13
F-test	**	**	**	**	N.S

5.10 Conclusions

Development of adaptability and stability in rice is a thought-provoking task that requires a comprehensive thoughtful of the various morphological, biochemical, physiological and molecular characters. Although remarkable progresses have been achieved through rice breeding, we still have several critical problems to overcome the effects of climate change on rice. Moreover, the complex nature and multigenic control of abiotic stress would be a major bottleneck for the current and coming future research in this respect. Maintenance of yield in rice under stress conditions is a multifaceted phenomenon governed by the cumulative effects of several traits. Transgenic approaches play a pivotal role in improving agronomic traits and yield characteristics of rice and it would be an efficient way to boost the rice-breeding program for stress tolerance and yield stability. In addition, appropriate agricultural techniques play a significant role in mitigating the impact of climate change on rice production of recent cultivars.

5.11   Recommendations

There are various strategies and options for adaptation to climate change in the rice sector production. The measures to take advantage of the positive impacts and mitigate the negative ones, including: (1) adjusting cropping patterns to expand rice cultivation; (2) breeding new rice varieties with short growing periods that are tolerant to environmental stress; (3) improving irrigation systems, and (4) improving pest and disease control, fertilizer application and mechanization.

References

Abdallah A, Gaballah M, Aml El-Saidy M, Ammar M (2014) Drought tolerance of anther culture derived rice lines. J Plant Prod 11:115–124

Anis G, Hassan H, Saneoka H, El Sabagh A (2019) Evaluation of new promising rice hybrid and its parental lines for floral agronomic traits and genetic purity assessment. Pak J Agric Sci 56:567–576

Anonymous (2020) Technical Recommendations for Rice Crop. Ministry of Agriculture and Land Reclamation, Agricultural Research Center, Field Crops Research Institute, Rice Research and Training Center, Egypt

Awaad HA (2021) Performance, adaptability and stability of promising bread wheat lines across different environments. In: Awaad HA, Abu-hashim M , Negm A (eds) Handbook of mitigating environmental stresses for agricultural sustainability in Egypt, Springer Nature Switzerland AG, pp 178–213

Bernardo R (2002) Breeding for quantitative traits in plants. Press, Woodbury, MN, USA, Stemma

Bin Rahman ANMR, Zhang JH (2016) Flood and drought tolerance in rice: opposite but may coexist. Food Energy Secur 5(2):76–88

Chukwu SC, Rafii MY, Ramlee SI, Ismail SI, Oladosu Y, Kolapo K, Musa I, Halidu J, Muhammad I, Ahmed M (2020) Marker-assisted introgression of multiple resistance genes confers broad spectrum resistance against bacterial leaf blight and blast diseases in PUTRA-1 rice variety. Agronomy 10:42

Cui M, Zhang WJ, Zhang Q, Xu ZQ, Zhu ZG, Duan FP, Wu R (2011) Induced over-expression of the transcription factor OsDREB2A improves drought tolerance in rice. Plant Physiol Biochem 49(12):1384–1391

Das CK, Bastia D, Naik BS, Kabat B, Mohanty MR, Mahapatra SS (2018a) GGE biplot and AMMI analysis of grain yield stability and adaptability behaviour of paddy (*Oryza sativa* L) genotypes under different agroecological zones of Odisha. ORYZA-An Int J Rice 55(4):528–542

Das CK, Bastia D, Swain SC and Mahapatra SS (2018b) Computational analysis of genes encoding for molecular determinants of arsenic tolerance in rice (*Oryza sativa* L) to engineer low arsenic content varieties. Oryza 55(2):248–259

Dash PK, Rai R, Rai V, Pasupalak S (2018) Drought induced signaling in rice: delineating canonical and non canonical pathways Front Chem 6:264

de Júnior AM, M, G A Aguiar, P H K Facchinello, P R R Fagundes, (2019) Genotypic performance, adaptability and stability in special types of irrigated rice using mixed models. Rev Ciênc Agron 50(1):66–75

Duan JZ, Zhang MH, Zhang HL, Xiong HY, Liu PL, Ali J, Li JJ, Li ZC (2012) *OsMIOX*, a myo-inositol oxygenase gene, improves drought tolerance through scavenging of reactive oxygen species in rice (*Oryza sativa* L) Plant Sci 196:143–151

El-Degwy IS and MM Kmara (2015) Yield potential, genetic diversity, correlation and path coefficient analysis in rice under variable environments. J Plant Product 6(5):695–714

El-Hashash EF, EL-Agoury RYA (2019) Comparison of grain yield based drought tolerance indices under normal and stress conditions of rice in Egypt. J Agric Vet Sci 6(1):41–5

FAO (2021) Production quantities of rice, paddy by country. Average 1994–2019. http://www.fao.org/faostat/en/#data/QC/visualize

Gaballah MM, Hassan HM, Shehab MM (2016) Stability parameters for grain yield and its components in some rice genotypes under different environments. In: 6th Field Crops conference, FCRI, AEC, Giza, Egypt, pp 3–14

Gaballah MM, Abdallah AA (2015) Effect of water irrigation shortage on some quantitative characters at different rice development growth stages. World Rural Obs 7:10–21

Gaballah MM, Metwally AM, Skalicky M, Hassan MM, Brestic M, EL Sabagh A, Fayed AM (2021) Genetic diversity of selected rice genotypes under water stress conditions. Plants 10, 27:1–19

Gars HG, Bhattacharya C (2017) Drought tolerance indices for screening some of rice genotypes. IJABR 7(4):671–674

Gauch HG (1992) Statistical analysis of regional yield data: AMMI analysis of factorial designs. Elsevier, New York, New York, 278 pages

Gupta A, Rico-Medina A, Caño-Delgado AI (2020) The physiology of plant responses to drought. Science 368:266–269

Hasan MJ, Kulsum MU, Hossain MM, Akond Z, Rahman MM (2014) Identification of stable and adaptable hybrid rice genotypes. SAARC J Agric 12(2):1–15

Hassan HM, El-Khoby WM, El-Hissewy AA (2013) Performance of some rice genotypes under both salinity and water stress conditions in Egypt. J Plant Product Mansoura Univ 4(8):1235–1257

Joshi R, Wani SH, Singh B, Bohra A, Dar ZA, Lone AA, Pareek A, Singla-Pareek SL (2016) Transcription factors and plants response to drought stress: current understanding and future directions. Front Plant Sci 7:1029

Karwa S, Bahuguna RN, Chaturvedi AK, Maurya S, Arya SS, Chinnusamy V, Pal M (2020) Phenotyping and characterization of heat stress tolerance at reproductive stage in rice (*Oryza sativa* L). *Acta Physiologiae Plantarum* 42(1):1–16

Khattab EA (2019) Performance evaluation of some rice varieties under the system of planting in Egypt. AJRCS 3(2):1–10

Kim H, Lee K, Hwang H, Bhatnagar N, Kim DY, Yoon IS, Byun MO, Kim ST, Jung KH, Kim BG (2014) Overexpression of *PYL5* in rice enhances drought tolerance, inhibits growth, and modulates gene expression. J Exp Bot 65(2):453–464

Kottegoda N, Sandaruwan C, Priyadarshana G, Siriwardhana A, Rathnayake UA, Berugoda Arachchige DM, Kumarasinghe AR, Dahanayake D, Karunaratne V, Amaratunga GA (2017) (Urea-hydroxyapatite nanohybrids for slow release of nitrogen. ACS Nano 11(2):1214–1221

Krishnamurthy SL, Sharma PC, Sharma SK, Batra V, Kumar V, Rao LVS (2016) Effect of salinity and use of stress indices of morphological and physiological traits at the seedling stage in rice. Indian J Exp Biol 54:843–850

Li HW, Zang BS, Deng XW, Wang XP (2011) Overexpression of the trehalose-6-phosphate synthase gene *OsTPS1* enhances abiotic stress tolerance in rice. Planta 234(5):1007–1018

Lia A, Ai K, Ilma H (2019) Stability and adaptability genotype of hybrid rice in the West Java province of Indonesia. Asian J Agric Rural Dev Asian Econ Soc Soc 9(2):204–215

Liu CT, Mao BG, Ou SJ, Wang W, Liu LC, Wu YB, Chu CC, Wang XP (2014) OsbZIP71, a bZIP transcription factor, confers salinity and drought tolerance in rice. Plant Mol Biol 84:19–36

Meena RS, Lal R (2018) Legumes and sustainable use of soils. In: Meena RS et al.) eds) Legumes for soil health and sustainable management. Springer. https://doi.org/10.1007/978-981-13-025 3-4_1

Meng TY, Wei HH, Li C, Dai QG, Xu K, Huo ZY, Wei HY, Guo BW, Zhang HC (2016) Morpho-logical and physiological traits of large-panicle rice varieties with high filled-grain percentage. J Integr Agric 15:1751–1762

Metwally TF, El-Malky MM, Glelah AA, Ghareb A (2012) Performance of elite aromatic rice varieties under different sowing dates under Egyptian condition. J Plant Product 3(2):111–333

Omar AEEA (2002) Rice fertilization under saline soil conditions. D, Argon Dept, Fac of Agric, Zagazig Univ, Egypt, Ph

Passioura JB (2007) The drought environment: Physical, biological and agricultural perspectives. J Exp Bot 58:113–117

Priya B, Amarendra K, Sanjeev K, Azad CS (2018) Impact of fungicides and nanoparticles on Ustilaginoidea virens causing false smut disease of rice. J Pharmacogn Phytochem 7(1):1541–1544

Putri FM, Suedy SWA, Darmanti S (2017) The effects of nano-silica fertilizer on the number of stomata, chlorophyll content and growth of black rice (*Oryza sativa* L Cv Japonica). www.ejo urnal.undip.ac.id2017

Rahman H, Ramanathan V, Nallathambi J, Duraialagaraja S, Muthurajan R (2016) Over-expression of a NAC 67 transcription factor from finger millet (*Eleusine coracana* L) confers tolerance against salinity and drought stress in rice BMC Biotechnol 16:35

Raman A, Verulkar S, B, N Mandal P, M Variar, V Shukla D, J Dwivedi L, B Singh N, O Singh N, P Swain, A Mall K, S Robin, R Chandrababu, A Jain, T Ram, Sh Hittalmani, S Haefele, H Piepho & A Kuma, (2012) Drought yield index to select high yielding rice lines under different drought stress severities. Rice 5(31):1–12

Rana BB, Kamimukai M, Bhattarai M, Koide Y, Murai M (2019) Responses of earliness and lateness genes for heading to different photoperiods, and specific response of a gene or a pair of genes to short day length in rice. Hereditas, 156. Article Num 36:1–11

Ravikumar G, Manimaran P, Voleti SR, Subrahmanyam D, Sundaram RM, Bansa KC, Viraktamath BC, Balachandran SM (2014) Stress-inducible expression of AtDREB1A transcription factor greatly improves drought stress tolerance in transgenic indica rice Transg Res 23(3):421–439

Sahebi M, Hanafi MM, Rafii MY, Mahmud TMM, Azizi P, Osman M, Abiri R, Taheri S, Kalhori N, Shabanimofrad M, Miah G, Atabaki N (2018) Improvement of drought tolerance in rice (*Oryza sativa* L): genetics, genomic tools, and the WRKY gene family. Biomed Res Int 1–20

Shamshudin NA, Swamy BM, Ratnam W, Cruz MT, Raman A, Kumar A (2016) Marker assisted pyramiding of drought yield QTLs into a popular Malaysian rice cultivar, MR219. BMC Genet 17:3

Sharifi P, Aminpanah H, Erfani R, Mohaddesi A, Abbasian A (2017) Evaluation of genotype × environment interaction in rice based on AMMI model in iran. Rice Sci 24(3):173–180

Shrestha J, Kushwaha UKS, Maharjan B, Kandel M, Gurung SB, Poudel AP, Karna MKL, Acharya R (2020) Grain yield stability of rice genotypes. Indonesian J Agric Res 03(02):116–126

Sitaresmi T, Susanto U, Pramudyawardani EF, Nafisah NY, Sasmita P (2020) Genotype x environment interaction of rice genotype. IOP Conf Ser: Earth Environ Sci 484, 012028

Sohn-Jaekeun K, Lee JJ, Kwon YS, Kim KK (2002) Varietal differences and inheritance of resistance to ozone stress in rice (*Oryza sativa* L). SABRAO J Breed Genet 34(2):65–71

Sorour FA, Etwally TF, EL-Degwy IS, Eleisawy EM, Zidan AA (2020) The effects of nano phos-phatic fertilizer application on the productivity of some Egyptian rice varieties (*Oryza sativa* L). Appl Ecol Environ Res 18(6):7673–7684

Stokstad E (2020) Rice genetically engineered to resist heat waves can also produce up to 20% more grain. Science. https://www.sciencemag.org/news/2020/04/rice-genetically-engineered-res ist-heat-waves-can-also-produce-20-more-grain

Suh JP, Won YJ, Ahn EK, Lee JH, Ha WG, Kim MK, Cho YC, Jeong EG, Kim BK (2014) Field performance and SSR analysis of drought QTL introgression lines of rice. Plant Breed Biotechnol 30:158–166

Sultan MS, El-Kassaby AT, El-Habashy MM, Taha AS (2013) Yield and yield components of hybrid one rice cultivar as affected by irrigation intervals, fertilization combinations and their interaction. J Plant Product Mansoura Univ 4(8):1149–1157

Uga Y, Sugimoto K, Ogawa S, Rane J, Ishitani M, Hara N, Kitomi Y, Inukai Y, Ono K, Kanno N, Inoue H, Takehisa H, Motoyama R, Nagamura Y, Wu JZ, Matsumoto T, Takai T, Okuno K, Yano M (2013) Control of root system architecture by deeper rooting 1 increases rice yield under drought conditions. Nat Genet 45:1097–1102

Upadhyaya H, Panda SK (2019) Drought stress responses and its management in rice. In: Hasanuz-zaman M, Fujita M, Nahar K, Biswas JK (eds) Advances in rice research for abiotic stress tolerance. Elsevier, UK, pp 177–200

Venuprasad R, Dalid CO, Del Valle M, Zhao D, Espiritu M, Sta Cruz MT, Amante M, Kumar A, Atlin GN (2009) Identification and characterization of large-effect quantitative trait loci for grain yield under lowland drought stress in rice using bulk-segregant analysis. Theoret Appl Genet 120:177–190

Vikram P, Swamy BPM, Dixit S, Cruz S, Ahmed HU, Singh AK, Kumar A (2011) qDTY1.1, a major QTL for rice grain yield under reproductive-stage drought stress with a consistent effect in multiple elite genetic backgrounds. BMC Genet 12:89

Wei SY, Hu W, Deng XM, Zhang YY, Liu XD, Zhao XD, Luo QC, Jin ZY, Li Y, Zhou SY, Sun T, Wang LZ, Yang GX, He GY (2014) A rice calcium-dependent protein kinase OsCPK9 positively regulates drought stress tolerance and spikelet fertility. BMC Plant Biol 14:133

Wu MY (2013) Effects of incorporation of nano-carbon into slow-released fertilizer on rice yield and nitrogen loss in surface water of paddy soil. Third international conference on intelligent system design and engineering applications, Hong Kong, pp 676–681

Wu S, Hu Z, Hu T, Chen J, Yu K, Zou J, Liu S (2018) Annual methane and nitrous oxide emissions from rice paddies and inland fish aquaculture wetlands in southeast China. Environ. https://doi.org/10.1016/j.atmosenv.2017.12.008

Wu-Shutu et al. (2000). Wu-Shutu YC, Kuo B, Thseng F, Wu S, Yu C, Kuo B, Thsenge F (2000) Diallel analysis of cadmium tolerance in seedling rice. SABRAO J Breed and Genet 32(2):57–61

Yu LH, Chen X, Wang Z, Wang SM, Wang YP, Zhu QS, Li SG, Xiang CB (2013) Arabidopsis Enhanced Drought Tolerance1/HOMEODOMAIN GLABROUS11 confers drought tolerance in transgenic rice without yield penalty. Plant Phys 162:1378–1391

Zhang Z, Xi-fu F, Lei SS, Xue-li C, Yu-juan L, Ming Z (2010) Effect of nano-fertilizers on rice growth [J]. Heilongjiang Agric Sci 8:19

Chapter 6
Approaches in Faba Bean to Mitigate Impact of Climate Change

Abstract Faba bean plants face environmental stress, mainly towards the end of the growing season. In recent years, Egypt has confronted a severe crisis in the faba bean production, which has declined significantly, due to the decrease in its cultivated area and exposure to unsuitable environmental factors. So, this chapter focused on the size and nature of the interaction between genotype × environment, and the determination of the adaptability and stability of genetic makeups under different conditions. Also, determined the genetic behavior of different characteristics, such as earliness, number of pods, number of seeds and seed weight, which are used as selection criteria for improving yield under various environments. Illustrations the role of the new developments in marker technology together with biotechnology in improving crop stability under stress environmental conditions. Also, some agronomic practices to sustain faba bean production and reduce the impact of environmental stress have been discussed.

Keywords Faba bean · Heat stress · Drought stress · Genotype × environment interaction · Stability · DNA markers · Gene transfer · Heritability · Stress measures · Mitigation

6.1 Introduction

Faba bean (*Vicia faba* L.) is grown as legume crop mainly as human food and animal feed (Crépon et al. 2010). Faba bean is generally considered as a good source of protein, starch, cellulose and minerals for humans in developing countries, it also has multiple advantages in the world and Egypt. The crop used as a foundation of food and feed with valuable cheap sources of protein as an accompaniment to cereals crops for the common of the poor mostly for those who cannot afford to use proteins from animal resources (Jarso and Keneni 2006). Then, faba bean is an important grain legume crop being used principally as a source of protein in human diets, as fodder and a forage crop for animals, and for available nitrogen in the biosphere (Rubiales 2010). Faba bean is also a source of income for farmers and provide the country's foreign currency (Keneni et al. 2006). Faba bean have tremendous value in improving soil fertility in crop rotation programs, and offering valuable ecological services in sustainable agriculture as it forms symbiotic association with nitrogen-fixing bacteria (Jensen et al. 2010). Therefore, faba bean contribute significantly to maintaining and enhancing the fertility of soil nitrogen through biological N_2-fixation. Its widespread world distribution demonstrates a basic good capability to adapt with diverse climates, however its low and unstable yields hamper its competitiveness (Arbaoui et al. 2008; Link et al. 2010; Duc et al. 2011). With climate change hazard, a breeding effort is essential for tolerance to various abiotic stresses i.e. freezing, heat, drought, salt and elevated CO_2 (Maalouf et al. 2015) and to related diseases and pests (Maalouf et al. 2016).

Worldwide, China, Italy, Spain, the United Kingdom, Egypt, Ethiopia, Morocco, Russia, Mexico, and Brazil are the main faba bean-growing countries (Fig. 6.1). The area of the world cultivated with faba bean fell from 5 million hectares in 1965 to less than half in 2007 due to biotic and abiotic stresses and then unstable yield (Rubiales 2010). Though, although the problems affecting the production of faba bean, the worldwide average yield has increased from 0.9 tons/ha between 1961 and 1964–1.86 tons/ha in 2016 (FAOSTAT 2018).

The deficiency of soil moisture is a major limiting factor for the production of faba bean (De Costa et al. 1997). Heat stress is also considered as detrimental environmental factor to the growth and yield of faba bean plants. Faba bean plants face heat stress when exposed to supra-optimal temperatures or water stress, mainly towards the end of the growing season (Stoddard et al. 2006). Katerji et al. (2011) registered a decrease in faba bean seed yield at soil salinity levels of ≥ 6.5 dS m^{-1}. Increased soil salinity affects growth and nitrogen fixation parameters in faba bean plants (Bulut et al. 2011), which are expressed moderately sensitive legume to salinity (Delgado et al. 1994).

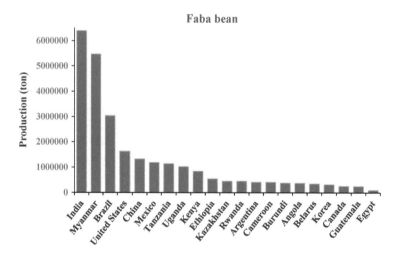

Fig. 6.1 Production/Yield quantities of faba bean worldwide (FAOSTATE 2019)

In recent years, Egypt has witnessed a severe crisis in the quantity of production of the faba bean production, which has declined significantly, due to the decrease in its cultivated area. The area cultivated with faba beans decrease from 350 thousand faddans before 2005–77,426 faddans in 2017 with an average productivity 9.4 ard./fad. While the total area in the world 5,864,239 faddan with an average productivity 5.32 ard./faddan which caused a deficit in the domestic needs of faba bean (FAO 2018). The decline in the area of beans in Egypt is due to the spread of the yellow spot disease in faba bean in the Delta, and the spread of hemp broomrape in lands cultivated with faba bean in Upper Egypt. Besides, the high costs of fertilizers, pesticides and seeds, and the low price of the crop have played a major role in farmers' abstention from growing faba bean. Egypt's imports of Faba bean throughout the first 10 months of 2018 reached to about 182 million dollars, rather than 173 million dollars through the same period in 2017, based on the General Organization for Export and Import Control (Zikrallah 2019). Faba bean global acreage declined from 3.7 to 2.1 million ha between 1980 and 2014, and yields are highly variable within specific countries (FAO 2017).

Thermal stress conditions are a specific factor for growing faba bean under modern sustainable farming systems, which necessitates the release of heat tolerant varieties. Bishop et al. (2016) subjected winter faba bean plants to heat stress regimes (18/10; 22/14; 26/18; 30/22; 34/26 °C day/night) for 5 days during floral development and anthesis. Sensitivity of faba bean to heat stress fluctuated between floral stages; flowers were most affected during initial green-bud stages. Plant yield showed a linear negative relationship to stress with high plasticity in yield allocation.

So, expanding the cultivation of faba bean under sandy stress conditions is faced by a number of yield limiting factors among them is the drought, heat stresses, and poor population of native *Rhyzobium leguminosarum* and hereafter an expected decrease in nodulation and nitrogen fixation. In addition, these soils are of a poor fertility level from macro and micronutrients. However, the extension of field crop cultivation in these soils has become a must to minimize the food gap which is wide due to the ever growing population. To face these yield limiting factors, seed inoculation was recommended to activate atmospheric N fixation (El-Karmity 1990). Also, Khan et al. (2010) showed the importance of cultivating resistant faba bean cultivars to ascochyta blight, particularly in areas exposed to terminal drought stress.

Faba bean is cultivated fundamentally as a cool-season crop. Fluctuations in sowing dates and shortage in precipitation expose faba bean crop to drought and heat stresses. Because of climate change, high temperatures are likely to exacerbate the adverse effects of hot and dry climates on faba bean growing. High heat stress is particularly damaging to faba bean during flowering, when the viability of pollen is critical for successful reproduction. Recent studies have shown that maintenance of protein homeostasis through synthesis of heat shock proteins plays a key role in the heat response of plants (Lavania et al. 2015). Furthermore, Siddiqui et al. (2015) subjected Egyptian and various faba bean genotypes to different levels of water stress. Water stress reduced all growth parameters i.e. plant height, fresh weight and dry weight/ plant, leaf area/plant; physiological and biochemical characters i.e. leaf relative water content, proline content, total chlorophyll content and activities of catalase, peroxidase and superoxide dismutase of all genotypes.

6.2 Genotype × Environment Interaction and Its Relation to Climatic Change on Faba Bean Production

The interaction between genotype × environment is mainly due to the geographical climate, experimental factors and environments studied. This gives an indication of the size and nature of the existing interaction, leads to the determination of the stability of genetic makeups under different conditions. Most often, plant breeders face an interaction between G × E when assessing breeding program materials across environments. To identify stable genotypes, breeders should divide the G × E interaction into the stability statistics assigned to each genotype evaluated across a range of environments (Tekalign 2018).

Under twelve environments which are the combinations of 3 seasons × 2 plant population densities × 2 locations, eleven faba bean varieties has been grown and tested by Awaad (2002). Stability analysis of variance (Table 6.1) showed that the mean squares of environments were highly significant suggesting that the environments under study were different. Genotypes mean squares were found to be highly significant, hereby faba bean varieties were different in genes controlling seed yield and leaf chlorophyll content. Highly significant G × E item was detected for both characters, provide evidence that the studied faba bean genotypes differed in their response to the environmental conditions. The regression approach partitions G × E interaction into two components i.e. heterogeneity and the remainder. It is evident that both heterogeneity and the remainder exhibited highly significant values in leaf chlorophyll content. The heterogeneity mean squares were highly significant when tested against the remainder mean square for seed yield/fad and leaf chlorophyll content, suggesting that there were differences in regression coefficients among the

Table 6.1 Joint regression analysis of seed yield (ard/fad) and leaf chlorophyll content characters (Awaad 2002)

Source	d.f	Seed yield (ard/fad)	Leaf chlorophyll content
Environments (E)	11	73.302**	32.263**
Reps within environ	24	9.896	2.099
Genotypes (G)	10	9.960**	55.099**
G × E	110	1.802**	8.977**
Heterogeneity	10	6.046**	4.428**
Remainder	100	1.377	9.431**
Pooled error	240	1.255	0.891

* Highly significant at 0.01 probability level

genotypes. Since, the major portion of differences in stability was due to the linear regression and not to the deviation from the linear relationship for these characters.

Sixteen faba bean genotypes under 13 environments were evaluated in Ethiopia during the main cropping season for three years of 2009–2011 by Temesgen et al. (2015). They computed combined analysis of variance for seed yield for 16 faba bean genotypes tried across 13 environments. The main effect differences among genotypes, environments, and the interaction effects were highly significant. Environment main effect accounted for 89.27% of the total variance of seed yield, however genotype and G × E interaction effects were negligible and contributed for 2.12 and 3.31% of the total variation, respectively. Hereby seed yield was highly affected by changes in environmental conditions, then G × E interaction and genotypic effects. They suggested that the highly significant environment effect and its high variance component might be attributed to the great variances between the studied locations in altitude and variances in both quantity and distribution of annual rainfall. There was a clear G × E interaction effect existing in faba bean different environments, reflecting in the presence of considerable variances in genotypic responses to the tested environmental conditions. Thus selection and commendation of recent cultivars should be hard at such situations, where the effects of a high G × E reaction lead to the masking effects of variable environments. Under divers environments in Ethiopia, Tekalign (2018) showed that mean square for G × E interaction was highly significance for faba bean seed yield, hereby it divided into environment linear, G × E (linear) interaction effects (bi) and unexplained deviation from linear regression (S^2di). Analysis of variance due to linear regression was highly significant among genotypes. Mean squares attributed to environment (linear) were highly significant, representative differences between environments. The G × E (linear) interaction was highly significant, suggesting differential response of genotypes to change in environments. The non-linear responses as estimated by pooled deviation from regression were highly significant, so, the changing in the performance of genotypes under various environments was partially unpredictable. According to the previous results, Pham and Kang (1988) stated that G × E interaction diminishes the usefulness of genotypes by confounding their yield performances. So, it is of necessary to evaluate in depth the yield levels, adaptation and stability patterns of faba bean entries in mult-environmental conditions.

Furthermore, under Egyptian conditions, at two environments, control and artificial infection by chocolate spot disease (*Botrytis fabae* Sardo), El-Abssi (2020) tested 21 faba bean genotypes for some economic characters (Table 6.2). Results revealed that mean squares due to genotypes, parents, crosses and parents *versus* crosses (P vs. C) were highly significant for chocolate spot resistance, seed yield/plant, protein and carbohydrates contents under both conditions. Hereby, high genetic variances have been existed between faba bean genotypes for those characters under both conditions, except P. versus C for chocolate spot resistance under the natural infection and seed yield/plant under the artificial infection condition. So, these results provide evidence for the presence of adequate amount of genetic variability valid for further biometrical assessments. Moreover, the highly significant mean squares

Table 6.2 Mean squares of chocolate spot resistance, seed yield/plant, protein and carbohydrate contents for parental faba bean genotypes and their F_1 crosses under control and artificial infection conditions (El-Abssi 2020)

Source of variation	d.f	Chocolate spot resistance		Seed yield/plant (g)		Protein content (%)		Carbohydrate content (%)	
		Natural	Artificial	Control	Artificial	Control	Artificial	Control	Artificial
Replicates	2	0.68	0.59	3.05*	1.24	0.08	0.001	0.004	0.001
Genotypes	20	6.52**	7.23**	6187.98**	2415.36**	7.03**	7.67**	14.00**	13.48**
Parents	5	14.22**	18.73**	7182.78**	4363.41**	9.22**	6.83**	19.21**	18.77**
Crosses	14	4.20**	3.24**	6105.47**	1891.98**	6.11**	7.53**	8.71**	8.76**
P. versus C	1	0.41	5.62**	2369.13**	2.54	8.98**	13.74**	62.01**	53.12**
Error	40	0.58	0.14	0.75	1.28	0.09	0.001	0.001	0.001

*, ** Significant at 0.05 and 0.01 probability levels, respectively

due to parents versus crosses (P vs. C) observed for previous characters, revealed the attainability of heterosis for both characters in the studied genotypes.

6.3 Performance of Faba Bean Genotypes in Response to Environmental Changes

The possibility of increasing the cultivated area of faba bean under environmental pressure is an important goal for breeders and crop producers. Thus increasing productivity through the development of new high-yield varieties prerequisit. Different characteristics, such as flowering duration, number of pods, number of seeds and seed size, are used as a selection criteria for improved yields (El-Hady et al. 1998). A successful selection depends upon the information on the genetic variability and correlation of morpho-agronomic characters with seed yield. Under different environmental conditions, Sharifi (2014) and Kumar et al. (2017) showed that seed yield could be improved through selecting each of pod length, number of seeds/pod and 100- seed weight in faba bean breeding programs.

Ahmed et al. (2008) tested seven genotypes of faba bean, one of them is Egyptian origin (Giza Blanca) and the rest six genotypes were delivered by ICARDA, based on tolerance to drought and salt. Faba bean plants were tested under three diverse irrigation water intervals, 5, 10 and 15 days and two concentrations of NaCl, 25 and 50 mM. Results indicated that control plants displayed higher growth and yield compared to genotypes exposed to water or salt stress. The seven faba bean genotypes arranged by yield in the following order; V5, V6, V3, V2, V1 and V4. Variety 4 exhibited significant stimulation of seed yield under salinity effect of the adopted concentrations, and classified as highest salt tolerance among the seven tested genotypes. Expose the faba bean genotypes (V1, V2, V4, V5, V6 and V7) to water stress caused significant increase in photosynthetic pigments after 45 days from sowing, but this effect took the opposite direction after 90 days. Salt stress reduces chlorophyll content however increase carotenoids in the leaves. Variety 3 showed increase in photosynthetic pigments under water and salt stresses. Salinized genotypes exhibited the maximum values of total soluble sugars, proline and total free amino acids. Varieties V5, V6 and V1 recorded the highest total soluble sugars among the seven genotypes of control plants. Higher estimates of total soluble sugars were verified for genotypes V1, V7 and V3 under 25 mM and V6, grown as legume crop mainly V5 and V1 under 50 mM NaCl. Genotype 3 presented the maximum quantities of proline under water and salt stresses. Genotypes 2 displayed the maximum total free amino acids under water stress and genotypes V2, V5 and V1 under 50 mM NaCl. Yield reduction percentage in the seed yield as a result to water and salt stresses, seemed that V4, V6 and V3 appeared to be more tolerant to drought, as well as varieties V4 and V3 for salt tolerance compared to the remaining genotypes.

Duc et al. (2010) shows that faba bean displays wide genetic variability which has been well characterized. Internationally, more than 38,000 accessions are involved in at least 37 registered collections varied in their characteristics, composition and tolerance to biotic and abiotic stresses. Alghamdi (2007) tested the genetic behavior of six faba bean genotypes grown in the central region of Saudi Arabia and registered differences between the genotypes in most studied traits containing number of pods and number of seeds per plant, seed weight per plant and seed yield.

Tavakkoli et al. (2012) noted significant variations in the level of salt tolerance in eleven faba bean genotypes. Concentrations in Na^+ and Cl^- at 75 mM NaCl were significantly related with biomass production under controlled conditions ($r = -0.97$ and -0.95) and classified faba bean genotypes for its grain yield under field conditions. Moreover, six parents and their 15 F_1 cross were sown in two adjacent fields to avoid the differences in the soil fertility and irrigated with two levels of soil moisture (60–70 and 40–50%) of field capacity in addition to the rainfall. Omar et al. (2014) indicated that environments (Env.) mean squares were significant for the considered characters, representative overall differences between the normal and moisture stress environments. Mean performance of genotypes of all characters under normal environment were significantly higher than those the corresponding stress one. Significant genotypes mean squares were obtained for all studied characters under both treatments and the combined analyses. This shows high degree of genetic diversity between the parental genotypes. Significant genotype × environment mean squares were attained for all yield characters, revealing that genotypes differed in their performance from environment to another. Significant mean squares due to interaction between parents and environments were observed for all yield characters except 100-seed weight, showing that the parents differed in their response to environments. The parental variety Nubaria 1 gave highest values for 100-seed weight, Sakha 2 for number of seeds/ pod; Giza 429 for number of pods/plant under both levels of soil moisture as well as the combined analyses. Sakha 1 gave the second values for 100-seed weight, while it produces reasonable values for other characters under the two soil moisture levels and the combined analyses. Giza 716 exhibited the lightest 100-seed weight with lowest number of seeds/pod under the two soil moisture levels and the combined analyses with moderate values for other characters. Giza 2 displayed uppermost values for plant height under both levels of soil moisture as well as the combined analysis with highest values for number of branches/plant and seed yield/plant under normal irrigation and the combined analyses.

The cross Sakha 1 × Giza 429 had the highest mean value of number of branches/plant, under two levels of soil moisture and the combined data, but the cross Giza 716 × Giza 2 provided the lowest values under normal irrigation and the combined results, and Nubaria 1 × Sakha 2 under water stress level. The cross Giza 429 × Giza 2 under normal irrigation and in the combined and Giza 429 × Giza 716 under stress irrigation produce highest values for number of pods/plant. The cross Sakha 1 × Giza 716 under stress irrigation and the combined and Nubaria 1 × Giza 429 under normal irrigation were registered maximum values for seed yield/plant. However, the crosses Sakha 2 × Giza 429 under stress environment and the combined as well as Nubaria 1 × Sakha 2 under normal irrigation produced the lowest seed yield/plant. Therefore, the promising faba bean crosses were important for improving yield and its components under drought environments. Moreover, sixteen faba bean genotypes were screened by Temesgen et al. (2015) under 13 environments. The average environmental seed yield across genotypes varied from lowest at 2.31 t /ha in Bekoji 2011 to the uppermost at 5.24 t /ha in Koffale 2009. Mean seed yield of faba bean genotypes across environments differed from 3.2 t/ha for genotype EK01024-1–1 to 3.88 t /ha for EK 1042–2-1with an overall environment mean of 3.58 t /ha. The highest seed yield ranged from 5.16 t /ha in genotypes EK01024-1–1 and EK 01,024–1-2 to 6.53 t /ha for the check cultivars Moti and EK 01,001–8-1. The lowest yield varied from 0.82 t /ha in genotype EK01001-8–1 to 2.04t /ha in EK01004-2–1. The minutest yield amplitude was attained from EK01004-2–1 (3.55 t /ha followed by EK01015-1–1 (3.62 t /ha).This indicating their steady performance through the evaluated environments. However the highest yield amplitude was registered for EK01001-8–1 (5.70 t /ha) followed by the check cultivar Moti (5.33 t /ha). This inconsistency yield rank of genotypes across the environments revealed that the G × E interaction effect was of the crossover type.

Migdadi et al. (2016) tested nine faba bean genotypes for seed yield and its components and free proline content in leaves under three water regimes viz well-watered, mild and severe drought. The results showed that drought caused clear negative impacts on yield and its components, however the effect was positive on proline content of leaves. Cultivar Hassawi 2 out yielded all genotypes at water treatments and was followed by Giza 843 and ILB 1814 under well irrigation and by Giza Blanca and Giza 843 under severe drought stress. Also Hassawi 2 and Giza Blanca revealed higher drought tolerance efficiency (42.3 and 39.5), less drought stress sensitivity index (0.6) and smallest decrease in seed yield 58.3 and 60.4%, respectively. Proline content ranged from 46.3 µg/g for Gazira 2 to 69.7 for ILB 1814 under well-watered and from 89.8 for Kamline to 264.0 for Gazira 1 under severe drought.

6.4 Adaptability and Yield Stability

An adaptation phenomenon is more complex process than just reduced growth and productivity (Conde et al. 2011). Evaluating genotypes in diverse environments is essential to obtain information regarding phenotypic stability, which is beneficial for the selection of promising genotypes and improvement programs. Most often, plant breeders face an interaction between G × E when assessing breeding program outcomes across environments. To identify stable genotypes, breeders should divide the G × E interaction into the stability statistics assigned to each genotype evaluated across a range of environments.

In this respect, Faba bean crop is sensitive to drought stress (Khan et al. 2007, 2010; Ammar et al. 2014). The response to environment as measured by the regression technique was found to be highly heritable and controlled by genes with additive action (Hayward and Lawrence 1970). The regression coefficient "b" is a measure of the linear response or the adaptability of a genotype to different environments (Langer et al. 1979). It the case of the insignificant "b" value, the deviation from regression "S^2_d" is considered most appropriate for measuring phenotypic stability, because it measures the predictability of genotypic reaction to various environments (Becker et al. 1982). Gulian Yue et al. (1990) reported that the deviation from regression seemed to be very important for estimating the stability. In this connection, Eberhart and Russel (1966) described the stable genotype which having high mean performance over environments, with "b" value approaching near unity and the deviation from regression as minimum as possible ($S^2_d = 0$).

In this respect, Darwish et al. (1999) emphasized that cultivar Giza Blanca proved to be more adapted under newly reclaimed sandy soil condition. Moreover, in order to determine adaptability and stability of faba bean seed yield, Awaad (2002) grown eleven varieties under twelve environments (the combinations of 3 seasons × 2 plant population densities × 2 locations). As shown in Table 6.3 "b" value varied from 0.542 (Giza 429) to 1.829 (Giza 461). The regression coefficient deviated significantly from unity (b > 1) in faba bean genotypes Giza 402 and Giza 461, indicating higher production potential in favourable environments. Otherwise, the "b" value

Table 6.3 Means and phenotypic and genotypic stability parameters for seed yield of the eleven faba bean varieties grown under twelve environments (Awaad 2002)

Variety	$\overline{X}$	b	S^2_d	α	λ	Degree of stability		
						0.99	0.95	0.90
1-Giza 1	9.316	0.799	0.162	−0.385†	0.809	+ + +	+ + +	+ + +
2-Giza 2	8.131	0.795	1.854*	−0.704†	3.088*	+	+	+
3-Giza 3	9.680	0.717	0.281	−0.976†	2.049	+ + +	+ + +	+ + +
4-Improved G. 3	10.778	1.087	0.653	−0.374	1.913	+ +	+ +	+ +
5-Giza Blanca	11.558	0.669*	0.985	0.687†	3.165*	+	+	+
6-Giza 402	10.746	1.683**	0.389	0.868†	19.070*	+	+	+
7-Giza 429	10.108	0.542**	1.534*	−1.003†	1.741	Nearly	Perfect	+ +
8-Giza 461	9.333	1.829**	1.189	−0.124	1.248	+ +	+ +	+ +
9-Giza 714	11.183	1.133	−0.219	0.212	1.812	+ +	+ +	+ + +
10-Giza 843	12.019	0.845	0.616	0.541†	1.213	+ + +	+ + +	+ +
11-Giza 957	11.691	0.893	0.743	0.389†	1.256	+ + +	+ + +	
General mean	10.419							
L.S.D$_{0.05}$	1.393							

* and ** denotes significant at 0.05 and 0.01 levels of probability, respectively * λ value greater than Fa value derived from F-table with n1 = 10, n$_2$ = 240 and a = 0.05
† α Value significantly from α = 0 at the 0.05 probability level
+ + + Genotypes with above average degree of stability
+ + Genotypes with average degree of stability
+ Unstable genotype

Fig. 6.2 Distribution of stability statistics for seed yield/fad of the eleven faba bean genotypes (Awaad 2002)

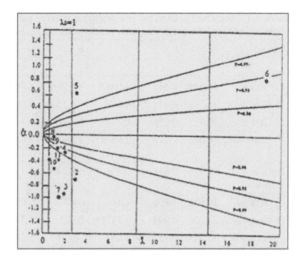

was deviated significantly from one and less than unity (b < 1) in Giza Blanca and Giza 429 which appeared to be more adapted to less favourable environments. The deviation from regression "S^2_d" was very small and did not deviate significantly from zero in Giza 1, Giza 3, Improved Giza 3, Giza Blanca, Giza 402, Giza 461, Giza 714, Giza 843 and Giza 957 varieties which showed stability for seed yield. Whereas, Giza 2 and Giza 429 appeared to be more sensitive to the fluctuating environmental conditions. A simultaneous consideration of the three stability parameters (X, b and S^2_d), evidenced that the most stable and high yielding genotype was Giza 843 followed by Giza 957, Giza 714 and Improved Giza 3. Moreover, genotypic stability (Table 6.3), evidenced that, the great variation in λ statistics suggested that the relatively unpredictable component of the G × E interaction variance may be much more important than the relatively predictable component of variation α, for the studied genotypes which showed different degrees of stability (Tai 1971). As illustrated in Fig. 6.2 the average stability area contained faba bean varieties improved Giza 3, Giza 461 and Giza 714. Among those, Improved Giza 3 and Giza 714 gave higher seed yield (10.778 and 11.183 ard/fad, respectively) than grand mean (X = 10.419 ard/fad). Giza 843 and Giza 957 exhibited high yield potentiality, however Giza 1 and Giza 3 were below the level of grand mean, but they showed above average stability as revealed by α estimates deviated significantly from zero with λ not deviate significantly from one. Giza 429 gave relatively low yield but showed nearly perfect stability. However, Giza 2, Giza Blanca and Giza 402 were unstable. So, it may be possible to select a high yielding cultivars which show relatively low level of instability such as Giza Blanca as a source of high yielding genes to be crossed with genetic makeup have stability genes and practice selection for genotypes with both high yield and good stability. However, it has to be kept in mind that stable genotypes are generally low yields than instable ones.

The cause of yield stability or instability is often unclear due to the diverse mechanisms of physiological, morphological and phonological aspects (Heinrich et al.

Table 6.4 Means and phenotypic and genotypic stability parameters for leaf chlorophyll content of the eleven faba bean varieties grown under twelve environments (Awaad 2002)

Variety	$\overline{X}$	b	S^2_d	α	λ	Degree of stability		
						0.99	0.95	0.90
12-Giza 1	35.625	0.347**	−0.009	−0.280	1.845	+ +	+ +	+ +
13-Giza 2	35.775	0.574**	0.114	−0.583†	0.861	+ + +	+ + +	+ + +
14-Giza 3	40.083	0.936	6.443**	1.294†	9.440*	+	+	+
15-Improved G. 3	38.692	2.122**	2.937**	1.380†	4.840*	+	+	+
16-Giza Blanca	35.958	0.671	0.738	−0.242	1.737	+ +	+ +	+ +
17-Giza 402	36.692	1.136	1.314	0.627†	0.094	+	+	+
18-Giza 429	39.200	0.999	3.837**	0.682†	3.342*	+	+	+
19-Giza 461	40.283	0.139**	0.833	−0.319	1.883	+ +	+ +	+ +
20-Giza 714	34.950	2.301**	2.648**	1.400	5.163*	+	+	+
21-Giza 843	40.467	1.038	1.367	0.604†	7.700*	+	+	+
22-Giza 957	39.600	0.149	0.068	1.012†	1.916	+	+	+
General mean	37.938							
L.S.D$_{0.05}$	1.511							

* and ** denotes significant at 0.05 and 0.01 levels of probability, respectively * λ value greater than Fa value derived from F-table with n1 = 10, n2 = 240 and a = 0.05
† α Value significantly from α = 0 at the 0.05 probability level
+ + + Genotypes with above average degree of stability
+ + Genotypes with average degree of stability
+ Unstable genotype

1983). Leaf chlorophyll content is an important physiological character contributing to seed yield (Hafiz and Abd El-Mottable 1998). Awaad (2002) noticed from phenotypic stability estimates for leaf chlorophyll content (Table 6.4) that the regression value varied from genotype to another. It ranged from 0.139 (Giza 461) to 2.122 (Improved Giza 3). The regression coefficients deviated significantly from unity (b > 1) in Improved Giza 3 and Giza 714 genotypes which showed good response to improved environments. However, Giza 1, Giza 2 and Giza 461 showed specific adaptability to poor environments (b < 1). Faba bean genotypes Giza 1, Giza 2, Giza Blanca, Giza 402, Giza 461, Giza 843 and Giza 957 appeared to be more stable as revealed by lowest and insignificant S^2_d values. Whereas, Giza 3, Improved Giza 3, Giza 429 and Giza 714 were unstable. It is likely to notice that, only two faba bean genotypes: Giza 843 and Giza 957 ranked as the most desired and stable genotypes and the remaining nine faba bean genotypes have been ranked as sensitive ones. Genotypic statistics of Fig. 6.3 illustrated that varieties Giza 1, Giza Blanca and Giza 461 were located in the area of average stability with high concentration

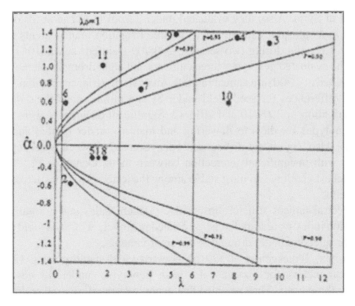

Fig. 6.3 Distribution of stability statistics for leaf chlorophyll content of the eleven faba bean genotypes (Awaad 2002)

of leaf chlorophyll for Giza 461 (40.283) than grand mean. Giza 2 could be classified as above average stability genotype as revealed by α value less significantly than unity. Giza 402 gave positive and significant α value, thus it could be classified as more responsive to the environmental conditions. Giza 957 was located in the area of below average stability. The other faba bean genotypes exhibited different degrees of instability (Table 6.4). Finally, based on phenotypic stability parameters, it could be concluded that Giza 402 characterized by high seed yield potentiality with greater number of pods/plant and Giza 461 gave relatively low yield. Both varieties showed high degree of stability and were adapted to be grown under Zagazig region as favourable environment. Whereas, Giza Blanca exhibited high mean values of seed yield/ fad and 100 seed weight and Giza 429 had relatively low yield. They classified as highly adapted to be grown under Khattara region as stress environment. Moreover, the most desired and stable varieties for seed yield/ fad were Giza 843, Giza 957, Giza 714 and Improved Giza 3 at both phenotypic and genotypic levels, therefore they could be grown under wide range of environments.

Water stress is a major limiting factor in faba bean (*Vicia faba* L.) production in the Mediterranean region, which is well-known for its irregular water distribution and modest moisture levels (~500 mm rainfall). In this study, Maalouf et al. (2015) valued faba bean genotypes for spectral indices, yield traits, rhizobium nodulation and yield stability under various environments, and their relations. Eleven faba bean genotypes were tested under two water regimes, rainfed and supplemental irrigation (SI), in 2008/09; and under three water regimes (rainfed, 50 and 100% soil water capacity irrigation) during three successive seasons (2009/10, 2010/11 and 2011/12), under

Tel Hadya in Syria. Also, they evaluated under Terbol in Lebanon for one season (2011/12) in a split-plot experiment with three irrigation managements and under Kfardan in Lebanon during two seasons, 2008/09 (rainfed) and 2010–11 (rainfed and SI). The evaluated genotypes three cultivars and eight drought-tolerant breeding strains selected at < 300 mm annual rainfall. Analysis of variance revealed significant genotypic differences for seed yield under SI and rainfed conditions in 2008/09, and full irrigation in 2009/10 and 2011/12. Significant differences were registered among genotypes for days to flowering and maturity under rainfed and irrigated conditions. Also, significant differences were recorded between water regimes and genotypes with insignificant interaction between them. Genotype FLIP06-010FB was the highest yielding and most stable among the tested genotypes across different environments.

The spectral indices structure-insensitive pigment index and normalized pheophytinisation index were found to be associated positively with seed yield and might be used for selection under drought stress environments.

In Ethiopia, Temesgen et al. (2015) assessed yield stability using 17 different stability parameters for sixteen faba bean genotypes under 13 environments throughout three years. They revealed that genotypic superiority index (Pi) and FT3 were very useful for selecting both high yield and stable faba bean genotypes. Twelve of the 17 stability parameters, comprising CVi, RS, α, λ, S^2di, bi, Si (2), Wi, σi^2, EV, P59, and ASV, were influenced by yield and stability. Although not any of the varieties exhibited consistently superior performance across the studied environments, the genotype EK 01,024–1-2 rated in the top third of the investigated genotypes in 61.5% of the test environments and was recognized as the most stable one, with type I stability. EK 01,024–1-2 also displayed a 17.0% seed size advantage over the standard genotypes and was wide production. Diverse stability parameters explained genotypic performance differently, irrespective of yield performance.

6.5 Additive Main Effects and Multiplicative Interaction

AMMI permitted the presentation of the adaptation of groups of genotypes to specific environments, which could not be perceived from the deviation from regression (S^2di) estimates. Link et al. (1996) stated that broadening agronomic adaptation will improve yield stability in the grain legume faba bean. The adaptation of European and Mediterranean genotypes to European and Mediterranean environments are involved. The materials included 20 genotypes (12 European and 8 Mediterranean lines) and 99 intra- and interpool-crosses in F1 crosses. Genotypes were evaluated under 9 environments which are two spring-sown Southern German environments (SGermE), and seven autumn-sown Mediterranean environments (MedE) in Sicily, Puglia, Andalucia and South Africa. Stability analyses and AMMI analyses were performed. Mean yield in F_1 was 257 g/row, exceeded the overall parental mean 144 g/row. Environmental means varied from 94 to 411 g/row. The average regression coefficient in F_1 valued $b_i = 1.07$, was significantly greater than their parents

($b_i = 0.68$). The opposite was true for the relative magnitude of the deviations from the regressions (S^2di), which were highly associated to the AMMI-PC1-results. The AMMI analysis obviously separated the SGermE from the MedE, as well as the germplasm pools. Though the supremacy of the F_1-crosses over their parents was apparent and their interactions with the environments accurately reflected on their parents. They recognize number of promising crosses as a nucleus of a broadly adapted faba bean gene pool. A strong pattern of specific adaptations of the 20 parental lines was recognized by the AMMI-based biplot. It caused from abundant positive interaction of the minor lines and the SGermE and the Mediterranean lines and the MedE (except Gouda). This interaction was strictly reflected by the respective minor and Mediterranean in trapool-hybrids. The maximum average yields were attained in two of the MedE (Cordoba and Catania in 1992) and in both SGermE. The Mediterranean group of lines and crosses produced the great average yields at the two MedE. While the European material caused the high average at the two SGermE. Then, large deviations form regression was occurred on these environments. The regression could describe only very few of this pattern, and the correlation between the (S^2di) estimates and the AMMI-scores was very high.

Moreover, Flores et al. (2012) compared field performance in autumn sowings of 15 faba bean winter-type cultivars were sown in two successive autumns in 12 climatically divergent sites in Austria, France, Germany, Spain and the UK. Crossover genotype × environment was greater and principally attributed to the geoclimatic area. GGE biplot permitted identification of three mega-environments, viz. continental, oceanic, and Mediterranean. As a result of the climatic diversity of the environments, no cultivar performed well in all environments. Cultivars Clipper, Castel, Target, Wizard, and Gabl-107 performed well in oceanic mega-environment. However cultivars Castel, HIX and Target gave a good performance under continental mega-environment. No one of the studied genotypes was suitable for Mediterranean environments, and only Irena cultivar was able to give some moderate seed yield under Cordoba. Wizard and Gabl-107 were the highest yielding cultivars being relatively stable over both oceanic and continental situations. On the contrary, the two cultivars Irena and Divine yielded poorly at all situations. The results support the specific breeding for each major geoclimatic zone based on distinct genetic bases and selection environments. The first two principal components give 74–96% of total genotype + genotype by environment interaction sum of squares. Hereby the first two principal components were significant factors in all subset of results, excluding the second principal component in the first subset. The much greater G × E interaction comparative to the genotype main effect produce heavy crossover genotype × environment interactions, as supported by the information that first environment principal component marks took different signs and the environments dropped in all quadrants. Faba bean genotypes that were farthest from the biplot origin viz. cvs. Castel, Clipper, Divine, and Irena formed the corners of the polygon. The GGE biplots based on each subset data separated the mega-environments. Clipper was found to be the winning cultivar for the first mega-environment (Bersée-03, Brosse-03, Brosse-04, Paris-03, Paris-04, Edgmond-03, and Edgmond-04). Castel was the winning cultivar for the second mega-environment (Dijon-03, Dijon-04, and Montbartier-03). Based on the

"ATC" of the biplot, subset 3, Wizard and Gabl-107 were the uppermost yielding cultivars on average and were relatively stable over the environments. The test environments, Paris-04, Brosse-04, Dijon-04, Bersée-03, and Edgmond-03 were the most differentiating. The mega-environments, Paris, La Brosse, Bersée, Edgmond, and Hohen-lieth locations tended to be grouped distinctly from the other locations, named as oceanic mega-environment. Dijon, Montbartier, Logroño, Göttingen, and Gleisdorf tended to be grouped, as a second mega-environment named continental. The GGE biplots revealed three groups of cultivars based on their response to the mega-environments. Group1 contained of five cultivars, viz. Clipper, Castel, Target, Wizard, and Gabl-107 that performed well under the oceanic mega-environment. Group 2 comprised of three cultivars i.e. Castel, HIX and Target that performed well under the continental mega-environment. Group 3 entailed Irena cv. performed better than the others in Mediterranean mega-environment, while its productivity was low. The poor adaptation in Mediterranean areas could be attributed to the little earliness of cycle and drought tolerance of these winter types relative to the Mediterranean germplasm.

Based on AMMI analysis, Tolessa (2015) showed that cultivars EH91016-5–1-1, CS20DK, MOTI, TUMSA, DOSHA and EH98086-2, exhibited IPC1 scores close to zero, have less response to the interaction and showed general adaptation to the test environments. Variety SELALE-KASIM revealed large positive IPC1 score and appeared to be better adapted to environment Bekoji 2007 as it illustrates larger and similar sign IPC1 score. While, varieties NC-58, BULGA-70 and MESSAY, with larger negative IPC1 scores were adapted to environments Asassa 2007 and Koffale 2009. Faba bean recent cultivars, TUMSA and DOSHA, and the oldest variety CS20DK were shared the highest mean yield over test environments with low IPC1 scores are deliberated as the most stable ones with relatively less variable yield performance across environments. Environments, Koffale 2007 and Koffale 2009, were joined larger main effects with larger interaction effects. Therefore, the relative ranking of cultivars were unstable at Koffale making it less predictable location for faba bean valuation and production rather the rest environments. Furthermore, Tadele Tadesse et al. (2018) evaluated fifteen faba bean genotypes with the improved and local checks under different environments. They utilized additive main effects and multiplicative interaction (AMMI) analyses. The principal components (IPCA1) and (IPCA 2) described 52.8 and 47.2% of the interaction, separately. Seed yield varied from 2.6 to 4.2t/ha among genotypes with an average of 3.3t/ha. Almost of the genotypes were highly responsive to the environment and adapted to improved environments. Genotype EH03014-1 and EKLS01013-1 provided greater seed yield compared to the checks, with regression coefficient near to one and deviation from regression close to zero, and revealed stable overall locations. Also, EH03014-1 and EKLS01013-1 appeared to be more resistant to major disease and superior in seed yield rather than the check.

6.6 Gene Action, Genetic Behavior and Heritability for Faba Bean Traits Related to Environmental Stress Tolerance

Genetic variations and heritability for faba bean genotypes in arid conditions of Saudi Arabia has been determined by Ghandorah and El-Shawaf (1993). They registered high broad sense heritability estimations and predicted genetic advances. Therefore, selection between the genotypes might be improved yield, yield components and plant height. Moreover, Alghamdi (2007) recorded highest estimates of heritability in broad sense for flowering date 0.986, No. of pods/plant 0.96, No. of seeds/plant 0.957 and days to maturity 0.905, respectively in the first season. While, in the second season the highest values of broad sense heritability were registered for days to flowering, No. of seeds/plant and days to maturity with 99.9, 94.2 and 91.0%, respectively.

A diallel cross set involving six parents and their 15 F1 crosses were tested under two adjacent fields to avoid the differences in the soil fertility productivity and irrigated with two levels of soil moisture (at 60–70 and 40–50%) of field capacity in addition to the rainfall by Omar et al. (2014). They indicated that mean squares associated with general GCA and specific SCA combining ability were highly significant for all yield characters except number of branches/plant under the two levels of soil moisture. Both additive and non-additive gene effects were involved in determining the performance of single cross progeny. GCA/SCA ratio was more than unity for all characters except plant height under stress irrigation, number of branches and number of seeds/pod under both environments, representative the prevalence of additive and additive × additive types of gene action in the genetics of such characters. The magnitude of additive and additive × additive types of gene action varied from irrigation level to another. Significant mean squares of interaction between SCA and irrigation levels were registered for all traits, suggesting that non-additive type of gene action was varied from irrigation level to another. Ratios for SCA × Env./SCA was much higher than ratios of GCA × Env./GCA for all characters except plant height and number of branches/plant. Therefore, non-additive effects were highly influenced by irrigation treatments than additive ones. The best general combining ability (ĝi) effects was attained by Nubaria 1 for 100-seed weight under the two levels of soil moisture, Sakha 1 for number of seeds/plant under stress condition and 100-seed weight under the two levels of soil moisture, Giza 429 for number of pods/plant, number of seeds/plant and seed yield/plant under the two levels of soil moisture, Giza 716 for plant height under stress condition, number of pods/plant and yield/plant under the two levels of soil moisture. Hence, Giza 716 could be considered as excellent combiner for improving drought tolerance using these characters. The highest desirable specific combining ability Ŝij effects were achieved in the crosses Nubaria1 × Giza 429, Sakha 1 × Giza 716 and Giza 429 × Giza 716 for seed yield/plant across two irrigation levels. Obiadalla-Ali et al. (2015) showed significant positive heterosis values over mid-parent for most yield traits. They detected highest value of heterosis over the mid-parent for total dry seed yield (128.8), while the lowest value (1.2%) of hybrid vigor was displayed by weight of 100-seeds. The highest value for heterosis was registered by cross Misr 2 × Giza 429 as the best for total dry seed yield and cross Giza 429 × Misr 1 for No. of branches/plant. Giza 429 is the best general combiner for utmost traits.

Furthermore, Soad Mahmoud et al. (2018) evaluated some selected Faba bean genotypes under two water, well-watered (100% from ETo) and severe water stress (60% from ETo) at the experimental farm of the Faculty of Agriculture, Suez Canal University, Ismailia, Egypt. They registered highly significant general (GCA) and specific (SCA) combining ability for morpho-physiological and yield characters under two water deficit, showing that GCA and SCA were important in the genetics of the yield traits. The majority of crosses exhibited highly significant heterosis over mid parents (MP). The values of heritability in narrow sense were 7.50 and 47.8% for Seed yield/plant; 23.3 and 56.8% for chlorophyll a; 41.6 and 57.5% for chlorophyll b as well as 1.68 and 41.9% for carotenoids under 100% ETo and 60% ETo water regimes, respectively. Therefore these characters are greatly influenced by dominance and environmental conditions.

In order to determine gene action and heritability controlling earliness, yield and chocolate spot disease in faba bean, El-Abssi et al. (2019) crossed six diverse parental genotypes in a half diallel crosses and evaluated under two environments, control and artificial infection by chocolate spot disease (*Botrytis fabae* Sardo) as illustrated in Table 6.5. The results indicated highly significant differences among faba bean genotypes for all the studied characters under both conditions. The additive (D) and dominance (H1 and H2) genetic components were significant for days to flowering and maturity, chlorophyll content (SPAD), seed yield/plant and resistance to chocolate spot under both conditions. The additive genetic component was higher in its magnitude as compared to the dominance ones for resistance to chocolate spot under the natural infection environment, resulting in average degree of dominance $(H1/D)^{0.5}$ less than the unity. Whereas, dominance component (H1 and H2) made up the most part of the total genetic variation as it was larger in its magnitude than the corresponding additive one for earliness characters, chlorophyll content and seed yield/plant under both conditions and resistance to chocolate spot under the artificial infection only. Thus, the average degree $(H1/D)^{0.5}$ was more than the unity for these characters. Narrow sense heritability (h2n) differed in its magnitude, due to the change in the genetic components from the control, natural to the artificial infection of chocolate spot disease. It was high for days to maturity under the artificial infection condition and resistance to chocolate spot under the natural infection one. And moderate for days to flowering under the control and the artificial conditions; days to maturity, chlorophyll content and seed yield/plant under the control condition as well as resistance to chocolate spot under the artificial infection one. Thus, selection based on phenotype could be effective to improve these characters. While, it was low for chlorophyll content and seed yield/plant under the control condition, suggesting that selection for both characters in early generations may not be useful and had to be delayed till late segregating generations. Hence, utilization of heterosis breeding could be rewarding for these characters.

Table 6.5 Additive (D), dominance (H) genetic variances and their derived parameters for faba bean characters under control, natural infection and artificial infection of chocolate spot disease conditions

Character component	Days to flowering (day)		Days to maturity (day)		Chlorophyll content (SPAD)		Seed yield/plant (g)		Chocolate spot disease	
	Control	Artificial	Control	Artificial	Control	Artificial	Control	Artificial	Control	Artificial
Genetic components										
D	385.381**	382.048**	14.146**	12.237*	49.429*	38.202*	2393.972	1454.045**	4.547**	6.189**
H_1	580.636*	618.680*	57.530**	55.864**	73.928**	103.883**	6669.924**	6165.212**	5.560**	3.525**
H_2	477.835*	507.279*	49.131**	48.440**	67.68**	82.103**	5670.133**	2646.000**	3.985**	2.771**
F	304.44	322.430	1.738	-6.485	28.191*	50.085*	1570.744*	1643.706	4.169**	4.159**
h^2	98.505	94.667	11.723	17.326	13.135	97.970**	511.690	0.313	-0.020	1.185**
E	1.360	1.037	0.021	0.063	0.229	1.063	0.287	0.426	0.196	0.053
Derived parameters										
$(H_1/D)^{0.5}$	1.227	2.017	1.223	1.137	1.223	1.649	1.669	1.475	1.106	0.755
$H_2/4H_1$	0.206	0.214	0.227	0.217	0.227	0.198	0.213	0.209	0.179	0.196
KD/KR	1.947	1.063	1.608	0.779	1.608	2.320	1.489	2.231	2.416	2.608
h(n.s)	43.2	45.8	45.3	51.8	45.3	18.6	39.1	20.4	45.0	65.1

*, **: Significant at 0.05 and 0.01 levels of probability, respectively. h(n.s): Heritability in narrow sense

6.7 Role of Recent Approaches

6.7.1 Biotechnology

The ICARDA faba bean MAGIC population comprising >2,200 F_4 lines is now under development, combining eight various parents with resources for heat, drought, ascochyta blight, chocolate spot, rust and *Orobanche* resistance.

The large genome size of faba bean (13,000 Mb), has hindered the speed of improvement programs (Rispail et al. 2010). Mitochondrial genome size of Broad Windsor is 580 Kb sharing 45% homology (Negruk 2013). Quantitative trait loci (QTLs) were recognized for resistance against numerous stresses in faba bean (Torres et al. 2006). To improve faba bean tolerance to drought and heat stresses, it is necessary to employ DNA Markers techniques and recognize stress resistance QTLs beside field experiments under different agro climatic regions. Marker-assisted selection (MAS) was utilized for improving faba bean against biotic stresses such as ascochyta blight, crenate broom rape, rust and against abiotic stresses like drought and frost (Torres et al. 2010). Detection the association between morpho physiological traits and drought tolerance in faba bean by using molecular tools is very important. Abdellatif et al. (2012) registered negative relationship between morphological traits and drought tolerance in faba bean. Thought SDS-PAGE examination, optical variances were detected between drought tolerant cultivar Giza 843 and drought sensitive Giza 3 in their protein patterns. Several protein bands were detected in cultivar Giza 843 that were absent in cultivar Giza 3. While, some protein bands were observed in Giza 3, which was absent in the protein pattern of Giza 843. Significant correlation was detected between morphological characters and biochemical markers.

The use of Random Amplified Polymorphic DNA (RAPD) method offers a simple, fast, efficient and inexpensive tool (Baheer-Salimia et al. 2012). Furthermore, it does not requirement knowledge of marker sequence and can produce abundant polymorphic DNA fragments (Kocsis et al. 2005; Achtak et al. 2009). Therefore, RAPD is a powerful and accurate tool for analyzing the genetic relatedness and diversity in many species. The DNA markers like RAPD, ISSR. AFLP and SSR were extensively used in assessing genetic diversity in faba bean (Wang et al. 2012).The analysis of faba been cultivars leads to collection of information about the genetic diversity at the genome level. The RAPD profile analysis can be useful to the selection of cultivar containing good information and properties in faba bean improvement program (Tahir 2015).

Yang et al. (2012) constructed and characterized a library with 125,559 putative SSR sequences for repeat type and length from a mixed genome of 247 spring and winter sown faba bean genotypes using 454 sequencing. A suit of 28,503 primer pair sequences were designed and 150 were randomly selected for validation. They found 94 produced reproducible amplicons were polymorphic amongst 32 faba bean genotypes chosen from different geographical localities. Number of alleles per locus varied from 2 to 8. The expected heterozygocities ranged from 0.0000 to 1.0000, and the observed heterozygosities ranged from 0.0908 to 0.8410. The justification by

UPGMA cluster study of 32 genotypes based on Nei's genetic distance, showed high quality and effectiveness of SSR markers developed via next generation sequencing technology in faba bean breeding efforts. Yahia et al. (2014) utilized SSR and RAPD markers to estimate the genetic diversity of 13 Tunisian faba bean genotypes and revealed that the polymorphic fragments % were 100% and 60.63% for SSR and RAPD markers, respectively. Maximum genetic similarity was verified between Sakha 4 and Wadi 1 (0 0.970), followed by Misr 3 and Nubaria 3 (0.919). The lowermost relationship was verified for cultivars Giza 3 and Giza 429 (0.619). The dendrogram resulted in two main clusters, one of them contained the cultivars Giza 3 and Giza 40, while the second cluster included the remaining cultivars. The second cluster was separated into two subclusters, one comprised Misr 3 and Nubaria 3, however the second cluster included Sakha 4, Wadi 1, Nubaria 1, Giza 429, Giza 2 and Giza 843. Obiadalla-Ali et al. (2015) showed that twelve arbitrary primers generated diverse degrees of genetic polymorphism amongst the parental faba bean genotypes. A total of 65 amplification products were scored polymorphic. Polymorphic bands % ranged from 33 to 100% with an average of 66.47%. The average of amplified bands was 5.42 polymorphic bands per primer. Khazaei et al. (2014) studied stomal properties related to drought stress tolerance in faba bean viz. density, length and conductance profoundly. They recognize eight QTLs governing stomatal features and the putative candidate genes within QTLs. Cui et al. (2008) indicated that expression of faba bean *VfPIP*1, gene in Arabidopsis enhanced drought tolerance in faba bean. Link et al. (2010) stressed that evidences on receptive genes for winter hardiness in faba bean, and their function is progressively remains untested in faba bean and HSP and HSF genes.

Naglaa et al. (2018) selected five primer of RAPD to identify genetic diversity among ten faba bean cultivars. The primers produce multiple bands, which ranged between 4 bands for primer D13 to 9 bands for primer C06. The total number of bands was 32, which 21 of them were polymorphic with 65.6% polymorphism. The highest level of polymorphism was observed in primer C19 which showed 75%, while the lowest polymorphism 50% was in primer D13 (Table 6.6 and Fig. 6.4). RAPD analysis produce four positive markers in Egyptian cultivar Giza 3, which present in the two primers (C06 and C19) at different molecular weight (MW) produce 2 bands, 1850 bp and 1200 bp in primer C06 and the other two at mw 2850 bp and 1850 bp in primer C19. Giza 40 present two positive markers one of them in primer B09 in MW 580 bp, while the other in MW 1210 bp at primer C19. Whereas, Giza 429 recorded one negative band at 450 bp in primer C19. Giza 843 has one positive band marker at 1020 bp in primer C19.

They added that similarity index and dendrogram across the ten faba been cultivars according to RAPD analysis (Table 6.7), revealed that the maximum genetic similarity was registered between Sakha 4 and Wadi 1 (0 0.970), then Misr 3 and Nubaria 3 (0.919). Otherwise, the lowest association was noted between cultivar Giza 3 and Giza 429 (0.619). The dendrogram (Fig. 6.5) resulted in two main clusters. One of them included the cultivars Giza 3 and Giza 40, whereas the second contained the remaining cultivars. The second cluster was divided into two subclusters, one

Table 6.6 Levels of polymorphism by the five RAPD primers across ten faba bean cultivars (Naglaa et al. 2018)

Primer name	MW(bp)	Number of monomorphic bands (nmb)	Number of Polymorphic bands (npb)	NB	P (%)	Genotypes	MM PM	NM
B09	2600:580	2	3	5	60	Giza 40	580	
C06	1850:310	3	6	9	66.7	Giza 3	1850	
							1200	
C08	2800:675	2	4	6	66.7			
C19	2850:450	2	6	8	75	Giza 3	2850	
							1850	
						Giza 40	1210	
						Giza 843	1020	
						Giza 429		450
D13	1220:450	2	2	4	50			
Total		11	21	32	65.6			

Where:

NB,	Number of bands	MM,	Molecular marker	P (%),	Polymorphism (%)
PM,	Positive marker	NM,	Negative marker		
MW,	Molecule weight				

Fig. 6.4 RAPD banding
patterns amplified with 5
primers a cross ten faba bean
cultivars (Naglaa et al. 2018)

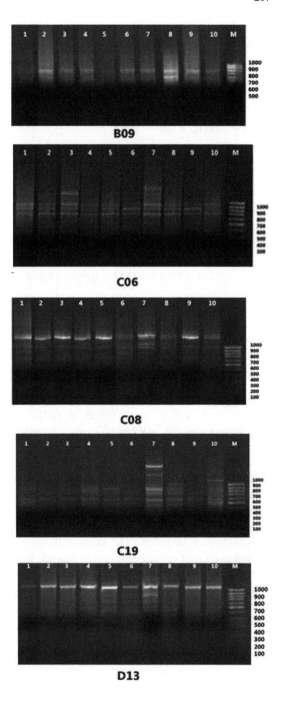

Table 6.7 Similarity matrix among the ten faba bean cultivars using RAPD analysis (Naglaa et al. 2018)

Cultivar	Misr 3	Sakha 4	Wadi 1	Nubaria 1	Nubaria 3	Giza 2	Giza 3	Giza 40	Giza 429
Sakha 4	0.778								
Wadi 1	0.757	0.970							
Nubaria 1	0.811	0.909	0.882						
Nubaria 3	0.919	0.848	0.842	0.882					
Giza 2	0.872	0.800	0.778	0.833	0.778				
Giza 3	0.809	0.651	0.682	0.682	0.727	0.783			
Giza 40	0.744	0.769	0.811	0.750	0.700	0.762	0.800		
Giza 429	0.686	0.902	0.875	0.875	0.750	0.824	0.619	0.737	
Giza 843	0.789	0.882	0.857	0.800	0.800	0.865	0.711	0.780	0.848

comprised Misr 3 and Nubaria 3, however the second included Sakha 4, Wadi 1, Nubaria 1, Giza 429, Giza 2 and Giza 843.

Rescaled Distance Cluster Combine

Agrobacterium-mediated transformation has been successfully achieved in faba bean (Hanafy et al. 2005). Also, Yan et al. (2010) successfully expressed the human cytomegalovirus (HCMV) pp150 gene in transgenic faba bean for providing new resources for the improvement of an edible vaccine against HCMV. Gene PR10a isolated from potato as pathogenesis-related was found to be increased tolerance to drought and salinity stresses in transgenic faba bean. Next-generation sequencing techniques and genotyping techniques have evolved as new tools to address crop genetics (Varshney et al. 2012).

Hanafy et al. (2005) used method for the production of transgenic faba bean by Agrobacterium-mediated transformation based upon direct shoot organogenesis after transformation of meristematic cells resultant from embryo axes. Explants were co-cultivated with *A. tumefaciens* strain EHA105/pGlsfa, which confined a binary vector of a gene encoding a sulphur rich sunflower albumin (SFA8) linked to the bar gene. The sequences of SFA8 and LysC genes were fused to seed specific promoters *Vicia faba* legumin B4 promoter (LeB4). They achieved seven phosphinothricin (PPT) resistant clones were recovered from genotypes Mythos and Albatross. Southern blot, PCR, enzyme activity analyzes and Western blot procedures were utilized for integration, genetics and expression of the transgenes. In continuous, in their reports Hanafy et al. (2013) revealed that amongst seed legumes, faba bean is principally sensitive to abiotic stress viz water stress or soil salinity and suffers from severe yield reduction. Pathogenesis-related proteins are expressed in crop plants in responsive biotic and abiotic stresses. They showed that expression of potato PR10a gene procures enhanced tolerance to drought and salinity in faba bean. The PR10a gene was selected and isolated from the potato cultivar Desiree and transferred into faba bean cultivar Tattoo through *Agrobacterium tumefaciens*-mediated transformation system based upon direct shoot regeneration after transformation of meristematic cells resulting from embryo axes. They obtained recovered fertile transgenic faba bean plants. Inheritance and expression of the foreign genes were proved by PCR, RT-PCR, Southern blot and monitoring of Luciferase activity. Under water stress situation, after preventing water for 3 weeks, the leaves of transgenic plants were remain green, however non-transgenic plants (WT) wilted. Twenty-four hours after re-watering, the leaves of transgenic plants still green, but WT plants did not recover. The transgenic lines exhibited higher tolerance to NaCl stress. Hereby, introgression a novel PR10a gene into faba bean might be considered a recent procedure to increase drought and salt tolerance capability.

The development of tissue culture, molecular biology and the biological techniques of faba bean are important in obtaining desirable regenerated plants. Bahgat et al. (2008) used a technique for inducing somatic embryos in faba bean callus for two Egyptian cultivars of Giza 2 and 24 Hyto. Embryos developed into plantlets and plants were regenerated. RAPD technique was implemented to examine the genetic stability of the obtained regenerated plants. They found that cultivar Giza 2 displayed more genetic stability compared to 24 Hyto. Added that regeneration system was recognized as appropriate for gene transformation and somaclonal mutants isolation will be used to improve the nutritional value of faba bean.

RNA-Seq has been utilized for recognizing stress-inducible transcripts and find out different systems that control plant system biology (Mizuno et al. 2010). AltafKhan et al. (2019) showed that de novo assembly of a total of 606.35 M high-quality pair-end clean reads produced 164,679 unigenes of faba bean tissue leaf. A total of 35,143 of them 12,805 upregulated and 22,338 downregulated unigenes and 28,892 of them 16,247 upregulated and 12,645 down-regulated genes were varied in their expressed under water stress situations in the vegetative and flowering stages, respectively. A total of 538 of them 272 were upregulated and 266 downregulated, and 642 of them 300 upregulated and 342 downregulated putative transcription factors in the vegetative and flowering stages, respectively, were recognized under diverse transcription factor families. Also, a considerable proportion of recognized genes were unique, so signifying a specific reaction to water stress in faba bean. They stated that RNA-seq results were validated by quantitative reverse-transcription PCR technique.

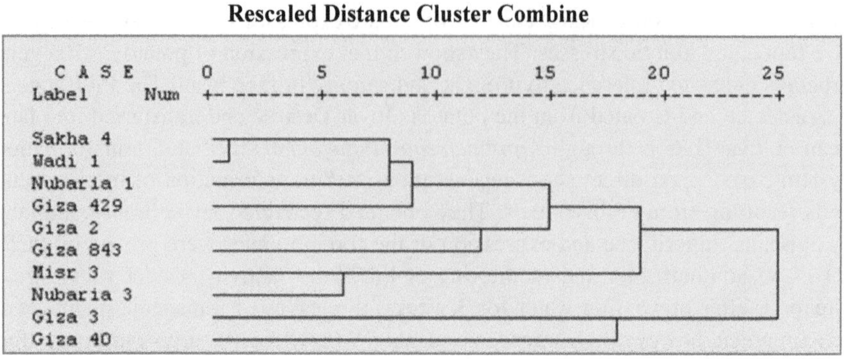

Rescaled Distance Cluster Combine

Fig. 6.5 Dendrogram of the genetic distances among the ten faba bean cultivars based on RAPD analysis (Naglaa et al. 2018)

6.7.2 Nano Technology

Due to of growing concerns about the negative impact of chemicals on the environment, agricultural practices involving organic and environmentally friendly compounds such as optimization are gaining global acceptance. Enhancing soil productivity, nutrition use efficiency and protection plants from environmental stress, can be achieved by Nanofertilizers. So, environmental friendly trend of producing high quality organic fertilizer (Abo-Sedera et al. 2016). Nanotechnology has received great attention to the progress and economic development of science and knowledge (Sherif 2009).

Almosawy et al. (2018) carried out experiment at Iraq, Karbala, Hussania during 2016–2017 season. The experiment comprised of optimum nanoparticles (T0 = control, T1 = 1 ml L^{-1}, T2 = 2 ml L^{-1}, T3 = 3 ml L^{-1}, T4 = 1 ml L^{-1}, magnetic, T5 = 2 ml L^{-1} magnetic and T6 = 3 ml L^{-1} magnetic) and four cultivars Mulch, Acudlage, Local and Luz de Otono and interaction between them. The highest plant height (66 cm), No of branches/plant (15.33) were noticed in T5 with Local cultivar, No of seeds/pod (5.07 and 5.05) was recorded in T5 + Mulch cultivar and T5 + Luz de Otono, respectively, seed weight (1.56 g) and seed yield (9346.86 kg.ha^{-1}) were recorded inT5 with Luz de Otono cultivar. Megahed et al. (2018) study the effect of foliar application with different concentrations of nanosilica and different salinity of irrigation water on some soil properties and productivity of faba bean at Sakha Agriculture Research Station Farm, Egypt. Results showed that there are no differences between soil salinity and foliar application with various concentrations of Nano-silica. Anatomical characters of faba bean roots showed increment (15.54%) with 300 mg L^{-1} Nano-Si more than the control. Similarly, the anatomical characters of roots were reduced by increasing of the salinity irrigation water and registered lowermost values up to T2 (2.45 dS m^{-1}) and without Nano-Si. T2 (2.45 dS m^{-1}) gave a significant decrease in seed yield (80.13%), and in straw yield (78.06%) compared to the control treatment, T1. Otherwise, T4 produced the maximum values 1.74 and

1.84 Mg fad^{-1} in seed and straw yield of faba bean, under foliar application with 300 mg L^{-1} of Nano-silica compared with other concentrations. Similar trend was detected for chlorophyll content, nitrogen uptake and nitrogen use efficiency of faba bean. Hereby, foliar application with 300 mg L^{-1} of Nano-silica is the appropriate concentration to alleviate the salt stress on faba bean plants.

6.8 How Can Measure Sensitivity of Faba Bean Genotypes to Environmental Stress?

6.8.1 Stress Sensitivity Measurements

Significant differences between six faba bean inbred lines in response to drought were recorded by Amede et al. (1999). They found that five high yielding inbred lines from Europe exhibited a yield sensitivity index SI more than 0 and therefore were relatively drought sensitive. However the low yielding group was collected from drought prone regions and attained SI less than 0 and expressed as drought resistant. The drought sensitivity index of the inbred line L6 was significantly lesser than Adriewaalse at water stress circumstances in the field. Ahmed et al. (2008) evaluated tolerance of seven faba bean varieties to drought and salt stresses. Faba bean plants were grown under three different irrigation water intervals, 5, 10 and 15 days and two concentrations of NaCl, 25 and 50 mM. Results showed that that seed yield of variety 5 having the top yield was 50 and 71% suppressed under water stress treatments of 5 and 10 days intervals, respectively. Variety 4 was the highest salt tolerance among the seven studied varieties. In the other varieties, the yields were suppressed under salt stress. The clustering analysis showed that the seven faba bean varieties as affected by water stress classified into two groups. Group (a) contains four varieties; V1, V5, V2 and V7, the yield and yield components of these varieties highly reduced as affected by water stress. Group (b) contains three varieties; V3, V6 and V4, the yield of these three varieties slightly reduced when compared to group (a). According to the reduction % in the seven faba bean varieties yield, clustering analysis classified the seven faba bean varieties in their response to salt stress to two groups (c and d). Group (c) represented by six varieties; V1, V6, V2, V7, V5 and V3, the yield of these varieties was reduced as a result of salt stress. While, group (d) represented only by V4. Abdellatif et al. (2012) found that sensitivity test for drought tolerance indicated that faba bean cultivar Giza 3 displayed the highest statistical significant sensitivity to water stress and can be expressed as sensitive one. However, Giza 843 was more tolerant to drought stress and provided moderate mean performance across the genotypes.

Subjected 11 faba bean cultivars to three levels of water deficit (90, 50 and 30% of field capacity FC) was done by Abid et al. (2017). They recorded significant cultivar differences in their responses to water shortage stress. 'Hara' give a moderate yield at 50% FC, presented the best yield at 30% FC and was found to be less affected concerning photosynthetic parameters and produced the maximum value of grain yield, with higher antioxidant enzyme activities, osmoprotectant accumulations, chlorophyll b and relative water content. Hence 'Hara' cultivar was found to be more tolerant to water deficit stress compared to the other cultivars. 'Giza 3' produces the maximum seed yield under control conditions, but its yield decreased intensely at 30% FC.

6.9 Agricultural Practices to Mitigate Environmental Stress on Faba Bean

The effects of **environmental stress** can be addressed through the following procedures:

– Cultivate the most tolerate genotypes to environmental stress at the appropriate time. In this respect, it is recommended to cultivate faba bean varieties Sakha 1, Sakha 2, Sakha 3, Improved Giza 3, Giza 716, Giza 843 in the northern and central Delta governorates. While cultivars Misr 1, Giza 40, Giza 429 and Wadi 1 in the provinces of South Delta and Central Egypt. As for the new land in the Nubaria region, Nubaria 1, Giza 716. While Wadi1 is cultivated in the New Valley. The planting date is considered an important to avoid environmental stress conditions. Faba bean is grown under Upper Egypt in the first half of October in Sohag, Qena, Aswan and New Valley. And in the second half of October and the first week of November in Assiut, Menia, Beni Suef, Fayoum and Giza governorates. While in the sea face governorates is grown in the first half of November (Anonymous 2020).
– Some agricultural practices reduce the impact of environmental stress on the productivity of field crops. In this regard, under Khattara region as sandy soil stress environment, Mokhtar (2003) found that seed yield/fad in the two faba bean cultivars Giza Blanca and Sakha1 was affected by N, P and K levels and their interactions. Giza Blanca had significantly higher seed yield/fad than Sakha1. The increase of N level from 10 to 20 kg N/fad secured a significant increase in each of seed yield/fad. For P increment, seed yield of Sakha 1 was increased from 5.60 to 9.70 ardab/fad and that of Giza Blanca was increased from 5.68 to 13.43 ardab/fad due to the increase of P level from zero to 31 kg P_2O_5/fad. Addition of 24 kg K_2O/fad produced a significant increase in each of seed yield/fad of the two faba bean cultivars and the combined analyses. Moreover, Kubure et al (2016) showed that application of phosphorus tended to produce marginally longer tap root than no Phosphorus. Dawood et al. (2019) found that addition of humic acid has had a positive effect on photosynthesis pigments, seed yield and its components as well as some biochemical components of faba bean seeds.
– Taking into account the interest of the critical periods of plant life and avoid heat stress through irrigation. Also, link crop irrigation dates with meteorological data to avoid the harmful effects of high heat waves. So, irrigation must be moderated

and irrigation should be stopped when the rainfall is sufficient. In central and upper Egypt, regular irrigation is received during the flowering and fruiting periods to resist the adverse effects of frost.

6.10 Conclusions

Reference studies show the importance of studying the interaction between genotype × environmental on faba bean crop production. This is useful in assessing the adaptation and stability of the genotypes and the outcome of breeding programs. In this chapter, the role of genetic differences in responding to different geographical and environmental conditions, the importance of planting date, irrigation, fertilization and biotechnology in mitigating the harmful impact of environmental stress on the faba bean yield was clearly demonstrated.

6.11 Recommendations

Because of the sensitivity of faba bean to climate change, it seems necessary to identify the size and nature of the interaction between the genotype × environment, and determine the extent of adaptability and stability of faba bean genotypes under different conditions by applying appropriate statistical models. Also study the genetic behavior of various characteristics that could be used as a selection criteria for improving yield under different environments. Focus on the role of DNA-markers and biotechnology in improving crop stability beside some agricultural practices to reduce the impact of environmental stress on faba bean production and achieve the sustainably.

References

Abdellatif KF, El-Absawy EA, Zakaria AM (2012) Drought stress Tolerance of Faba Bean as studied by morphological traits and seed storage protein pattern. J Plant Stud 1(2):1–8

Abid KH, Marwa A, Ibtissem A, Jean-Pierre B, Yordan M, Guy M, Khaled S, Myriam M, Fatma S, Moez J (2017) Agro-physiological and biochemical responses of faba bean (Vicia faba L var 'minor') genotypes to water deficit stress. Ghassen Biotechnol Agron Soc Environ 21(2):146–159

Abo-Sedera FA, Shams1 AS, Mohamed1 AHM, Hamoda (2016) Effect of organic fertilizer and foliar spray with some safety compounds on growth and productivity of snap bean. Ann Agric Sci Moshtohor 54(1):105–118

Achtak HA, Oukablio MA, Santoni S (2009) Microsatellite markers as reliable tools for fig cultivar identification. J Ame Soc Hort Sci 134:624–631

Ahmed AK, Tawfik KM, Zinab AA, El-Gawad (2008) Tolerance of seven Faba Bean varieties to drought and salt stresses. Res J Agric Biol Sci 4(2):175–186

Alghamdi SS (2007) Genetic behavior of some selected faba bean genotypes. Afr Crop Sci Conf Proc 8:709–714

Almosawy AN, Alamery AA, Al-Kinany FS, Mohammed HM, Alkinani LQ, Jawad NN (2018) Effect of optimus nanoparticles on growth and yield of some broad bean cultivars (*Vicia faba* L). Int J Agric Stat Sci 14 (2):1–4

AltafKhan M, Alghamdi SS, Ammar MH, Sun Q, Teng F, Migdadi HM, Al-Faifi SA (2019) Transcriptome profiling of faba bean (*Vicia faba* L) drought-tolerant variety hassawi-2 under drought stress using RNA sequencing. Electron J Biotechnol 39:15–29

Amede T, Kittlitz EV, Schubert S (1999) Differential drought responses of Faba Bean Vicia faba L Inbred Lines. J Agron Crop Sci 183:35–45

Ammar MH, Anwar F, El-Harty EH, Migdadi HM, Abdel-Khalik SM, Al-Faifi SA, Farooq M, Alghamdi SS (2014) Physiological and yield responses of faba bean (*Vicia faba* L) drought in a Mediterranean-type Environment. J Agro Crop Sci 201:280–287

Anonymous (2020) Recommendations techniques in Faba bean cultivation. Agricultural Research Center, Giza, Egypt

Arbaoui M, Balko C, Link W (2008) Study of faba bean (*Vicia faba* L) winter hardiness and development of screening methods. Field Crops Res 106:60–67

Awaad HA (2002). Phenotypic and genotypic stability of some faba bean (*Vicia faba* L) varieties. Egyptian J Plant Breed. 6(1):1–15

Baheer-Salimia R, Awad RM, Ward J (2012) Assessment of biodiversity based on molecular markers and morphological traits among westbank, Palestine fig genotypes (*Ficus caria* L). Am J Plant Sci 3:1241–1251

Bahgat S, Shabban OA, El-Shihy O, Lightfoot DA, El-Shemy HA (2008) Establishment of the regeneration system for *Vicia faba* L. Curr Issues Mol Biol 11(Suppl 1):47–54

Becker HC, Geiger HH, Morgen Stern K (1982) Performance and phenotypic stability of different hybrid types in winter rye. Crop Sci 22:340–344

Bishop J, Potts SG, Jones HE (2016) Susceptibility of Faba Bean (*Vicia faba* L) to heat stress during floral development and anthesis. J Agron Crop Sci 202(6):508–517

Bulut F, Akinci S, Eroglu A (2011) Growth and uptake of sodium and potassium in broad bean (*Vicia faba* L) under salinity stress. Commun Soil Sci Plant Anal 42:945–961

Conde C, Estrada F, Martínez B, Sánchez O, Gay C (2011) Regional climate change scenarios for México. Atmosfera 24(1):125–140

Crépon K, Marget P, Peyronnet C et al (2010) Nutritional value of faba bean (*Vicia faba* L) seeds for feed and food. Field Crops Res 115:329–339

Cui X-H, Hao F-S, Chen H, Chen J, Wang X-C (2008) Expression of the *Vicia faba* VfPIP1gene in Arabidopsis thaliana plants improves their drought resistance. J Plant Res 121:207–214

Darwish DS, El-Metwally EA, Shafik MM, H.H. EL-Hinnawy, (1999) Stability of faba bean varieties under old newly reclaimed lands. Poc. Pl. Poc PI Breed Cronf Giza Egypt J Plant Breed 3:365–377

Dawood MG, Abdel-Baky YR, M E E-Awadi, G Sh Bakhoum, (2019) Enhancement quality and quantity of faba bean plants grown under sandy soil conditions by nicotinamide and/or humic acid application. Bull Nat Res 43(28):2–8

De Costa WAJM, Dennett MD, Ratnaweera U, Nyalemegbe K (1997) Effects of different water regimes on field-grown determinate and indeterminate faba bean(*Vicia faba* L). I. Canopy growth and biomass production. Field Crops Res 49:83–93

Delgado MJ, Ligero F, Lluch C (1994) Effects of salt stress on growth and nitrogen fixation by pea, faba bean, common bean, and soybean plants. Soil Biol Biochem 26:371–376

Duc G, Bao S, Baum M et al (2010) Diversity maintenance and use of *Vicia faba* L genetic resources. Field Crops Res 115:270–278

Duc G, Link W, Redden RJ, Stoddard FL, Torres AM, Cubero JI (2011) Genetic adjustment to changing climates: faba bean. In: Crop adaptation to climate change, 1st edn. Wiley

Eberhart SA, Russel WW (1966) Stability parameters for comparing varieties. Crop Sci 6:36–40

El-Abssi MG (2020) Breeding faba bean (*Vicia faba* L) for high yield and disease resistance. MSc thesis, Agron Dept Fac of Agric Zagazig Univ, Egypt

El-Abssi MG, Rabi HA, Awaad HA, Qabil N (2019) Performance and gene action for earliness, yield and chocolate spot disease of faba bean. Zagazig J Agric Res 46(6A):1–11

El-Hady M, Omar MA, Nas SM, Ali RA, Esso MS (1998). Gene action of seed yield and some yield components in F_1 and F_2 crosses among five faba bean (Vicia faba L) genotypes. Bull Fac Agric Cairo Univ 49(3):369–388

El-Karmity AE (1990) Yield and its components of faba bean as affected by herbicide, inoculation and nitrogen fertilization. Minia J Agric Res Dev 12(3):1303–1314

FAO (2017) FAOSTAT database. Food and agriculture organization of the United Nations. www.fao.org/faostat/

FAO (2018) Production year book, 54. FAO, Rome

FAOSTAT (2018) http://www.fao.org/faostat/en/#data/QC

FAOSTATE (2019) Food and agriculture organization of the United Nations, Production/Yield quantities of crops worldwide. http://www.fao.org/faostat/en/#data/QC

Flores F, Nadal S, Solis I, Winkler J, Sass O, Stoddard FL, Link W, Raffiot B, Muel F, Rubiales D (2012). Faba bean adaptation to autumn sowing under European climates. Agron Sust Dev 32:727–734. Springer, France

Ghandorah MO, El-Shawaf IIS (1993) Genetic variability, heritability estimates, and predicted genetic advance for some characters in Faba Bean (Vicia faba L). J King Saud Univ Agri Sci 5:207–218

Gulian Yue SK, Perng TL, Walter CE, Wassom GHL (1990) Stability analysis of yield in maize, wheat and sorghum and its implications in breeding programs. Plant Breed 104:72–82

Hafiz SI, Abd EL-Mottaleb HM (1998) Response of some fabe bean cultivars to foliar fertilization under newly reclaimed sandy soil. Proceedings of the 8th Conference on Agron, Suez Canal University, Ismalia, Egypt, pp 307–316

Hanafy M, El-Banna A, Schumacher HM, Hassan F (2013) Enhanced tolerance to drought and salt stresses in transgenic faba bean (Vicia faba L) plants by heterologous expression of the PR10a gene from potato. Plant Cell Rep 32:663–674

Hanafy M, Thomas P, Heiko K, Hans-Joerg J (2005) Agrobacterium-mediated transformation of faba bean (Vicia faba L) using embryo axes. Euphytica 142(3):227–236

Hayward MD, Lawrance T (1970) The genetic control of variation in selected population of Lolium pernne L. Can J Cytol 12:806–815

Heinrich GM, Francis CA, Eastin JD (1983) Stability of grain sorghum yield components across diverse environments. Crop Sci 23:209–212

Jarso M, Keneni G (2006). Vicia faba L. In: Brink M, Belay G (eds) Cereals and pulses (Plant Resources of Tropical Africa 1). Backhuys Publishers, Netherlands, pp 195–199. ISBN-13: 9789057821707

Jensen ES, Peoples MB, Hauggaard-Nielsen H (2010) Faba bean in cropping systems. Field Crops Res 115:203–216

Katerji N, Mastrorilli M, Lahmer FZ, Maalouf F, Oweis T (2011) Faba bean productivity in saline-drought conditions. Eur J Agron 35:2–12

Keneni G, Jarso M, Wolabu T (2006). Faba Bean (Vicia faba L) genetics and breeding research in Ethiopia: a review. In: Ali K, Keneni G, Ahmed S, Malhotra R, Beniwal S, Makkouk K, Halila MH (eds) Food and Forage Legumes of Ethiopia: progress and prospects, ICARDA, Aleppo, Syria

Khan HR, Paull JG, Siddique KHM, Stoddard FL (2010) Faba bean breeding for drought-affected environments: a physiological and agronomic perspective. Field Crop Res 115:279–286

Khan HR, Link W, Hocking TJ, Stoddard FL (2007) Evaluation of physiological traits for improving drought tolerance in faba bean (Vicia faba L). Plant Soil 292:205–217

Khazaei H, O'Sullivan DM, Sillanpa¨ä MJ, Stoddard FL, (2014) Use of synteny to identify candidate genes underlying QTL controlling stomatal traits in faba bean (Vicia faba L). Theor Appl Genet 127:2371–2385

Kocsis ML, Jaromi P, Kozma P (2005) Genetic diversity among twelve grape cultivars indigenous to the Carpathian basin revealed by RAPD markers. Vitis—Geilweilerhof 1(44):87–91

Kubure TE, Raghavaiah CV, Hamza I (2016) Production potential of Faba Bean (*Vicia faba* L) genotypes in relation to plant densities and phosphorus nutrition on vertisols of central Highlands of West Showa Zone, Ethiopia, East Africa. Adv Crop Sci Tech 4:214

Kumar P, Das RR, Bishnoi SK, Sharma V (2017) Inter-correlation and path analysis in faba bean (*Vicia faba* L). Electronic J Plant Breed 8 (1):395–397

Langer S, Frey KJ, Baily T (1979) Associations among productive, production and stability indexes in oat varieties. Euphytica 28:17–24

Lavania D, Siddiqui MH, Al-Whaibi MH, Singh AK, Kumar R, Grover A (2015) Genetic approaches for breeding heat stress tolerance in faba bean (*Vicia faba* L). Acta Physiol Plant 37:1–9

Link W, Balko C, Stoddard FL (2010) Winter hardiness in faba bean: physiology and breeding. Field Crops Res 115:287–296

Link W, Bruno S, Ernst K (1996) Breeding for wide adaptation in faba bean. Euphytica 92(1–2):185–190

Maalouf F, Ahmed S, Shaaban K et al (2016) New faba bean germplasm with multiple resistances to Ascochyta blight, chocolate spot and rust diseases. Euphytica 211:157–167

Maalouf F, Miloudi N, Ghanem ME, Murari S (2015) Evaluation of faba bean breeding lines for spectral indices, yield traits and yield stability under diverse environments. Crop Pasture Sci 66(10):1012–1023

Megahed M, Amer F, El- Emary A (2018) Impact of foliar with Nano-silica in mitigation of salt stress on some soil properties, crop-water productivity and anatomical structure of maize and faba bean. Environ Biodiv Soil Secur 2:25–38

Migdadi HM, El-Harty EH, Salamh A, Khan MA (2016) Yield and proline content in faba bean genotypes under water vstress treatments. J Animal Plant Sci 26(6):1772–1779

Mizuno H, Kawahara Y, Sakai H et al (2010) Massive parallel sequencing of mRNA in identification of unannotated salinity stress-inducible transcripts in rice (*Oryza sativa* L). BMC Genomics (1):683

Mokhtar TS (2003) Effect of some agronomic practices on faba bean under newly cultivated lands. MSc thesis, Agron Dept, Fac of Agric Zagazig University, Egypt

Naglaa Q, Helal AA, Rasha YS, Abd E-K (2018) Evaluation of some new and old faba bean cultivars (*Vicia faba* L) for earliness, yield, yield attributes and quality characters. Zagazig J Agric Res 45(3):821–833

Negruk V (2013) Mitochondrial genome sequence of the legume *Vicia faba*. Front Plant Sci 4:1–11

Obiadalla-Ali HA, Mohamed NEM, Khaled AGA (2015) Inbreeding, outbreeding and RAPD markers studies of faba bean (*Vicia faba* L) crop. J Adv Res 6(6):859–868

Omar SA, Belal AH, Eman IE-S, El-Metwally MA (2014) Genetic studies on breeding faba bean for drought tolerance: 2-Heterosis and combining ability. J Plant Produc Mansoura Univ 5(11):1897–1914

Pham HN, Kang MS (1988) Interrelationships among repeatability of several stability statistics estimated from international maize trials. Crop Sci 28:925–928

Rispail N, KaloʹP, Kiss GB, et al (2010) Model legumes contribute to faba bean breeding. Field Crops Res 115:253–269

Rubiales D (2010) Faba beans in sustainable agriculture. Field Crops Res 115:201–202

Sherif AM (2009) Nanotechnology, half a century between dream and reality. Arab magazine—Ministry of Information Kuwait, 607:152–159. In: 4th international conference on biotechnology applications in agriculture (ICBAA), Benha University, Moshtohor and Hurghada, 4–7 Apr 2018, Egypt

Sharifi P (2014) Correlation and path coefficient analysis of yield and yield component in some of broad bean (*Vicia faba* L) genotypes. Genetika 46 (3):905–914

Siddiqui MH, Al-Khaishany MY, Al-Qutami MA, Al-Whaibi MH, Grover A, Ali HM, Al-Wahibi MS, Bukhari NA (2015) Response of Different Genotypes of Faba Bean Plant to Drought Stress. Int J Mol Sci 16(5):10214–10227

Soad Mahmoud A, Amal M, Abd E-M, Enas SI (2018) Combining ability, heritability and heterosis estimates in faba bean (*Vicia faba* L) under two water regimes. Egypt. J Agron 40(3):261–284

Stoddard FL, Balko C, Erskine W et al (2006) Screening techniques and sources of resistance to abiotic stresses in cool-season food legumes. Euphytica 147:167–186

Tadele Tadesse , Behailu Mulugeta, Gashaw Sefera, Amanuel Tekalign (2018). Genotypes by environment interaction of Faba Bean (*Vicia faba* L) Grain yield in the highland of Bale Zone, Southeastern Ethiopia. Plant 5(1):13–17

Tahir NA (2015) Identification of genetic variation in some faba bean (*Vicia faba* L) genotypes grown in Iraq estimated with RAPD and SDS-PAGE of seed proteins. Indian J Biotechnol 14:351–356

Tai GCC (1971) Genotypic stability analysis and its application on potato regional traits. Crop Sci 11:184–190

Tavakkoli E, Paull J, Rengasamy P, McDonald GK (2012) Comparing genotypic in faba bean (*Vicia faba* L) in response to salinity in hydroponic and field experiments. Field Crops Res 127:99–108

Tekalign A (2018) Genotype × environment interaction for grain yield of faba bean (*Vicia faba* L) varieties in the highlands of Oromia region, Ethiopia. MSc thesis, Haramaya University, Haramaya

Temesgen T, Keneni G, Sefera T, Jarso M (2015) Yield stability and relationships among stability parameters in faba bean (*Vicia faba* L) genotypes. The Crop J. 3:258–268

Tolessa TT (2015) Application of AMMI and Tai's stability statistics for yield stability analysis in Faba bean (*Vicia faba* L) cultivars grown in central Highlands of Ethiopia. J Plant Sci 3(4):197–206

Torres AM, Avila CM, Gutierrez N et al (2010) Marker-assisted selection in faba bean (Vicia faba L). Field Crop Res 115:243–252

Torres AM, Román B, Avila CM et al (2006) Faba bean breeding for resistance against biotic stresses: towards application of marker technology. Euphytica 147:67–80

Varshney RK, Kudapa H, Roorkiwal M et al (2012) Advances in genetics and molecular breeding of three legume crops of semi-arid tropics using next-generation sequencing and high-through-put genotyping technologies. J Biosci 37:811–820

Wang HF, Zong X, Guan JP, Yang T, Sun XL, Ma Y (2012) Genetic diversity and relationship of global faba bean (*Vicia faba* L) germplasm revealed by ISSR markers. Theor Appl Genet 124:789–797

Yahia Y, Hannachi H, Ferchichi A (2014) Genetic diversity of (*Vicia faba* L) based on random amplified polymorphic DNA and simple sequence repeat marker. Acta Botanica Gallica: Botany Lett 161(2):151–158.

Yan H, Yan H, Li G et al (2010) Expression of human cytomega-lovirus pp150 gene in transgenic Vicia faba L. and immunogenicity of pp150 protein in mice. Biological 38:265–272

Yang T, Bao S-Y, Ford R et al (2012) High-throughput novel microsatellite marker of faba bean via next generation sequencing. BMC Genom 13:602. https://doi.org/10.1186/1471-2164-13-602

Zikrallah A (2019) Current Crisis in Egypt's Faba Beans Crop. Egyptian Insttute For Studies. https://en.eipss-eg.org/current-crisis-in-egypts-faba-beans-crop/

Chapter 7
Approaches in Sesame to Mitigate Impact of Climate Change

Abstract Environmental stress adversely affected growth and yield parameters of sesame. Components of genotype by environment interaction are an important step in any breeding strategies to sustain sesame productivity. Extensive variations in seed yield due to genotype, location, and genotype × location interaction were observed, demonstrating that genotypes performed differently across locations and the essential for adaptability and stability analyses. Many investigators registered different degrees of tolerance to drought, heat and salt in sesame genotypes. Hence, genotypes can be utilized in breeding programs to improve seed yield stability in sesame. At the genome-wide level 287 MYB genes are highly active in growth and adaptation to the major abiotic stresses in sesame. Extensive variations in seed yield due to genotype, location, and genotype × location interaction were recorded, demonstrating that genotypes performed differently across locations and the essential for adaptability and stability analysis. Hereby, pyramiding the favorable alleles is important for enhancing stress tolerance in sesame. The negative effects of severe drought stress on quantitative and qualitative yield can be compensated by foliar application of iron Nano-chelates and organic compounds such as fulvic acid. So, this chapter highlights on genotype × environment interaction and its relation to climatic change, adaptability and yield stability as well as role of recent approaches in this respect.

Keywords Sesame · Environmental changes · Performance · Adaptation · Nano-technology · Heritability · Agricultural practices

7.1 Introduction

Sesame (*Sesamum indicum* L.) is known as orphan crop. Currently, though world demand for its seeds is interestingly increasing due to its good quality oil (50%), protein (25%). Sesame oil is rich in polyunsaturated fatty acids and natural antioxidants like sesamin, sesamolin and tocopherol homo-logues (Anilakumar et al. 2010). Beside these nutritional advantages, sesame cropping has many agricultural benefits, it grows well in tropical to temperate environments. Over the past decade, sesame seed production has enlarged more than twice and its selling price has nearly

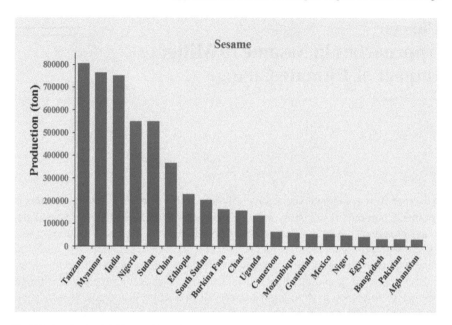

Fig. 7.1 Production/yield quantities of sesame worldwide (FAOSTAT 2017)

tripled (FAOSTAT 2017). Nevertheless, sesame seed yields are generally low (300–400 kg/ha) in most of the arid and semi-arid regions (Islam et al. 2016). The cultivated area is about 75,000 feddans (Feddan = 4200 m^2), mostly in the governorates of Upper Egypt, Upper Egypt and Ismailia in the coastal region. The crop ranges from 4.5 to 5.0 Ardab per feddan. Figure 7.1 show production/yield quantities of sesame worldwide.

Adaptability, stability and biotechnology are important indicators in the light of climate change in the world, especially in the Mediterranean region and in Egypt. Also, environmental stress adversely affected growth parameters of sesame including plant height, fresh weight, dry weight, number of leaves and leaf area. Also, caused significant decrease in endogenous phytohormones including salicylic acid and kinetin (Hussein et al. 2015). However, severe drought adversely impairs the plant normal growth and development (Anjum et al. 2011) and the reproduction and yield (Sun et al. 2010).

Sesame can be grown on soil moisture stored without the need for precipitation or irrigation, and be grown in pure stands with little input. Sesame crop has great potential as an industrial crop, but its production faces challenges due to drought and salt pressures (Li et al. 2018). Sesame is moderately tolerant to drought stress (Eskandari et al. 2009). However, severe drought is detrimental to natural plant growth and development (Anjum et al. 2011) and the reproduction and productivity (Sun et al. 2010). Hussein et al. (2015) showed that drought stress adversely affected plant growth of sesame including plant height, fresh weight, dry weight, number of leaves and leaf area, also caused significant decrease in endogenous phytohormones i.e. salicylic acid and kinetin. Hence, breeding sesame varieties with higher tolerance to abiotic pressures is an important target for sesame breeders (Dossa et al. 2017). Ayoubizadeh et al. (2019) found that severe drought stress significantly reduced leaf area, number of leaves, number of capsules, number of seeds, 1000-seed weight, seed yield and biological yield compared to non-stress conditions.

7.2 Genotype × Environment Interaction and Its Relation to Climatic Change on Sesame Production

Components of genotype by environment interaction are an important step in any breeding strategies to improve productivity. Significant variances were attained for genotypes (G) respecting seed yield and its contributing characters overall environments. Awaad and Aly (2002) indicated that pooled analyses of variance for sesame genotypes over environments (Table 7.1) provide evidence for highly significant environmental effects on seed yield and seed oil content as well as relevant characters. Partitioning the environmental effects into years (Y), locations (L) and plant population densities (D), and their interactions, revealed that they were significant in all the studied characters, except (L), (Y × D) and (L × D) for seed oil contents as well as (Y × L) and (L × D) for 1000-seed weight which were insignificant. Thus, the combination of environmental components (Y), (L) and (D) were sufficient to obtain reliable information about the studied genotypes regarding both seed oil content and 1000-seed weight. The factors under study accounted, generally for 62.25% (plant population densities), 13.11% (genotypes), 12.64% (locations) and 11.72% (years) from the total variance of sesame seed yield. First order interactions of genotypes with each of years, locations and plant population densities were significant in all characters. Also, highly significant second (G × Y × L), (G × Y × D) and (G × L × D) as well as third (G × Y × L × D) order interactions for sesame seed yield and its contributing characters. This implying different response of genotypes over year, location and plant population density combinations. Highly significant mean squares of sesame genotypes for seed yield and its contributing characters, revealing that sesame genotypes were genetically different for genes controlling these characters. Highly significant environment + (genotype × environment) component and environment "linear" mean squares were recorded for all characters, thus the studied characters were highly influenced by the combination of environmental components (seasons, locations and plant population density). Highly significant genotype × environment "linear" interactions was shown for seed yield and its contributing characters, suggesting that sesame genotypes differed in their response to the environmental variation (Table 7.2). The non-linear responses as measured by pooled deviations from regressions were highly significant, indicating that differences in linear response among genotypes across environments did not account for all the G × E interaction effects. Therefore, the fluctuation in performance of genotypes grown in various environments was not fully predictable. Hereby it is necessity of evaluating

Table 7.1 Pooled analyses of variance for seed yield and its contributing characters of 12 sesame genotypes grown under 3 plant population densities in two seasons and two locations (Awaad and Aly 2002)

Source of variance	d.f	Seed yield (ard./fad)	Seed oil content (%)	Fruiting zone length	No. of branches/plant	No. of capsules/plant	No. of seeds capsule	1000-seed weight
Years (Y)	1	0871**	7.857*	14,489.823**	2.654*	16,088.266**	1657.690**	0.921**
Locations (L)	1	1.271**	0.704	1721.925**	106.446.**	98,684.927**	2504.570**	0.553**
Year × Location (Y × L)	1	12.040**	15.618**	833.000**	43.092**	12,706.762**	1640.185**	0.004
Reps in (Y × L) combined	8	0.014	1.021	4.938	0.424	13.890	15.591**	0.013
Plant population density (D)	2	45.380**	10.413**	6599.887**	48.259**	21,600.494**	5314.034**	15.684**
Y × D	2	0.092*	1.714	2530.931**	11.0411**	1283.385**	1029.519**	0.079*
L × D	2	0.65*	0.148	184.004**	2.316**	305.643**	268.894**	0.038
Y × L × D	2	2.360**	3.667**	214.955**	2.534**	1136.226**	799.025**	0.149**
Genotypes (G)	11	7.320**	89.878**	879.767**	9.391**	988.903**	431.073**	1.764**
G × Y	11	0.873**	13.799	1780.998**	3.112**	1544.238**	130.169**	0.146**
G × L	11	1.894**	6.706**	308.412**	6.205**	1201.394**	205.837**	0.135**
G × D	22	0.725**	17.838**	381.589**	1.285*	257.259**	51.736**	0.399**
G × Y × L	11	0.574**	3.987**	501.201**	6.096**	1150.757**	145.034**	0.486**
G × Y × D	22	1.031**	2.245**	429.693**	1.369**	444.486**	36.692**	0.112**
G × L × D	22	0.433**	2.739**	47.424**	1.568**	404.389**	78.404**	0.107**
G × Y × L × D	22	0.663**	2.096**	69.528**	1.058**	539.903**	32.575**	0.132**
Error	280	0.016	0.587	3.052	0.228	5.894	1.405	0.023

*, ** denote significant at 5% and 1% levels of probability, respectively

Table 7.2 Mean squares of stability analysis of seed yield and its contributing characters in sesame (Awaad and Aly 2002)

Source	Mean squares							
	d.f	Seed yield/ fed	Seed oil content (%)	Fruiting zone length	No. of branches/ plant	No. of capsules/ plant	No. of seeds/ capsule	1000-seed weight
Genotypes	11	7.320**	89.878**	879.767**	9.391**	988.903**	431.304**	1.764**
Environment + (genotype × environment)	132	0.554**	2.206**	215.823**	1.739**	1235.304**	76.662**	0.725**
Environment (linear)	1	44.624**	18.630**	12,151.120**	94.452**	81,364.359**	6884.820**	10.393**
Genotype × environment (linear)	11	0.584**	3.693**	296.939**	2.012**	2489.942**	32.814**	1.271**
Pooled deviation	120	0.184*	1.933**	108.927**	0.612*	452.554**	23.946**	0.595**
Pooled error	264	0.014	0.768	3.223	0.336	6.820	1.704	0.019

*, ** denote significant at 5% and 1% levels of probability, respectively

the studied sesame genotypes under multi environments in order to identify the best genetic make-up to be grown under particular environments.

At El Saff, Giza governorate and Nubaria, Behaira governorate, Egypt, six experiments were conducted under two locations through 2005, 2006 and 2007 years using seven sesame genotypes by Abd El-Mohsen (2008). A standard multi-factor analysis of variance for seed yield (kg/plot) revealed that main effects due to years and locations were highly significant. The main effects for genotypes, first order interaction genotype × location and genotype × year and the second order interaction (genotype × location × year) were highly significant. Furthermore, under, Sagadividi Farm, Junagadh, Gujarat, Parmar et al. (2018) evaluated 56 sesame genotypes and two checks, G.Til-2 andG.Til-3 at four environments during diverse time and location for sowing of summer 2016. Stability analysis showed the existence of significant G × E interaction for all the characters when tested against pooled error accepts days to flowering, days to maturity, width of capsule and number of capsules/leaf axil. Mean squares of E + (G × E) were significant when tested against the pooled error for all the characters. Check Bhuva-2 × G.Til-10 ($\bar{X}i$ = 6.49) had greater tolerance to environmental changes and thus this hybrids would had specific adaptability to poor yielding environments. Furthermore, during three growing seasons (2013–2015) under a rainfed condition, Baraki and Berhe (2019) found that the genotype, year and genotype × year interaction components were statistically significant variation for days to 50% flowering, days to 50% maturity, plant height, seed yield (kg/ha) traits which clearly confirms the presence of genotype × year interaction. The highest combined mean grain yield (906.3 kg/ha) was attained from Hirhir followed by Serkamo white (756.5 kg/ha). The growing seasons were varied from one another and the agronomic traits, except thousand seed weight, were statistically different across the three growing seasons.

7.3 Performance of Sesame Genotypes in Response to Environmental Changes

Environmental stress is one of the most imperious constraints on the growth and productivity of sesame plants in arid and semi-arid areas. Performance of sesame genotypes in their response to environmental conditions varied from genotype to another. Seven sesame genotypes i.e. Mutation-8, F5 generation, Giza 32, Local Sharkia, B_{10}, B_{11} and B_{35} were evaluated for seed yield and oil yield characters under two plant population densities (56 and 112 thousand plants/fad.) by Awaad and Basha (2000) under sandy soil stress conditions, Khattara region, Sharkia Governorate, Egypt. Significant differences were detected among the seven sesame genotypes respecting yield characters in both seasons and their combined (Table 7.3). B_{10} genotype was the tallest with longer fruiting zone and higher number of seeds/capsule than the rest of genotypes. Local Sharkia was the shortest genotype but has higher number of seeds/capsule and branches/plant. However, Mutation 8 out yielded the

Table 7.3 Effect of genotype and plant population density on seed yield (ton/fad.), seed oil content and oil yield (kg/fad) of sesame (Awaad and Basha 2000)

Main effects and interaction	Seed yield (ton/ fad.)			Seed oil content (%)			Oil yield (kg/fad.)		
	1 st season	2 nd season	Combined	1 st season	2nd season	Combined	1 st season	2nd season	Combined
Genotype "G"									
Mutation 8	0.340ab	0.282a	0.311a	48.78a	46.84a	47.81a	165.852a	132.089a	148.971a
F5 generation	0.209c	0.229b	0.219bc	43.85d	43.25d	43.55e	91.647cd	99.043c	95.345d
Giza 32	0.202c	0.182c	0.192c	43.58d	45.62b	44.60d	88.031d	83.028d	85.529d
Local Sharkia	0.296b	0.211bc	0.254b	43.60d	43.20d	43.40e	129.056b	91.152 cd	110.104c
B-10	0.352a	0.207bc	0.279ab	44.70c	46.90a	45.80bc	157.978a	97.124c	127.551b
B-11	0.204c	0.275ab	0.239bc	44.29cd	42.11e	43.20e	90.351 cd	115.803b	103.077cd
B-35	0.212c	0.274ab	0.243b	46.75b	44.27c	45.51c	99.110c	121.299b	110.205c
F. test	**	**	**	**	**	*	**	**	**
Plant population density "D"									
D1	0.246b	0.196b	0.221b	44.700	44.850	44.850	111.333b	88.415b	99.874b
D2	0.273a	0.278a	0.276a	45.389	44.199b	44.494	123.531a	122.882a	123.206a
F. test	*	**	*	N.S	*	N.S	*	*	**
Interaction									
G × D	**	**	*	**	**	N.S	**	**	**

*, ** denote significant at 5% and 1% levels of probability, respectively

other genotypes due to its superiority in number of capsules/plant, 1000-seed weight and seed oil content and thus recorded the highest seed and oil yields. Doubling the plant population density increased significantly plant height, seed and oil yields but decreased the components. Significant genotype × plant population density interactions were recorded regarding seed and oil yields/fad. and some of their attributes. Sesame genotypes varied significantly in their response to doubling the plant population density. The highest increase of seed yield (54.651%) and oil yield (59.129%) was achieved by genotypes F_5 generation and local Sharkia, respectively.

Under stress condition, performance of three sesame varieties i.e., Giza 32, Local Sharkia and B_{10} were verified at three harvesting dates i.e., 90, 100 and 110 days from planting (Basha and Awaad 2000). The results revealed significant differences among sesame varieties based on combined data (Table 7.4). Performance of sesame genotypes was varied from season to another. B_{10} variety surpassed the other tested varieties in seed and oil yields/fad. However, Giza 32 variety gave the lowest values in both characters. Harvesting sesame plants after 110 days from planting gave higher values in all studied characters compared with the other two dates (90 and 100 days). Meanwhile, the highest oil percentage was obtained at duration of 100 day. Therefore, premature harvesting produced lighter sesame seeds with lower oil content. The high reduction of seed yield from early harvesting (90 days) might be due to a high level of immature seeds, while the slight reduction form late harvesting (120 days) might be due to capsules shattering. Therefore, sesame plants are usually harvested as soon as the lower capsules become brown and the seeds take normal color.

Interaction effect based on the combined analyses showed clearly a highly significant interaction between three tested varieties and harvesting date. Therefore, the studied varieties differed significantly in their response to harvesting dates in both seed and oil yields/fad (Tables 7.5 and 7.6). Under all tested varieties, late harvest (110 days) produced the highest seed and oil yields/fad. particularly with genotype B10.

The performance of eight different genotypes of sesame including (Gorgan, Shiraz, Markazi, Birjand, Arzoieh, Sirjan, Ardestan and Safiabad) under different irrigation regimes were tested by Afshari et al. (2014). Drought tolerance was assessed at the base of field capacity (FC) (100, 80 and 50%) in greenhouse conditions. The maximum and minimum values of dry weight were detected in Arzoieh and Gorgan genotypes, respectively. The results of water stress have significant effect on all traits, except for the number of capsule/plant. The maximum (1.15 ton/ha) and lowermost (0.24 ton/ha) estimates of seed yield were detected by genotypes Arzoieh and Gorgan, respectively. Horacek et al. (2015) indicate that black sesame seed genotypes might be to some extent less sensitive to drought stress and better adapted than white sesame genotypes, which might be useful information for the selection and breeding of the optimal cultivar of sesame for a region with limited water resources. They applied nuclear SSRs and cpSSRs, and no significant genetic variation was detected between black and white seeded genotypes. Li et al. (2018) registered an extensive variation for drought and salt responses in the sesame genotypes and most of the accessions were moderately tolerant to both stresses. Tariq and Shahbaz (2020) appraised two

Table 7.4 Performance of seed oil percentage, seed and oil yields (kg/fad.) of three sesame varieties under different harvesting dates (Basha and Awaad 2000)

Main effects and interaction	Seed oil percentage			Seed yield (kg/fad)			Oil yield (kg/fad)		
	1 st season	2nd season	Comb	1 st season	2nd season	Comb	1st season	2nd season	Comb
Varieties (V)									
Giza 32	48.9	48.9	48.9	175.880b	151.040	163.460b	85.949b	74.003	79.976b
Local Sharkia	49.0	50.1	50.1	171.960b	169.133	170.547b	84.391b	85.098	84.744b
B$_{10}$	49.2	49.6	49.6	317.997a	175.740	246.868a	155.801a	87.479	121.640a
F. test	N.S	N.S	N.S	*	N.S	*	**	N.S	**
Harvesting date (H)									
90	48.5	48.6b	48.6b	151.978b	116.582b	134.284c	73.762b	56.868b	65.315c
100	49.3	50.0a	50.5a	185.400b	179.127a	182.263b	91.646b	89.595a	90.620b
110	49.2	50.0a	50.0a	328.450a	200.204a	264.327a	160.732a	100.117a	130.425a
F. test	N.S	**	**	**	**	**	**	**	**
Interaction									
V × H	N.S	N.S	*	**	N.S	**	N.S	N.S	**

*, ** denote significant at 5% and 1% levels of probability, respectively

Table 7.5 Interaction effect between varieties and harvesting date on seed yield (kg/fad.) of sesame as combined date (Basha and Awaad 2000)

Varieties	Harvesting date (DAS)		
	90	100	110
	C	B	A
Giza 32	114.017 b	169.993 a	206.370 b
	B	A	A
Local Sharkia	135.193 ab	173.850 a	202.597 b
	C	B	A
B_{10}	153.643 a	202.947 b	384.015 a

Table 7.6 Interaction effect between varieties and harvesting date on oil yield (kg/fad.) of sesame as combined date (Basha and Awaad 2000)

Varieties	Harvesting date (DAS)		
	90	100	110
	C	B	A
Giza 32	55.112 b	84.615 b	100.202 b
	C	B	A
Local Sharkia	65.819 ab	85.856 b	102.558 b
	C	B	A
B_{10}	75.014 a	101.391 a	188.515 a

sesame genotypes (TS-5 and TH-6) against salt stress (70 mM NaCl) and revealed that TS-5 showed comparatively better salt tolerance than TS-6.

Abbasali et al. (2017) examined response to drought of eight genotypes of sesame in Iran, along with cultivars Oltan and Darab1, under normal irrigation and restriction of irrigation since beginning of flowering as stress situation. Results revealed that, genotypes KC50662 and Oltan were superior in yield and yield components under both conditions. Comparing three diverse sesame populations contains of 80 sesame lines was done based on agronomic traits by Mohamed and Bedawy (2019). Results revealed largest amount of variation for plant height, first capsule height, fruiting zone length, days to 50% flowering, capsules number, capsule length, 1000-seed weight and seed yield. Sesame lines classified into three main clusters, the lines of Pop2 were gathered in cluster 1, while the lines of Pop1 were grouped into cluster 2. The lines in Pop3 were distributed among the different clusters, whereas 6, 5 and 7 lines were located to Cluster 1, 2 and 3, respectively.

Furthermore, Ayoubizadeh et al. (2019) showed that the maximum leaf area index was attained under control treatment in Halil (4.46) and Dashtestan (2.76) genotypes compared to under water stress in the first and second years. The highest palmetic acid content was achieved from Dashtestan genotype (12.09 and 11.41%) in first and second years, respectively under non application of fertilizer.

7.4 Adaptability and Yield Stability

Adaptation is the process of change by which an organism or species becomes better suited to its environment. In this regard, Awaad and Aly (2002) evaluated 12 sesame genotypes for seed yield and seed oil content (%) grown under 12 environments. Significant differences were obtained for genotypes (G) respecting the studied characters overall environments. The regression coefficients (Table 7.7 and Figs. 7.2 and 7.3) varied from 0.429 in Venezulea 7 to 1.639 in Mutant 48 for seed yield/fed. and from -2.084 in Venezulea 7 to 3.478 in Line 5/91 for seed oil content. It is evident that "b" value deviated significantly from unity (b > 1) in genotypes Giza 25, Mutant 48 and B-11 for seed yield/fed as well as Giza 32, B-35 and Line 5/91 for seed oil content; indicating that these genotypes were highly adapted to favorable environments. Otherwise, the "b" value deviated significantly from unity and was less than one (b < 1) in genotypes, Local Sharkia and Venezuela 7 for seed yield/fed. The deviation from regression "S^2d" showed that seven genotypes, i.e., Giza 25, Mutant 8, B-10, B-11, Line 5/91, F5 generation and Venezuela 7 displayed stability for seed yield/fed as well as four i.e. Local Sharkia, Mutant 48, F5 generation and Venezuela 7 for seed oil content. However, the remaining genotypes were sensitive to the environmental changes. Stability parameter termed ecovalence (Wi) indicate that a genotype with Wi = 0 is regarded as stable. Thus, the most stable sesame genotypes were Giza 32, Local Sharkia, Mutant 8, Mutant 48, B-10 and Line 5/91 for seed yield/fed as well Giza 25, Local Sharkia, Mutant 48, B-10, B-35 and F5 generation for seed oil content. The coefficient of determination (r^2i) as a measure of the predictability of the estimated response (stability) was a best measure of phenotypic stability because its value lies between zero and one, and is strongly related to S^2di. Among the studied sesame genotypes, "r^2" values accounted for 40.3 to 92.5% of the variation in seed yield/ fed. Significant values have been recorded in ten genotypes (Giza 25, Giza 32, Mutant 8, Mutant 48, B-10, B-11, Line 5/91, F5 generation, Adnan hybrid and Venezuela 7) for seed yield/fed; eight (Giza 32, Local Sharkia, Mutant 48, B-10, B-35, Line 5/91, F5 generation and Adnan hybrid) for seed oil content. There were great similarities between (bi), (S^2di), (Wi) and (r^2i) in sesame genotypes Giza 32 and Line 5/91 for seed yield/fed. with specific superiority for high yield potentiality. Meanwhile, F5 generation was the same for stability statistics but exhibited low oil content. The most desired and stable sesame genotypes based on four stability parameters (bi, S^2di, Wi and r^2i) accompanied with mean performance (Xi) were; Mutant 8, Giza 32 and Line 5/91 for seed yield/fed; Mutant 48 for seed oil content. Positive and significant relationship was detected between (bi) and (S^2di) for seed oil content (bi) and (Wi) for seed oil content. Negative and significant correlation was recorded between (Xi) and (Wi) for seed yield/fed. Hereby, the high performing genotypes tend to be ranked as less stable and vice versa (Table 7.8).

Six experiments were investigated under two locations through three years 2005, 2006 and 2007 using seven sesame genotypes by Abd El-Mohsen (2008). Phenotypic stability assessed using Eberhart and Russell model revealed that, the most desired and stable genotypes were Mutant 10, Mutant 47, Mutant 50 and Mutant 56 for

Table 7.7 Stability parameters for seed yield and its contributing characters of 12 sesame genotypes grown under 12 environments (Awaad and Aly 2002)

Character	Seed yield (ard./fed)					Seed oil content (%)				
Genotype	X_i	b_i	S^2d_i	W_i	r^2	X_i	b_i	S^2d_i	W_i	r^2
1. Giza 25	2.987	1.606*	2.641*	0.674*	0.504*	48.376	0.509	1.185**	0.771	0.338
2. Giza 32	3.124	0.998	0.439	0.201	0.664*	51.533	2.207*	1.840**	8.173*	0.513*
3. Local Sharkia	3.076	0.575*	2.189**	0.138	0.403	49.005	−0.526	0.982	0.001	0.891**
4. Mutant 8	3.652	0.684	0.594	0.024	0.788**	52.290	0.839	2.278**	3.466*	0.230
5. Mutant 48	3.349	1.639**	1.517**	0.102	0.685*	50.103	0.711	0.056	0.223	0.800**
6. B-10	2.733	0.738	0.312	0.008	0.925*	49.568	1.796	1.629**	0.196	0.633*
7. B-11	2.984	1.508*	0.471	0.731*	0.736**	47.473	1.269	1.962**	1.852*	0.387
8. B-35	2.670	1.070	1.801**	0.458*	0.460	49.245	2.761*	2.844**	0.189	0.573*
9. Line 5/91	2.919	1.014	0.052	0.099	0.483*	49.123	3.478**	3.841**	4.002*	0.619*
10. F5 generation	**3.159**	1.018	0.531	0.572*	0.775**	48.943	1.510	0.720	0.078	0.983**
11. Adnan hybrid	2.169	0.658	1.341**	1.501**	0.793**	46.476	−0.490	1.899**	6.492*	0.230
12. Venezuela 7	2.033	0.429*	0.223	0.511*	0.667*	46.068	−2.084	0.483	3.334*	0.678*

*, ** denote significant at 5% and 1% levels of probability, respectively

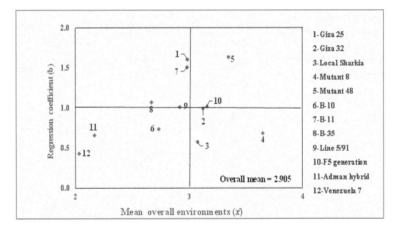

Fig. 7.2 Diagram of phenotypic stability for seed yield (Ardab/fad) (Awaad and Aly 2002)

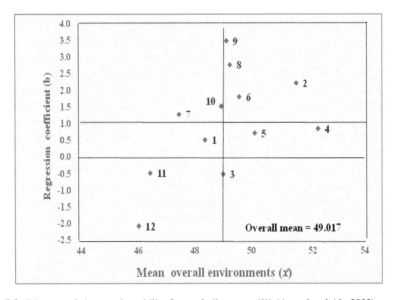

Fig. 7.3 Diagram of phenotypic stability for seed oil content (%) (Awaad and Aly 2002)

seed yield as they exhibited regression coefficients near to unit with lower deviations from regression and high mean yield. Also, their coefficients of determination r^2 values were high and valued 84, 78, 76 and 80%, confirming their stability. Moreover, Chemeda et al. (2014) tested ten sesame genotypes in four sites during 2011 and 2012 growing seasons. They used nine statistical methods to determine stability of the sesame seed yield. The results showed significant variations in seed yield due to genotype, location, and genotype × location interaction. Mean and cultivar superiority performance (Pi) showed high association with seed yield. The associations

Table 7.8 Correlation coefficients between pairs of different stability statistics for the studied sesame characters (Awaad and Aly 2002)

Statistics correlated	Seed yield /fed	Seed oil content (%)
$X_i - bi$	0.413	0.520
$X_i - S^2di$	0.075	0.188
$X_i - Wi$	−0.588*	0.055
$X - r2i$	−0.023	−0.004
$b_i - S2di$	0.318	0.682*
b_i-Wi	0.021	0.602*
$b_i - r2i$	−0.145	0.013
$S^2di - Wi$	0.218	0.327
$S^2di- r^2i$	−0.547	−0.434
$Wi - r^2i$	0.130	−0.510

between stability parameters S^2di, Wi, σ^2i, ASV, S_1 and S^2 were positive and significant. Sesame genotypes, EW002 and BG006, were recognized as stable with high seed yield and could be recommended for western Ethiopia. At the same time, Sedeek et al. (2014) evaluated twenty sesame genotypes were grown at three locations, i.e. Shandaweel (Sohag), El-Mataana (Luxor) and Assiut, Egypt on May 25th and May 20th in 2011 and 2012 summer seasons, respectively. Results indicated that genotypes, also genotype × environment (G × E) interaction was highly significant for all traits. Hybrid 137 recorded the highest seed yield/fed, while Shandaweel 3 produced the lowest value. In the meantime, genotypes No. 1, 3, 15, 16 and 17 were the most stable genotypes for seed yield. In advanced study, Abd-Elsaber et al. (2018) evaluated 16 sesame genotypes (15 new sesame lines and one commercial cultivar) during two successive seasons (2015 and 2016) at three locations; El-Mtaana, Shandaweel and Ismailia. Analysis of variance showed significant differences among genotypes for almost all studied traits. The joint regression analysis of variance showed that {E + (E × G)} was significant for all considered traits. The variance due to (E × G) interaction was highly significant for all traits, showing heritable variation among the genotypes. The most stable genotypes were No. 2, 7, 9 and 14 for plant height; Genotypes No 4, 11 and 14 for fruiting zone length; Genotypes No 6, 7, 8, 9 and 16 for No of branches/plant; genotypes No 2 for No. of capsules/plant; Genotypes No 4 and 8 for 1000-seed weight as well as genotype No. 16 for seed yield/plant. Furthermore, at four environments during different times and location during summer season of 2016, Parmar et al. (2018) evaluated 56 genotypes and two checks, G.Til-2 and G.Til-3. Both hybrids AT-322 × G.Til-10 and AT-319 × AT-285 displayed below average stability and well adapted to favorable environments. Sesame genotypes AT-319 for length of capsule, G.Til-3 for width of capsule, AT-319 for number of capsules/plant, IS-209 for number of seeds/capsule, G.Til-1 for 1000-seed weight and G.Til-3 for seed yield/plant appeared to be as average stable for environments. Check Bhuva-2 × G.Til-10 (X = 6.49) was better tolerance to environmental fluctuations and so expressed as specific adaptability to less favorable environments. These, genotypes can be utilized in breeding programs to improve seed yield stability in sesame.

7.5 Additive Main Effects and Multiplicative Interaction

The use of genotype main effect (G) plus genotype-by-environment (GE) interaction (G + GE) biplot analysis by plant breeders and other agricultural researchers has increased dramatically during the past 5 yrs. for analyzing multi-environment trial data (Yan et al. 2007). Recently, however, its legitimacy was questioned by a proponent of Additive Main Effect and Multiplicative Interaction (AMMI) analysis. They compare GGE biplot analysis and AMMI analysis on three aspects of genotype-by-environment data (GED) analysis, namely mega-environment analysis, genotype evaluation, and test-environment evaluation. They concluded that both GGE biplot analysis and AMMI analysis combine rather than separate G and GE in mega-environment analysis and genotype evaluation. The GGE biplot is superior to the AMMI1 graph in mega-environment analysis and genotype assessment as it describes more G + GE and has the inner-product property of the biplot. The discriminating power vsersus representativeness view of the GGE biplot is operative in assessing test environments, which is not possible in AMMI analysis. In this connection, six experiments were conducted using seven sesame genotypes under two locations through 2005, 2006 and 2007 years using by Abd El-Mohsen (2008). Principal components axis (PCA) model, by GGE biplot, showed that environment (E) described 19.25% of the total (G + E + GE) variation, while G and GEI accounted 68.32 and 12.42%, respectively. The first two principal components (PC1 and PC2), described 88.9 and 7.9% of GGE sum of squares, correspondingly. Additive main effects and multiplicative interaction (AMMI) fashion indicated that the first two principal component axes (PCA1 and PCA2) were highly significant and cumulatively contributed to 82.4% of the total GEI. Genotypes, cultivar Shandaweel 3, Mutant 56, Mutant 50, Mutant 45, Mutant 47 and Mutant 10 were the best stable overall the six evaluated environments. Mutant 10 and Mutant 47 being the most prominent and stability, they graded the first and second in seed yield.

Chemeda et al. (2014) showed that mean squares due to AMMI analysis of variance showed significant variations between genotypes, environments and their interaction of seed yield. The GEI was highly significant and accounting for 47% of the sum of squares approximately double that of the genotypes. The GEI was divided into two interaction principal component axes (IPCA). The IPCA 1 score was highly significant, explanation 60.78% of the variability due to GEI. The IPCA 2 was significant, explained of 26.16% of the variation. Based on AMMI Stability Value the most stable genotypes were BG006 followed by EW0011-4 and then EW002. Furthermore, Misganaw et al. (2017) evaluated twelve sesame genotypes in five locations under eight environments during 2010 and 2011 seasons. Genotypes cc.00047, NN-0143 and Borkena produced the highest oil yields valued 339.2, 306.0 and 287.5 kg ha^{-1}, respectively. Highly significant differences were recorded among genotypes, environments and GEI, demonstrating that genotypes performed differently across locations and the essential for stability analysis. Proportion of variance contributed by environment was 49.6%, genotypes 13.8 and GEI 32.1% of the total variation. The amounts of the GEI sum of squares were 2.3 folded of the genotypes sum of squares. The AMMI2 model, the Interaction Principal Component Axes (IPCA1 and IPCA2) exhibited highly significant and described 46.3 and 281.8% totally accounted 75.1% of the GEI variation. Based on AMMI Stability Value (ASV) model, genotypes Borkena, Local variety and Acc.202340 has been stable, however genotypes Acc.00047, Acc.00044 and Acc.00046 were unstable. Furthermore, Baraki and Berhe (2019) found that PCA1 was accounted for 38.3% of the variation, and positively associated with seed yield, branches per plant, length of the pod-bearing zone, plant height, number of pods per plant, number of seeds per pod, and 1000-seed weight. Otherwise, PCA2, which accounted for 19.7% of the variation, was positively associated with days to 50% flowering and days to 50% maturity.

7.6   Gene Action, Genetic Behavior and Heritability for Sesame Traits Related to Environmental Stress Tolerance

For breeding to water stress tolerance in sesame in Sudan with high yield capacity, Kuol (2004) using generation mean analysis. A wide genetic variability was registered in the verified materials which reflected in a broad heritability estimates of characters. Additive, dominance and three types of non-allelic gene interactions were detected for seed yield and yield components in almost crosses. Different types of gene action were governed the inheritance of oil, protein and components of fatty acid with great role due to dominance effects in most cases. He noticed negligible reduction in heritability for yield under water stress ($h^2 = 0.70$) compared to yield under normal irrigation ($h^2 = 0.73$). Stress sensitivity index was moderately heritable (0.69 and 0.59), whereas Stress Tolerance Index was highly and moderately heritable (0.80 and 0.51) in parental genotypes and progenies, respectively. Whereas, under Shandaweel Research Station, Sohag, Egypt as stress region, Sedeek (2015) recorded high broad sense heritability for days to 50% flowering, plant height, seed yield/plant and its components. The observed realized response after two cycles of pedigree line selection for number of capsules plant/plant were 29.9, 32.9 and 27.7% as measured from the best parent, bulk sample and check, respectively. Moreover, the highest values of correlated response were recorded for seed yield/plant 14.3, 21.6 and 29.5%, respectively. In continuous, under stress environment, Sedeek and Hassan (2015) revealed highly significant differences among genotypes, parents, crosses and parents *versus* crosses for yield and its components, indicating wide diversity among the parental materials. Mean squares due to both general combining ability GCA and specific combining ability SCA were highly significant for days to 50% flowering, plant height, seed yield/plant and its components. The ratio of GCA/SCA was less than one in all traits indicating that non-additive types of gene effects are playing the major role in the inheritance of these traits. Two parental genotypes (B-35 and Tusky-1) could be considered the best general combiners for seed yield/plant. However, the cross Hybrid family-1 × Intro. 640 had the most desirable

SCA effects for seed yield/plant and most studied traits. The F1 cross Sohag-78 ×
Intro. 640 showed the maximum heterosis percentage of seed yield/plant over the
better parent.

Moreover, at Experimental farm of Agricultural Research Center, Ismailia, Egypt
as stress environment, Rehab Abdel-Rhman (2018) showed that mean squares due
to both general and specific combining ability were significantly greater for yield
characters. The GCA/SCA ratios were less than unity for plant height, capsules
weight/plant, 1000-seed weight and seed yield/plant. This indicates presence of additive gene effect governed the previous characters. However, for fruiting zone, No. of
branches/ plant, No. of capsules/plant, leaf chlorophyll content and seed oil %, the
magnitude of SCA was higher than GCA. This indicates presence of non-additive
gene effect governed the previous characters. The parents Line 363, Shandaweel -3,
and Sohag could be considered as the best general combiners for seed yield. But, the
crosses H133A4 × Sohag, Line 363 × Shandaweel-3, and Line 363 × Sohag had
the most desirable SCA effects for seed yield/plant and the most studied traits, also
the previous crosses exhibited the uppermost heterosis % over the mid-parents for
seed yield.

7.7 Role of Recent Approaches?

7.7.1 Biotechnology

The SDS-PAGE detection of protein indicated some changes in amount and number
of protein bands and also new protein bands appeared in both stressed and phyto-
hormonal treated plants. Hussein et al. (2015) added that exogenous application of
drought stressed sesame plants with salicylic acid or kinetin might induce resistance
to drought stress by improvement of growth and different physiological processes
which are adversely affected by drought stress. On the other hand, the MYB gene
family constitutes one of the biggest transcription factors (TFs) modulating different
biological processes in plants. Ali Mmadi et al. (2017) gave the potential roles of
MYB proteins in the regulation of gene expression in response to abiotic environmental stresses. A survey of a genome-wide of this gene family in sesame was
done. They identified at the genome-wide level, 287 MYB genes and revealed that
the MYB proteins are highly active in growth and adaptation to the major abiotic
stresses namely drought and waterlogging in sesame.

A wide variant was detected to drought and salt responses in the population and
most of the genotypes were moderately tolerant to both stresses. Li et al. (2018)
showed that a total of 132 and 120 significant Single Nucleotide Polymorphisms
(SNPs) resolved to nine and 15 Quantitative trait loci (QTLs) has been identified for
drought and salt stresses, respectively. Only two common QTLs for drought and salt
responses were located on linkage groups 5 and 7, respectively. This means that the
genetic foundations of drought and salt responses in sesame are diverse. A total of

13 and 27 potential candidate genes were uncovered for drought and salt tolerance indices, respectively, encoding transcription factors, antioxidative enzymes, osmoprotectants and responsible in hormonal biosynthesis, signal transduction or ion sequestration. The recognized SNPs and potential candidate genes represent important sources for practical characterization for producing new cultivars more tolerant to drought and salt. Ali Al-Somain et al. (2017) showed that sequence-related amplified polymorphism (SRAP) markers were used to measure the genetic diversity amongst a collection of 52 sesame accessions representing diverse geographical environments, including eight Saudi landraces. SRAP analysis showed a high degree of genetic polymorphism in sesame accessions with a weak association between geographical origin and SRAP patterns. The wide genetic variation would be taken into consideration for sesame breeding programs.

Dossa et al. (2019) conducted genome-wide association study for characters associated to drought tolerance using 400 diverse sesame landraces and modern cultivars. Ten stable QTLs explanation more than 40% of the phenotypic variation and found on four linkage groups were associated significantly with drought tolerance characters. Accessions from the tropical area carrying higher numbers of drought tolerance alleles at the peak loci and were found to be more tolerant than those from the northern area, representing a long-term genetic adaptation for water stress environments. They identified at the genome-wide level 287 MYB genes and revealed that the MYB proteins are highly active in growth and adaptation to the major abiotic stresses namely drought and waterlogging in sesame. Therefore, sesame has essential alleles controlling drought, which explains relative tolerance to drought conditions. Hereby pyramiding the favorable alleles is important for enhancing drought tolerance in sesame. In addition, our results highlighted two important pleiotropic QTLs harboring known and unreported drought tolerance genes such as SiABI4, SiTTM3, SiGOLS1, SiNIMIN1 and SiSAM. They revealed that SiSAM gives drought tolerance by modulating polyamine levels and ROS homeostasis, and a missense mutation in the coding region gives the natural variant of drought tolerance in sesame.

7.7.2 Nano Technology

The negative effects of severe drought stress on quantitative and qualitative characters can be compensated by foliar application of iron nano-chelates and organic compounds such as fulvic acid. Mostafa et al. (2016) evaluated the effects of drought stress and foliar application of iron oxide nanoparticles on sesame plants. They tried drought stress treatments; $W_1 = 7$, $W_2 = 12$ and $W_3 = 17$ irrigation time, and four concentrations of foliar application of iron oxide nanoparticles, $F_1 = 0$, $F_2 = 0.5$, $F_3 = 1$ and $F_4 = 1.5$ kg at 1000 1 water. Results showed that the interaction between drought stress and iron oxide nanoparticles gave significant effect on seed yield, biomass, leaf stomatal conductance, chlorophyll "b" and carotenoids content in leaves and seed nitrogen. The uppermost seed yield and biomass were achieved under W_1F_3, stomatal conductance under W_1F_2, chlorophyll "b" under W_1F_4, carotenoids under W_2F_3 and seed nitrogen under the W_3F_3 treatments. Moreover, the number of branches/plant and potassium concentration in seeds were significantly affected by the two treatments. However, phosphorus in seeds, and potassium and phosphorous in leaves were unaffected by treatments. Number of seeds and seed weight/capsule reduced around 12.3 and 27.7%, respectively due to drought condition from W_1 to W_3. Hereby, under drought stress situation, foliar application of iron oxide nanoparticles about 1 kg per thousand liters of water increases physiological characters of sesame and seed yield.

Ayoubizadeh et al. (2019) examine the effects of drought stress (full irrigation as control, irrigation up to 50% seed ripening and flowering) and foliar application of iron nano-chelate and fulvic acid compounds on yield and quality characters of two sesame genotypes Dashtestan and Halil. Results showed significant effects of drought stress, foliar application, and genotype on number of capsules, number of seeds, 1000-seed weight, seed yield, biological yield, leaf area index, number of leaf, seed oil%, oil yield, and fatty acid components. Co-application of iron nano-chelate and fulvic acid improved number of capsule, number of seed, 1000-seed weight, seed yield, biological yield, leaf area index and number of leaf rather than the control. Interaction effects of foliar application × genotype produce the highest seed yield in Halil genotype under co-application of iron nano-chelate and fulvic acid in the first year (2507.2 kg. /ha) and foliar application of iron nano-chelate in the second year (2712.7 kg.ha^{-1}). Under drought stress, oleic content, linoleic content, linolenic content and palmitic content, were enhanced during two years of experiment, while oil % and oil yield were decreased. Under moderate stress level was observed the maximum oil % (48.29 and 53.27%) through the first and second year, respectively and seed oil yield (998.6 kg.ha^{-1}) through the second year. Co-application of iron nano-chelate and fulvic acid enhanced oleic and linolenic acid contents.

7.8 How Can Measure Sensitivity of Sesame Genotypes to Environmental Stress

7.8.1 Stress Sensitivity Measurements

Stress sensitivity measurements has been proposed by many authors of them, Rosielle and Hamblin (1981) described tolerance index (TOL) as yield difference under stress (Ys) and non-stress or potential (Yp) conditions. Likewise, they revealed Mean Productivity (MP) as mean production in stress and non-stress circumstances. Fischer and Maurer (1978) presented Stress Sensitivity Index (SSI). Fernandez (1992) proposed alternative resistance measurement called by Stress Tolerance Index (STI) which is utilized to determine genotypes with high yields in both stress and normal conditions. Clarke et al. (1992) applied Stress Sensitivity Index SSI to evaluating drought resistance in wheat genotypes. Guttieri et al. (2001) utilizing SSI and suggested that higher values than 1, shows more sensitivity and lower ones show less sensitivity to water deficit stress. Ramirez-Vallejo and Kelly (1998) revealed that GMP and SSI indices are mathematical derivatives of yield and genotype selection based on both indices can be more applicable criterion to valuation drought tolerance in common bean. In wheat, SSI and grain yield indices were utilized as genotype resistance parameters and identification of tolerant genotypes (Bansal and Sinha, 1991). Tantawy et al. (2007) indicated that the sesame had decreased by up to 6.42% when the number of irrigation decreased from seven to five irrigations. The most sensitive variety to water stress was Giza 32 compared to Toshky1, Shandaweel 3 and Sohag 3.

Molaei et al. (2012) found that the sesame cultivar Oltan with having of the largest Stress Tolerance Index STI and lower value of Stress Susceptibility Index SSI and Tolerance Index TOL recognized as a drought tolerance cultivar in compare with two other cultivars Hendi and Hendi 14.

Evaluation of 27 sesame genotypes under water deficit based on drought tolerance indices were done by Hassanzadeh et al. (2009). Results revealed that both Varamin 2822 and Hendi 12 genotype under stress situations attained the highest yield stability about tolerance (TOL) and Mean Productivity (MP) indices, respectively. Concerning, Geometric Mean Productivity (GMP), Karaj 1, Oltan and Naz takshakheh were at highest level. According to Stress Sensitivity Index (SSI), Varamin 237, Naz takshakheh, Naz chandshakheh, Oltan, Hendi 12, J-1, Panama genotypes and Jiroft line, were mid-resistant, whereas Zoodrass genotype classified as sensitive one. Stress Tolerance Index (STI) classified Varamin 2822 as mid resistant one. Therefore, Karaj 1, Naz takshakheh, Varamin 237 and Varamin 2822 exhibited highest rates for drought indices and are suitable for cropping under drought stress conditions. Boureima et al. (2016) used seven drought tolerance indices including stress susceptibility index (SSI), stress tolerance index (STI), mean productivity (MP), geometric mean productivity (GMP), tolerance (TOL), yield index (YI) and yield stability index (YSI) based on yield under drought and yield in optimal conditions. Biplot and factor analysis evidenced that genotypes LC 164, LC 162, BC 167, EF 147 and MT 169 had the highest grain yield under both DS and NS environments in 2010, whereas in 2011 the best performers in both environments were HC 108, 32–15, HB 168 and 38–1-7. Factor analysis and the mean rank method discriminated 32–15 as the highest drought-tolerant genotype.

Abbasali et al. (2017) examined response of eight Iranian sesame genotypes to drought, with cultivars Oltan and Darab1 under normal irrigation and limit of irrigation since beginning of flowering as stress situation. Both accessions KC50658 and KC50321 were the most tolerant and sensitive, respectively, and accession KC50662 and cultivar Oltan give high potentiality under normal and more yield under stress situations, and exhibited high STI index. The principal component analysis recognized the variation explanation characters and showed that genotype KC50662 and cultivar Oltan were suitable to improve the tolerance to environmental conditions. The regression results for Stress Tolerance Index STI, displayed that genotypes with higher index (STI) had more seed weight/capsule, higher capsules number/plant, and longer petiole in lower leaf in normal condition.

7.9 Agricultural Practices to Mitigate Environmental Stress on Sesame

Cultivate appropriate cultivar such as Giza 32, Shandawil 3, Toshka 1 or Sohag 1 in suitable date during Mid-April until the end of May, are considered important to increase yield and avoid unfavorable conditions.

Under newly cultivated sandy soil conditions, application of compost and mineral fertilizers can mitigate environmental stress and increased seed yield in sesame (Ghada; Abd Elaziz and Abo Elezz 2018).

It is preferable to add superphosphate and organic fertilization under sandy soils before planting. Also, applied 50 kg potassium sulphate 48% after thinning. Nitrogen fertilization with 45 units must be added at three doses. Sesame is considered to be a sensitive crop for irrigation. Therefore, it is necessary to avoid waterlogging or water stress, and irrigation at regular intervals during the growing season, and irrigation prevention during the afternoon due to high temperatures that help spread the fungus disease wilt (Anonymous 2021).

Under sandy soil stress conditions, add gypsum to the sesame plant, and spraying by humic acid and/or amino acids produced the highest seed yield, protein content, oil content and proteins yield and oil yield (Eisa 2011).

7.10 Conclusions

Measuring the adaptation and stabilization of genotypes is an important strategy in sesame crop production. In this chapter, the role of varieties in responding to different environmental changes, and the genetic behavior of different traits associated with stress tolerance, have been illustrated. Besides the importance of agricultural treatments and biotechnology in improving the tolerance of sesame varieties to environmental stress.

7.11 Recommendations

Extensive variations in seed yield due to genotype, location, and genotype x location interaction were verified, so demonstrating that genotypes performed differently across locations and the essential of analyzing adaptability and stability of genotypes. Hereby, it is of important to pyramiding the favorable alleles for enhancing stress tolerance in sesame. The results of previous studies have recommended that negative effects of severe climate change on quantitative and qualitative characters could be compensated by applying proper agricultural procedures, foliar application of Nano-chelates and organic compounds in order to achieve sustainable production of sesame.

References

Abbasali M, Gholipouri A, Tobeh A, Kh Sima NA, Ghalebi S (2017) Identification of drought tolerant genotypes in the Sesame (*Sesamum indicum* L) collection of national plant gene bank of Iran. Iran J Field Crop Sci 48(1): 275–289

Abd El-Mohsen AA (2008) The use of statistical methods for describing genotype-environment interaction and stability in sesame yield trials. In: The 33rd International conference for statistics, computer science and it's applications. 6–17 April 2008, pp. 1–24

Abd-Elsaber A, Ahmed HK, Teileb WML (2018) Performance and stability of some new sesame genotypes under various environments in Egypt. In: The seventh field crops conference. 18–19 Dec 2018, Giza, Egypt, (abstracts) P 7

Afshari F, Golkar P, Mohammadi-Nejad Gh (2014) Evaluation of drought tolerance in sesame (*Sesamum indicum* L) genotypes at different growth stages. https://www.researchgate.net/pub lication/312128552_Evaluation_of_drought_tolerance_in_sesame_Sesamum_indicum_L_geno types_at_different_growth_stages

Ali Al-Somain BH, Migdadi HM, Al-Faifi SA, Alghamdi SS, Muharram AA, Mohammed NA, Refay YA (2017) Assessment of genetic diversity of sesame accessions collected from different ecological regions using sequence-related amplified polymorphism markers. Biotech 7(1):82

Ali Mmadi M, Dossa K, Wang L, Zhou R, Wang Y, Cisse N, Oureye Sy M, Zhang X (2017) Functional characterization of the versatile myb gene family uncovered their important roles in plant development and responses to drought and waterlogging in sesame. Genes 8(12):362

Anilakumar KR, Pal A, Khanum F, Bawas AS (2010) Nutritional, medicinal and industrial uses of sesame *(Sesamum indicu*m L) seeds. Agric Conspec Sci 75:159–168

Anjum AH, Xie XY, Wang LC, Saleem MF, Man C, Le W, Morphological, (2011) Physiological and biochemical responses of plants to drought stress. Afr J Agric Res 6:2026–2032

Anonymous (2021) Recommendations techniques in sesame cultivation. Agricultural Research Center, Giza, Egypt

Awaad HA, Aly AA (2002) Genotype x environment interaction and interrelationship among some stability statistics in sesame (*Sesamum indicum* L). Zagazig J Agric Res 29(2): 385–403

Awaad HA, Basha HA (2000) Yield potentiality and yield analysis of some sesame genotypes grown under two plant population densities in newly reclaimed sandy soils. Zagazig J Agric Res 27(2):239–253

Ayoubizadeh N, Laei G, Dehaghi MA, Sinaki JM, Rezvan S (2019) Seed yield and fatty acids composition of sesame genotypes as affecet by foliar application of iron Nano0chelate and fulvic acid under drought stress. Appl Ecol Environ Res 16(6):7585–7604

Bansal KC, Sinha SK (1991) Assessment of drought resistance in 20 accessions of *Triticum aestivum* and related species I. total dry matter and grain yield stability. Euphytica 56(1): 7–14

Baraki F, Berhe M (2019) Evaluating performance of sesame (*Sesamum indicum* L) Genotypes in different growing seasons in northern Ethiopia. Int J Agron Article ID 7804621, p 7

Basha HA, Awaad HA (2000) Effect of harvesting date on three sesame varieties under newly cultivated sandy soil conditions. Zagazig J Agric Res 27(1):31–41

Boureima S, Diouf M, Amoukou AI, Van Damme P)2016(Screening for sources of tolerance to drought in sesame induced mutants: assessment of indirect selection criteria for seed yield. Int J Pure App Biosci 4(1): 45–60

Chemeda D, Ayana A, Zeleke H, Wakjira A (2014) Association of stability parameters and yield stability of sesame (*Sesamum indicum* L) genotypes in western Ethiopia. East Afr J Sci 8 (2): 125–134

Clarke JM, DePauw RM, Townley-Smith TM (1992) Evaluation of methods for quantification of drought tolerance in wheat. Crop Sci 32:728–732

Dossa K, Diouf D, Wang L, Wei X, Yu J, Niang M, Fonceka D, Yu J, Mmadi MA, Yehouessi LW et al (2017) The emerging oilseed crop *Sesamum indicum* enters the "Omics" era. Front Plant Sci 8:1154

Dossa K, Li D, Zhou R, Yu J, Wang L, Zhang Y, You J, Liu A, Mmadi MA, Fonceka D, Diouf D, Cisse N, Wei X, Zhang X (2019) The genetic basis of drought tolerance in the high oil crop *Sesamum indicum*. Plant Biotechnol J 17(9):1788–1803

Eisa SAI (2011) Effect of amendments, humic and amino acids on increases soils fertility, yields and seeds quality of peanut and sesame on sandy. Res J Agric Biol Sci 7(1):115–125

Eskandari H, Zehtab-Salmasi S, Ghassemi-Golezani K, Gharineh MH (2009) Effects of water limitation on grain and oil yields of sesame cultivars. J Food Agric Environ 7:339–342

FAOSTAT (2017) Food and Agriculture Organization statistical databases. http://faostatfao.org/

Fernandez GCJ (1992). Effective selection criteria for assessing plant stress tolerance. In: Adaptation of food crops to temperature and water stress, Kuo CG (Ed) AVRDC Publication, Shanhua, Taiwan, ISBN: 92–9058–081-X, 257–270

Fischer RA, Maurer R (1978) Drought resistance in spring wheat cultivars I Grain yield responses. Aust J Agric Res 29:897–912

Ghada B, Abd Elaziz, Abo Elezz AA (2018) Effect of applied fertilizer and planting density on growth and yield of two sesame cultivars under sandy soil conditions. In: The seventh field crops conference, 18–19 Dec 2018, Giza, Egypt (abstracts) P 16

Guttieri MJ, Stark JC, O'Brien K, Souza E (2001) Relative sensitivity of spring wheat grain yield and quality parameters to moisture deficit. Crop Sci 41:327–335

Hassanzadeh M, Asghari A, Jamaati-e-Somarin Sh, Saeidi M, Zabihi-e-Mahmoodabad R, Hokmalipour S (2009) Effects of water deficit on drought tolerance indices of sesame (*Sesamum indicum* L) genotypes in Moghan region. Res J Environ Sci 3:116–121

Horacek M, Hansel-Hohl K, Burg K, Soja G, Okello-Anyanga W, Fluch S (2015) Control of origin of sesame oil from various countries by stable isotope analysis and DNA based markers—a pilot study. PLoS ONE 10(4):1–12

Hussein Y, Gehan Amin, Adel Azab, Hanan Gahin (2015) Induction of drought stress resistance in sesame (*Sesamum indicum* L) plant by salicylic acid and kinetin. J Plant Sci 10(4): 128–141

Islam F, Gill RA, Ali B, Farooq MA, Xu L, Najeeb U, Zhou W (2016) Sesame. In: Gupta SK (ed) Breeding Oilseed Crop for Sustainable Production: Opportunities and Constraints. Academic Press, Cambridge, MA, pp 135–147

Kuol BG (2004) Breeding for drought tolerance in sesame (*Sesamum indicum* L) in Sudan. https://cuvillier.de/de/shop/publications/2951

Li D, Dossa K, Zhang Y, Wei X, Wang L, Zhang Y, Liu A, Zhou R, Zhang X (2018) Differential genetic bases for drought and salt tolerances in sesame at the germination stage. Genes 9:87

Misganaw M, Mekbib F, Wakjira A (2017) Study on genotype x environment interaction of sesame (*Sesamum indicum* L) oil yield. AJAR 2(5): 1–12

Mohamed NE, Bedawy IMA (2019) Comparative and multivariate analysis of genetic diversity of three sesame populations based on phenotypic traits. Egypt J Plant Breed 23(2):369–386

Molaei P, Ebadi A, Namvar A, Bejandi TK (2012) Water relation, solute accumulation and cell membrane injury in sesame (*Sesamum indicum* L) cultivars subjected to water stress. Ann Biol Res 3(4):1833–1838

Mostafa H, Maryam G, Hadi G, Mahdi BF (2016) Effect of drought stress and foliar application of iron oxide nanoparticles on grain yield, ion content and photosynthetic pigments in sesame (*Sesamum indicum* L). Iran J Field Crop Sci (Iran J Agric Sci)Winter, 46(4): 619–628

Parmar RS, Chovatia VP, Barad HR, Sapara GK, Rajivkumar (2018) G×E Interaction for seed yield and its components traits in summer sesame (Sesamum indicum L). Int J Curr Microbiol App Sci 7(12) 1921–1941

Ramirez-Vallejo P, Kelly JD (1998) Traits related to drought resistance in common bean. Euphytica 99:127–136

Abdel-Rhman Rehab HA, Okasha SA, Elareny IM (2018) Correlation, path coefficient analysis and genetic variability for assessment of yield and its components in F1 hybrid population of sesame (*Sesamum indicum* L). Int J Agric Environ Res 05(1): 130–147

Rosielle AA, Hamblin J (1981) Theoretical aspects of selection for yield in stress and non-stress environment. Crop Sci 21:943–946

Sedeek FSH (2015) Selection criteria associated with seed yield of sesame population using path analysis. Egyptain J Plant Breed 19(1):195–214

Sedeek FSH, Hassan MS (2015) Estimation of combining ability and heterosis in sesame. Egyptain J Plant Breed 19(1):111–123

Sedeek F SH, Mahmoud MW SH, El-Shaayi REA (2014) Mean performance and stability of some sesame genotypes under six different environments. In: The fifth field crops conference, 2014 (Abstracts) P 22

Sun J, Rao Y, Le M, Yan T, Yan X, Zhou H (2010) Effects of drought stress on sesame growth and yield characteristics and comprehensive evaluation of drought tolerance. Chin J Oil Crop Sci 32:525–533

Tantawy MM, Ouda SA, Khalil FA (2007) Irrigation optimization for different sesame varieties grown under water stress conditions. J Appl Sci Res 3(1):7–12

Tariq A, Shahbaz M (2020) Glycinebetaine induced modulation in oxidative defense system and mineral nutrients sesame (*Sesamum indicum* L.) under saline regimes. Pak J Bot 52:775–782

Yan W, Kang MS, Ma B, Woods S, Cornelius PL (2007) GGE biplot versus AMMI analysis of genotype-by-environment data. Crop Sci 47:641–653

Chapter 8
Approaches in Sunflower to Mitigate Impact of Climate Change

Abstract Adaptability and stability of yield is one of the most important requirements in sustainable agriculture. The ideal sunflower genotype would produce high yields when environmental conditions are abundant and had better small reduction in yields under stress condition. Phenotypic, genotypic and AMMI stability parameters for seed yield, oil content and other economic traits becomes prerequisite that incorporate to select superior genotypes. This chapter focused on genotype × environment interaction and its relation to climatic change on sunflower production, adaptability and yield stability, role of RECENT technologies, inheritance for traits associated with stress tolerance, sensitivity measurements beside agricultural practices to alleviate environmental pressure on sunflower production.

Keywords Sunflower · Response to environments · Performance · Adaptability · Heritability · Gene manipulation · Stress sensitivity · Mitigate environmental stress

8.1 Introduction

Sunflower (*Helianthus annuus* L) is considered one of the three most imperative annual oil crops in the world following after soybean and groundnut (Moghaddasi et al. 2011). Sunflower oil has found well-known acceptance as a high quality, edible oil, rich in the unsaturated fatty acids, oleic and linoleic acids, vitamin E and contains around 25% proteins. From a healthy point of view, sunflower oil is in the topmost of healthy vegetable oils. This vision might be due to its higher content of an essential unsaturated fatty acid, linoleic (ω6), lower content of long chain fatty acids, and lower saturated fatty acid content (Turhan et al. 2010). So, sunflower oil contains polyunsaturated fatty acids which are known to reduce the risk of cardiac related problems (Monotti 2004; Iocca et al. 2016).

Egypt's edible vegetable oil production suffers from many problems due to the decrease in domestic production of oil crops, which has led to the failure to meet domestic consumption needs (Hassan and Sahfique 2010). There is already a gap in oil production in Egypt amounted 97%, where the total domestic consumption reaching 2.6 million tons, while the production reaching only 400,000 tons, so

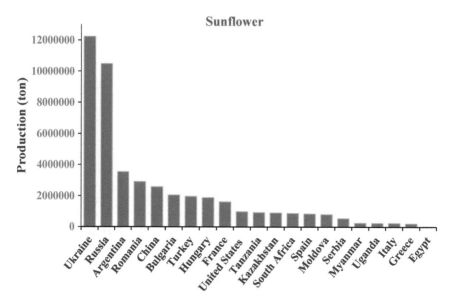

Fig. 8.1 Production quantities of sunflower worldwide (FAOSTAT 2019)

approximately 2.1 million tons annually are imported. Thus expanding the cultivation of the sunflower crop helps reduce the oil gap (Anonymous 2020). Sunflower has become an imperative oil crop in the world. The total area in 2018 reached about 24.05 million hectares worldwide with average production 1.84 ton/ha. gave total production 47.84 million metric tons. In the meantime, in Egypt, the area of sunflower cultivation is 0.01 million hectares with average production 2.38 ton/ha. gave total production 0.02 million metric tons (USDA 2019). Figure 8.1 shows production quantities of sunflower worldwide (FAOSTAT 2019).

The frequency, duration and severity of heat and drought stress events are predicted to increase in the future, devising severe effects for agricultural productivity and food safety (Killi et al. 2017). Sunflower is classified as moderately tolerant to water stress, but its production is significantly influenced by water stress (Pasda and Diepenbrock 1990).

Climate change is described by higher temperatures, elevated atmospheric CO_2 concentrations, risky climatic threats, and less water available for crop production. Sunflower could be more vulnerable to the direct effect of heat stress at anthesis and drought during its life cycle, both factors resulting in severe yield loss, oil content decrease, and fatty acid alterations. Adaptation of crops is a key factor to reduce the effect of climate change (Lobell et al. 2008). Hence, adaptations through breeding, crop management, and genetic engineering are important to partially cope with the negative impacts of climate changes. In addition, sunflower crop could contribute to the alleviation as a low greenhouse gas emitter compared with cereals and oilseed rape.

Water stress is a severe adversative feature that restrictions sunflower growth and their productivity. Environmental stress induces a variety of physiological and biochemical reactions in sunflower genotypes. Similarly, Bosnjak (2004) studied water availability over a period of 39 years and found drought to be a predominant feature in summer. He explained that moisture stress can lead to loss of sunflower yield of up to 50% which can be obtained in irrigated conditions. He also predicted that severe drought stress might be predictable in the future by worldwide climatic changes. In continuous, amongst several biotic and abiotic stresses constraining sunflower productivity, water stress is most predominant and important. Santhosh et al. (2017) showed that stress affects different physiological characters and processes, it causes up to 50% decrease in biomass and more than 50% lessening in yields. So, the genetic improvement of plant cultivars for water shortage tolerance and ability of a crop plant to produce maximum production under a wide range of stress and non-stress conditions, are expressed a major objective of stability breeding programs for a long time (Chachar et al. 2016).

High degree of genetic variability have been detected among sunflower genotypes for different physio-biochemical characteristics and can be useful in sunflower breeding programs for producing tolerant cultivars to water stress (Sarvari et al. 2017).

8.2 Genotype × Environment Interaction and Its Relation to Climatic Change on Sunflower Production

Sunflower breeders have consequently devoted effort to develop superior cultivars for seed yield and adaptation to the various stress effects. High yields are the major goal of any breeding program. Although high yields and drought tolerance are two different mechanisms, they often oppose each other. To achieve this purpose, a genetic database of germplasm and information on the performance and behavior of parents within the breeding program must be available. Therefore, we should review on the sunflower crop to avoid the environmental factors that lead to the reduction in yield and quality. Many investigators estimated performance and stability of oil crop genotypes and found significant genotype × environment for seed yield and oil content with different degrees of adaptability and stability among sunflower genotypes (Ali et al. 2006; Ghafoor et al. 2005; Tabrizi 2012; Awaad et al. 2016).

To assess G × E interaction, twelve sunflower genotypes (L38, L20, L11, L8, Giza 102, Sakha 53, L19, L235, L350, L990, L770 and L460) were examined by Salem et al. (2012) for seed yield and oil content under six diverse environments, which are the combination between three water regimes (control supplemented by 3000 m^3, moderate drought 2000 m^3 and severe drought 1000 m^3) during two summer seasons of 2009 and 2010 at El-Khattara Farm, Faculty of Agriculture, Zagazig University, Egypt with drip irrigation system. Results presented in Table 8.1 showed that sunflower genotypes (G), growing years (Y), levels of water regime (I) and

Table 8.1 Combined analyses of variance for seed yield and seed oil content (%) of 12 sunflower genotypes across 6 environments (Salem et al. 2012)

Source of variation	d.f	Seed yield (ton/fed.)	Seed oil content (%)
Reps (Env.)	18	0.016	4.928**
Genotype (G)	11	1.331**	245.176**
Year (Y)	1	0.377**	37.942**
Irrigation (I)	2	8.310**	141.704**
G × Y	11	0.004	19.453**
G × I	22	0.135**	1.622**
Y × I	2	0.006	18.769**
G × Y × I	22	0.005	3.645**
Error	198	0.017	0.736

*, ** Significant at 0.05 and 0.01 levels of probability, respectively

genotypes × levels of water regime (G × I) interaction revealed highly significant (P > 0.01) values for both seed yield and oil content. Higher values of mean squares for seed yield due to the levels of water regime show an abundant influence of water stress on sunflower seed yield. The analysis of variance for stability for seed yield (ton/fed) and seed oil content (%) through the six environments (Table 8.2) indicated that each of genotypes, environments and their interaction showed highly significant values for both characters. The partitioning of mean squares (environment + genotype × environment) showed that environments (linear) varied significantly on the performance of sunflower genotypes for both seed yield and oil content. The greater magnitude of mean squares due to environments (linear) rather than genotype × environment (linear) revealed that linear response of environments accounted for

Table 8.2 Analysis of variance for stability of 12 sunflower genotypes for seed yield (ton/fed) and seed oil content (%) across six environments (Salem et al. 2012)

Source of variation	d.f	Seed yield (ton/fed.)	Seed oil content (%)
Model	71	0.122**	11.789**
Genotype (G)	11	0.333**	61.294**
Environment (E)	5	0.850**	17.944**
G × E	55	0.014**	1.329**
E + G × E	60	0.084**	2.713**
Env. (linear)	1	4.252**	89.722**
G × E (linear)	11	0.054**	0.913**
Pooled deviation	48	0.004	1.313
Pooled Error	216	0.004	0.184

*, ** Significant at 0.05 and 0.01 levels of probability, respectively

the major part of total variation of the two characters. The significance of mean squares due to genotype × environment (linear) component against pooled deviation for seed yield and oil content advised that genotypes were different for their regression response to change in the environmental variations.

Hamza and Safina (2015) showed that combined analyses of variance illustration that the differences between years reached to the level of significance for leaf area index, 1000-seed weight, seed weight/plant, daily seed weight /plant and oil yield/fed. These differences could be due to the differences between years in climatic factors. Similarly, highly significant differences were established amongst planting dates for the studied traits. Mean squares due to cultivars were also highly significant for all studied traits. Mean squares due to planting dates × cultivars interaction were highly significant for all traits, excluding 1000-seed weight, seed and oil yields/fed were insignificant. Coefficients of variation (C.V) for the evaluated traits were varied from 0.1 to 5.2%.

Awaad et al. (2016) computed analysis of variance of sunflower genotypes i.e., parents and their F_1 crosses and between the three levels of water regime and genotypes × water regime interaction for leaf water content, leaf chlorophyll content, transpiration rate (mg $H_2O/cm^2/h$), plant height, head diameter, 100-achene weight, achene yield/plant and achene oil content (%). The results indicated highly significant differences among parental sunflower genotypes and their F_1 crosses in the three levels of water supply and genotypes × water regime interaction for the studied characters. This confirms the existence of substantial amount of genetic variability among the genetic materials. The sunflower genotypes fluctuated in their performance from adequate water supply, moderate to severe water stress. Furthermore, Ahmed and Abdella (2017) evaluated nineteen locally improved sunflower hybrids and introduced one (Hysun 33) under two irrigated locations, New Halfa and Rahad, Sudan during two successive seasons (2003/04 and 2004/05). They revealed that mean square due to genotype × environment interaction was highly significant, guiding that the yield of genotypes was inconsistent in different environments. Whereas, worldwide, Gonzalez-Barrios et al. (2017) showed that the experimental design and the number of locations, replications, and years are important aspects of multi-environment trial (MET) network optimization. They evaluated GEI in the network by delineating mega-environments, estimating genotypic stability and identifying relevant environmental covariates and used sunflower cultivars of Uruguay in a period of 20 years. They defined three mega-environments, explained mainly by different management of sowing dates. Late sowings dates gave the less performance in seed yield and oil production, due to higher temperatures before anthesis and fewer days to grain filling. Accordingly, the GEI and appropriate agronomic practices strategies have a positive impact on sunflower yield and selection of superior cultivars. Variation due to the location effect amounted (29.0%) followed by the location × year interaction (16.9%). Variance attributed to cultivar was small (2.6%), but was larger than the cultivar × location and cultivar × year interactions. The GEI accounted for a small part of the total variance (10.1%). The analysis of the 2000's period indicated a similar behavior of the 1990 period, wherever a greater number of cultivars were tested. Whereas, the highest proportion of variability was attributed to the year (29%), followed by the location × year interaction (18.2%) and GEI (8.7%) in the 2000 period.

8.3 Performance of Sunflower Genotypes in Response to Environmental Changes

Significant differences were distinguished between sunflower parents and their F_1 crosses under different environments for earliness, morpho-physiological characters by Herve et al. (2001) and Ibrahim et al. (2003) and for yield contributing characters by Rauf and Sadaqat (2007) and Iqbal et al. (2013).

Great differences were recorded among planting dates in number of days to maturity, plant height, number of leaves/plant, leaf area index, head diameter, 1000-seed weight, seed yield/plant and seed yield/ ha. (Abd El-Mohsen 2013 and Baghdadi et al. 2014). Moreover, Hussain et al. (2014) showed that sunflower genotypes display a huge variation in osmotic adjustment in response to water stress, as important in producing tolerant sunflower hybrids.

For earliness, morpho-physiological and biochemical characters, on sandy loam soil, Santhosh et al. (2014) screened four sunflower genotypes viz., GMU-337, GMU-437, EC-602063 and DRSF-113 under control and water stress in India. Water stress imposed from 30–65 DAS resulted in decreased days to 50% flowering, days to maturity, plant height, leaf number, leaf area index, crop growth rate, SPAD values, stomatal conductance, fluorescence, photosynthetic rate, chlorophyll stability index, membrane stability index and seed yield. While, proline content was increased under moisture stress. The genotypes DRSF-113 and EC-602063 recorded better values for morphological, physiological characters and yield attributes over other genotypes both under moisture stress and control conditions. Genotypes DRSF-113 and EC-602063 showed better drought tolerance as compared to GMU-337 and GMU-437. In advanced study, Santhosh et al. (2017) found that water stress resulted in decreased SPAD values, stomatal conductance, fluorescence, and photosynthetic rate. Significant variability was observed for physiological attributes and seed yield. The genotype DRSF-113 presented maximum values for SPAD chlorophyll meter readings, stomatal conductance, fluorescence photosynthetic rate and seed yield. In this context, Sarvari et al. (2017) studied physio-biochemical changes and antioxidant enzymes activities of six sunflower lines subjected to normal and irrigation at 60 and 40% of the field capacity. The results showed significant differences among sunflower lines for different characteristics such as relative water content, chlorophyll a and b contents, carotenoid and proline contents, lipids peroxidation and accumulation of malon dialdehyde, as well as activity of antioxidative enzymes like guaiacol peroxidase, ascorbate peroxidase, catalase and glutathione reductase under drought stress. The two lines C104 and RHA266 were best tolerant to water stress and can be useful in sunflower breeding programs for producing tolerant cultivars to water stress.

Performance of sunflower genotypes showed a wide range in achene yield and its attributes as well as quality characters, Taha et al. (2010) reported that Sakha-53 surpassed Giza-102 in plant height, head diameter, 1000-seed weight, seed weight plant, seed oil%, seed and oil yields/fed. Enas (2012) showed that among six sunflower genotypes, Bani Suef, Aswan and Beharia genotypes produced high proportion of polyunsaturated fatty acids compared to the other sunflower genotypes.

Under sandy soils at a wide range of planting dates in Egypt, Hamza and Safina (2015) investigated field experiments in the Desert Exp. Sta. Fac. Agric., Cairo Univ., Wadi ElNatroon, El-Beheira Governorate, Egypt under drip irrigation system in 2013 and 2014 years. Sakha-53 and Giza 102 sunflower cultivars were monthly planted, from January up to December. Results refers that there is a possibility of planting sunflower from February up to August in new reclaimed sandy soil under Egyptian conditions. Seed yield/fed significantly varied among planting date. The highest seed and oil yields/fed were obtained from 1st March planting (1038.3 and 400.1 kg/fed) followed by April planting (911.7 and 352.1 kg fed.). Sakha-53 surpassed Giza-102 in all studied traits. Fatty acids composition indicated that increasing growing temperatures resulted in more oleic acid and total unsaturated fatty acids as well as, less linoleic acid and total saturated fatty acids. Planting sunflower of March, April, May and June had roots longer than those of the other planting dates. While, August planting produced shorter roots. Planting of sunflower on February, March, April and June produce greater number of leaves plant than the other planting dates. Such rise on February and March might be attributed to the increase in plant height and the increase in temperature during those periods. Seed yield/fed varied significantly among planting dates. The highest seed yield/fed was achieved from planting on March (1038.8 kg fed) followed by April, February and May planting with values of 911.7, 868.2 and 838.4 kg/ fed, respectively. This superiority of the planting on March and April may be attributed to the considerable increase in seed yield/plant and 1000-seed weight. Sunflower cultivars exhibited differences for growing degree days accumulation during planting dates. Sakha-53 accumulated GDD more than Giza-102 through all planting dates. The maximum GDD (1328) and (1242) were attained by the interaction of Sakha-53 cv. × planting on December and Giza-102 cv. × planting on November, respectively. While, the minimum GDD (847 and 896) were obtained from Giza-102 cv. × planting on March and Sakha-53 cv. × planting on April, respectively (Fig. 8.2). The unsaturated fatty acids USFA were major components and the highest ratio in sunflower oil (oleic, linoleic and linoleinc) were attained by planting Sakha-53 and Giza-102 from March up to August. This means that sunflower oil is considered a good edible vegetable oil due to its lower content of saturated fatty acids which were obtained the highest values from September to December for both cultivars. Sakha-53 followed by Giza-102 cultivars are useful for development of healthier oils with high 18:2 (ω6) and low SFA. So, it can be increase 18:2 (ω6) in seed oil content by planting sunflower from September to November. Also, Schulte et al. (2013) indicated that increasing growing temperatures from 10 to 40 °C will result in more monounsaturated oils and less polyunsaturated oils.

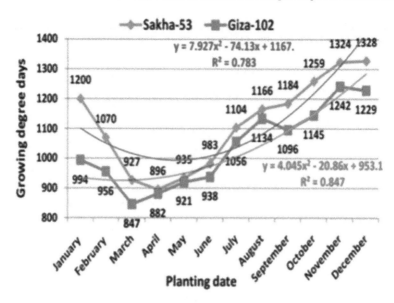

Fig. 8.2 Growing degree days for each planting date of sunflower cultivars (Hamza and Safina 2015)

Comprised 7 sunflower inbred lines (L38 and L11 from Egypt and L350, L460, L990, L770 and L235 from Bulgaria) and their the resultant 21 hybrids were evaluated under three levels of water regimes (adequate water supply with 7140 m^2, moderate 4760 m^2 and severe stress 2380 m^3/ha.). A Field experiments were carried out during the two successive seasons 2009 and 2010 at El-Khattara Agriculture Research Stations, Faculty of Agriculture, Zagazig University, Egypt by Awaad et al. (2016). The results indicated highly significant differences observed between parents and their F_1 crosses under the three levels of water regimes for various physiological characters (Table 8.3). The parental sunflower genotypes P_3 (L11) and P_4 (L460) exhibited the highest values for leaf water content, as well as their F_1 crosses $P_3 \times P_4$ and $P_4 \times P_7$ under adequate water supply; P_2 (L350) and P_4 (L460) as well as F_1 crosses $P_1 \times P_2$, $P_1 \times P_7$, $P_2 \times P_3$, $P_2 \times P_4$, $P_2 \times P_5$, $P_3 \times P_4$, $P_3 \times P_5$ and $P_4 \times P_7$ under moderate as well as P_1 and P_2 and their F_1 crosses $P_1 \times P_2$, $P_1 \times P_4$, $P_2 \times P_4$, $P_2 \times P_6$ and $P_4 \times P_7$ under severe stress. The highest leaf chlorophyll content were registered by P_1 (L38) and P_4 (L460) as well as F_1 crosses $P_1 \times P_2$, $P_1 \times P_4$, $P_3 \times P_4$ under adequate water supply and moderate as well as P_1 (L38), P_4 (L460) and P_7 (L235) and F_1 crosses $P_1 \times P_4$, $P_2 \times P_3$, $P_2 \times P_4$, $P_3 \times P_4$, $P_4 \times P_6$ and $P_4 \times P_7$ under severe stress. Lower transpiration rates were registered by P_1 (L38) and P_4 (L460) as well as the F_1 crosses $P_1 \times P_3$, $P_1 \times P_4$, $P_1 \times P_7$, $P_2 \times P_4$, $P_3 \times P_4$, $P_3 \times P_7$ and $P_4 \times P_7$ under adequate water supply; P_1 and P_3 as well as the F_1 crosses $P_1 \times P_2$, $P_1 \times P_3$, $P_1 \times P_7$, $P_2 \times P_7$, $P_3 \times P_7$ and $P_4 \times P_7$ under moderate, as well as P_1 and P_7 and the F_1 crosses $P_1 \times P_6$, $P_1 \times P_7$, $P_2 \times P_7$, $P_3 \times P_6$, $P_3 \times P_7$, $P_5 \times P_7$ and $P_6 \times P_7$ under severe stress. The general mean of transpiration

Table 8.3 Mean performance for physiological characters of sunflower genotypes of half-diallel analysis under three environments (Awaad et al. 2016)

Character Water supply Genotype	Leaf water content (%)			Leaf chlorophyll content (SPAD)			Transpiration rate (mg $H_2O/cm^2/h$)		
	Adequate water supply	Moderate stress	Severe stress	Adequate water supply	Moderate stress	Severe stress	Adequate water supply	Moderate stress	Severe stress
P1 (L38)	87.73	78.05	70.02	41.18	39.03	36.60	0.68	0.32	0.30
P2 (L350)	91.05	89.35	70.53	38.38	36.85	35.43	0.94	0.72	0.45
P3 (L11)	91.40	77.15	66.58	38.20	36.55	35.88	0.86	0.65	0.35
P4 (L460)	90.25	82.10	68.13	40.38	40.85	40.53	0.78	0.70	0.58
P5 (L990)	89.10	80.23	64.88	36.98	34.58	35.45	1.12	0.80	0.40
P6 (L770)	90.18	79.25	64.98	36.98	37.00	35.70	1.08	0.87	0.32
P7 (L235)	91.20	80.63	65.85	38.15	35.73	36.93	0.88	0.57	0.25
P1 × P2	94.60	85.40	72.60	39.45	39.25	37.50	0.90	0.65	0.40
P1 × P3	90.03	80.43	70.25	37.40	39.68	36.83	0.80	0.62	0.36
P1 × P4	91.13	82.23	73.35	38.45	39.05	39.18	0.82	0.68	0.51
P1 × P5	89.98	79.43	69.60	37.95	38.15	35.50	0.98	0.75	0.38
P1 × P6	92.33	79.45	72.28	37.00	38.50	36.05	0.96	0.80	0.35
P1 × P7	87.83	80.53	71.25	36.48	39.03	35.03	0.85	0.52	0.30
P2 × P3	92.00	84.65	69.80	36.80	40.80	38.80	0.92	0.70	0.42
P2 × P4	92.30	86.53	69.15	38.33	38.38	39.35	0.88	0.71	0.54
P2 × P5	90.25	87.58	69.80	36.00	37.00	37.03	1.02	0.78	0.43
P2 × P6	92.53	85.35	70.50	36.98	38.05	36.10	1.05	0.85	0.42
P2 × P7	90.18	83.68	69.48	35.88	39.03	35.30	0.93	0.60	0.35
P3 × P4	92.73	82.30	69.28	39.65	40.00	41.63	0.84	0.70	0.50
P3 × P5	89.60	87.28	66.80	36.18	38.08	36.05	0.99	0.76	0.38

(continued)

Table 8.3 (continued)

Character Water supply Genotype	Leaf water content (%)			Leaf chlorophyll content (SPAD)			Transpiration rate (mg H_2O/cm^2/h)		
	Adequate water supply	Moderate stress	Severe stress	Adequate water supply	Moderate stress	Severe stress	Adequate water supply	Moderate stress	Severe stress
P3 × P6	90.80	85.58	65.98	36.95	36.18	36.73	0.96	0.84	0.34
P3 × P7	89.40	80.53	66.58	36.00	40.95	36.00	0.89	0.65	0.32
P4 × P5	92.68	85.35	67.05	37.03	37.15	37.00	0.95	0.76	0.49
P4 × P6	93.53	81.30	69.95	41.48	37.48	38.98	0.90	0.82	0.45
P4 × P7	93.08	82.30	70.83	36.18	36.15	38.20	0.85	0.62	0.40
P5 × P6	89.20	80.48	61.53	34.98	37.13	36.20	1.00	0.83	0.35
P5 × P7	91.70	81.63	65.38	36.83	37.15	35.10	0.99	0.72	0.32
P6 × P7	91.38	80.30	67.33	36.80	36.50	35.40	1.02	0.77	0.30
Mean	91.00	82.46	68.56	37.61	38.01	36.94	0.92	0.71	0.39
L.S.D $_{0.05}$ (G)	3.42	6.28	3.71	2.10	3.50	3.41	0.08	0.07	0.06
L.S.D $_{0.05}$ (W)	0.87			0.57			0.01		
L.S.D $_{0.05}$ (G × W)	4.60			3.04			0.07		

rate and leaf water content tended to decrease from adequate water supply, moderate and severe stress. The interaction between sunflower genotypes and water supply treatments was significant, hereby the studied materials are effected by water stress applications. The grand mean of leaf water content and transpiration rate tended to decrease from adequate water supply, moderate to severe stress. This result could be due to drought acclimation affected the partitioning of water between the apoplastic and symplastic fractions (Maury et al. 2006), they also found differential responses of three sunflower genotypes to water stress for leaf water parameters i.e. predawn leaf water potential, photosynthetic rate and stomatal conductance.

Regarding achene yield/plant and achene oil % (Table 8.4), Awaad et al. (2016) added that, the best genotypes had high performance for achene yield/plant were P_4 (L460) followed by P_7 (L235) and P_6 (L770) along with the F_1 crosses $P_1 \times P_3$, $P_1 \times P_6$, $P_2 \times P_4$, $P_2 \times P_5$, $P_4 \times P_5$ and $P_4 \times P_6$ under adequate water supply and P_4 and P_5 as well as the F_1 cross $P_1 \times P_6$ under moderate and severe stress. Achene yield/plant tended to decrease by decreasing water supply from 7140 to 4760 and 2380 m^2/ha. with values of 80.05, 67.46 and 54.56 g, respectively. Furthermore, the parental sunflower genotypes P_1 (L38) followed by P_5 (L990) and P_7 (L235) gave the highest achene oil content and their F_1 crosses $P_1 \times P_3$, $P_1 \times P_5$, $P_1 \times P_7$, $P_3 \times P_5$ and $P_5 \times P_7$. The value of achene oil content showed relative stable under the three levels of water supply in parental sunflower genotypes P_2 (L350) and P_7 (L235) and the F_1 crosses $P_2 \times P_3$, $P_2 \times P_4$, $P_3 \times P_5$, $P_3 \times P_7$, $P_5 \times P_7$ and $P_6 \times P_7$. This result could be discussed on the basis that achene oil content was more heritable character and less influenced by the environmental conditions. These results could be discussed with reference to the connection between transpiration efficiency and plant growth accompanied by changes in plant growth characteristics and the effect of drought stress on partitioning of current assimilates between reproductive and non-reproductive organs as indicated by Virgona et al. (1990).

Under diverse environments, González-Barrios et al. (2017) evaluated seed yield and oil content of eleven sunflower cultivars amounted 2260 and 1031 kg/ha, respectively. Genotype 'INIA Butia' displayed the lowest seed yield (1988 kg/ha) and 'Pannar Pan 7355' the uppermost (2579 kg/ha). Oil content exhibited less variability, with an average of 45.7%. Otherwise, 'Pannar Pan 7355' gave the lowest content of oil (42.6%), while 'Dekasol 3810' exhibited higher values with 47.8%.

Significant differences of the mean values due to groups of sunflower parents and hybrids under normal and water stress environments were recorded for 12 morpho physiological, biochemical and yield traits by Tyagi et al. (2018). Parental genotypes and their hybrids performed well under normal conditions than under water stress. Also, hybrids exhibited more leaves, higher harvest index, seed yield, and oil content compared to their parental genotypes. Parents and hybrids overlapped with each other for harvest index and proline content under water stress. It was noted that the proline content was about three folded higher in the genotypes under water stress compared to normal condition, however, photosynthetic efficiency stayed unaffected at both environments.

To shows sensitivity to heat stress, Khan et al. (2017) evaluated 63 single cross hybrids under heat stress situation during two years and compared with the two

Table 8.4 Achene yield/plant, achene oil % and drought susceptibility index (DSI) of unflower genotypes under three environments (Awaad et al. 2016)

Character	Achene yield/plant (g)					Achene oil content (%)		
Water supply Genotype	Adequate water supply	Moderate stress	Severe stress	(DSI)		Adequate water supply	Moderate stress	Severe stress
				Moderate stress	Severe Stress			
P1 (L38)	50.33	43.80	40.28	0.82	0.63	36.10	35.40	34.90
P2 (L350)	76.78	67.23	48.73	0.79	1.15	23.40	23.10	23.00
P3 (L11)	74.85	62.80	53.18	1.02	0.91	29.50	27.90	27.10
P4 (L460)	85.73	74.68	55.10	0.82	1.12	25.40	24.80	24.50
P5 (L990)	79.10	75.63	61.85	0.28	0.68	30.50	29.50	28.80
P6 (L770)	80.20	68.05	53.28	0.96	1.05	27.40	26.20	25.60
P7 (L235)	81.30	61.40	51.13	1.56	1.17	30.40	29.80	29.10
P1 × P2	81.55	75.50	54.93	0.47	1.03	32.50	31.20	30.40
P1 × P3	84.15	73.40	52.93	0.81	1.17	35.20	33.50	32.60
P1 × P4	79.83	67.53	54.53	0.98	1.00	32.10	30.10	28.53
P1 × P5	64.48	51.83	48.20	1.25	0.79	36.40	35.40	34.10
P1 × P6	84.50	76.45	60.43	0.61	0.89	30.50	29.20	27.20
P1 × P7	77.73	67.30	56.55	0.85	0.86	38.30	34.40	32.20
P2 × P3	83.23	68.40	59.30	1.13	0.90	28.40	27.00	28.40
P2 × P4	83.95	69.20	53.90	1.12	1.12	26.80	26.10	25.20
P2 × P5	97.10	68.15	56.25	1.90	1.32	28.60	27.50	25.60
P2 × P6	77.93	72.33	51.45	0.46	1.07	28.80	25.40	24.20
P2 × P7	77.18	62.03	55.73	1.25	0.87	31.20	29.00	28.80
P3 × P4	78.78	69.58	58.40	0.74	0.81	29.90	27.40	26.10
P3 × P5	77.80	68.23	55.90	0.78	0.88	31.20	29.10	29.20
P3 × P6	74.80	63.50	52.15	0.96	0.95	29.40	27.80	28.30
P3 × P7	82.03	62.63	54.55	1.50	1.05	30.30	30.00	29.50
P4 × P5	87.60	75.70	55.95	0.86	1.13	30.30	29.90	28.00
P4 × P6	93.80	70.58	60.58	1.57	1.11	26.70	25.00	24.60
P4 × P7	81.85	67.45	53.38	1.12	1.09	30.40	29.10	27.30
P5 × P6	82.70	73.25	51.78	0.73	1.17	31.10	30.20	28.00
P5 × P7	80.70	67.23	57.88	1.06	0.89	30.60	29.10	30.10
P6 × P7	81.58	64.95	59.55	1.30	0.85	29.80	29.60	28.60
Mean	80.05	67.46	54.56			30.40	29.03	28.21

(continued)

Table 8.4 (continued)

Character	Achene yield/plant (g)					Achene oil content (%)		
Water supply Genotype	Adequate water supply	Moderate stress	Severe stress	(DSI)		Adequate water supply	Moderate stress	Severe stress
				Moderate stress	Severe Stress			
L.S.D $_{0.05}$ (G)	13.94	10.94	10.81			0.91	1.17	1.39
L.S.D $_{0.05}$ (W)	2.24					0.22		
L.S.D $_{0.05}$ (G × W)	11.86					1.16		

commercial hybrids in Pakistan. Results indicated that numerous single hybrids gave greater seed yield rather than standard check. Also, seed yield/plant was associated with pollen viability %, showing that seed yield was the produce of high gametophytic fertility at heat stress. Hybrids exhibited high seed yield production at heat stress had lesser cell membrane damage. Moreover, adaptation to high temperature mitigates the impact of water deficit during combined heat and drought stress in sunflower. To evaluate the impression on plant gas exchange, physiology and morphology, Killi et al. (2017) evaluated drought tolerant and sensitive varieties of sunflower under circumstances of elevated temperature for 4 weeks prior to subjection of water stress. The drought tolerant sunflower varieties retained ribulose-1,5-bisphosphate carboxylase/oxygenase (RubisCO) activity under heat stress to a greater extent than its drought sensitive ones. The drought tolerant sunflower varieties exhibited greater root-systems, permitting better uptake of the available soil water. Furthermore, heat stress tolerance of four traits in cultivated and wild sunflower for adaptation to heat stress was evaluated by Hernández et al. (2018). Flowering heads were covered with white (control) and black (HS) paper bags during seven successive days. Also, biogeographic tools were utilized to test the adaptation. Heat stress increased air temperature on black bags rather than the white ones by 9.4 °C on average and strongly decreased seed number and yield. They found large variability for heat stress tolerance, mainly in seed number and yield. The invasive group surpassed the cultivated and native groups in both years. Biogeographic analysis reveals variation in heat stress tolerance, populations from wetter environments were more tolerant to heat stress.

8.4 Adaptability and Yield Stability

Adaptation and stability of a genotype across environments are ordinarily tested by its degree of interaction with diverse growing environments. High mean yield should not be the only measure for stability of genotype without its high performance is stable over various environmental conditions (Sial and Ahmad 2000).Various statistical techniques have been developed to identify systematic variation in individual genotype response. Among these, Eberhart and Russell (1966) model which has been widely used in studies of adaptability and phenotypic stability of plant materials. Genotypic stability parameters have been proposed by Tai (1971) to provide information on the real response of genotype to environment. With regard to the genotypic stability, genotype $\times$ environment (G $\times$ E)-interaction effect of a genotype was partitioned into two components; i.e. "α" statistic which measures the linear response to environmental effects and "λ" statistic determines the deviation from linear response. A perfectly stable variety has $\alpha = -1$, $\lambda = 1$. However the average stable variety has $\alpha = 0$, $\lambda = 1$. Whereas, the above average stable genotype should have an estimate of $\alpha < 0$ and $\lambda = 1$, however the value $\alpha > 0$ and $\lambda = 1$ described as below average stable one. Ghafoor et al. (2005) tested stability of fifteen sunflower genotypes across eight locations in Pakistan for oil yield (kg/ha). According to six different stability parameters, sunflower genotypes viz SF-187, SMH-269, SC-110 and PSH-21 were stable regarding oil yield (kg/ha). Ward's fusion approach showed that genotypes SMH-32 and SMH-112 are different from rest genotypes over all environments, however in the largest group-7 genotypes, Hysun-33, SF-270 and SMH-269 have similar response pattern to oil yield over all environments. Among eight locations, there was a similarity in the behavior of genotypes NARC and Sariab, Kot. Diji and Dunya Pur, Tando Jam and Faisalabad and D.I.K. and Tarnab to oil yield over all genotypes. The results displayed that genotypes group-1 (Hysun-341 and NK-265) and genotypes group-2 (SF-187 and NK-277) steadily performed well over Kot Diji, Dunya Pur, Tando Jam and Faisalabad environments. SMH-32 was the best in performance at NARC and Sariab locations.

In this concern, Salem et al. (2012) tested twelve sunflower genotypes grown under six environments for seed yield and seed oil content % (Table 8.3). The regression " b" value deviated significantly from unity (b >) in sunflower genotypes L19, L990, L8 and Giza 102 for seed yield/fed as well as L235, L11, L460 and Giza 102 for seed oil content % (Fig. 8.3). Therefore, these genotypes could be grown under favorable environments. Otherwise, the "b" value was significantly less than unity (b < 1) in L38, L350, L11, L770 and Sakha 53 for seed yield/fed as well as L770 and L8 for seed oil contents %. These genotypes are suitable for drought stress environment. The deviation from linear regression (S^2d) was very small and insignificant in all sunflower genotypes for the studied characters, except genotypes L38 for seed yield (ton/fed), and L38, L19, L235, L11 and L460 for seed oil content %, which exhibited significant values of S^2d. Thus, these genotypes were more stable. A simultaneous consideration of the three stability parameters, ($\overline{X}$, b and S^2d) indicated that the most desired and stable genotypes were L20, L235 and L460 for seed yield/fed and L20, L350 and Sakha 53 for seed oil content. Genotypic stability parameters (Tai 1971) given in Table 8.5 and Fig. 8.4 showed that the most tested genotypes were stable and insignificant for linear response to environmental effects (α) and the deviation from linear (λ). The average stability area contained sunflower genotypes L20 for seed yield /fed and L350, L770, L460 and Sakha 53 for seed oil content, they recorded an statistic $\alpha = 0$ and $\lambda = 1$. Meanwhile, the below average stability area

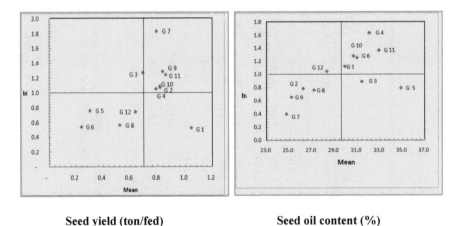

Seed yield (ton/fed)   **Seed oil content (%)**

Fig. 8.3 Illustration of stability parameters for seed yield (ton/fed) and oil content (%) of sunflower genotypes using Eberhart and Russell (1966) model (Salem et al. 2012)

contained sunflower genotypes L19, L990 and Giza 102 for seed yield (ton/fed). These genotypes exhibited an estimate of $\alpha > 0$ and $\lambda = 1$. Moreover, the above average stable genotypes were L38, L350, L770 and Sakha 53 for seed yield/fed and L350 and L770 for seed oil content %. They exhibited $\alpha < 0$ and $\lambda = 1$. Also, strong agreement has been detected between Eberhart & Russell and Tai procedures for measuring stability in sunflower genotypes L20 and L460 for seed yield (ton/fed) as well as L350 and Sakha 53 for seed oil content.

Nineteen improved sunflower hybrids and introduced one (Hysun 33) were screened by Ahmed and Abdella (2017) under two irrigated locations during two seasons in Sudan. According to stability parameters, the average ranking of the twenty hybrids, showed that Salih was the first ordered hybrid followed by Ka99 × 7, Ka99 × 29, Ka99 × 13 and Shambat 6, while Hysun 33 was the last grade one. Hybrids Salih and Shambat 6 were adapted to improved environments, however Ka99 × 25–2 was adapted to stress conditions. The graph for scatter points for yield illustrated that Salih and Shambat 6 were stable, however Hysun 33 was unstable and its yield was increased under improved conditions. Furthermore, under different mega-environments, González-Barrio et al. (2017) recorded differences in terms of sunflower seed yield stability. Cultivars 'Agrobel 972' (S = 1.0189) and 'Dekasol 4050' (S = 0.9882) were the most stable. Cultivars attained higher slopes as 'Dekasol 4040' or 'Macon RM' were adapted to the best environment. Whereas, both cultivars 'INIA Butia' and 'NK 55 RM' were less stability and poor average performance in all environments.

Table 8.5 Means for yield (ton/fed) and seed oil content (%) and various stability measurement of 12 sunflower genotypes evaluated across six environments (Salem et al. 2012)

Genotype	Seed yield (ton/fed)						Seed oil content (%)					
	$\bar{X}$	b_i	S^2d_i	α	λ	ASV	$\bar{X}$	b_i	S^2d_i	α	λ	ASV
1-L 38	1.043	0.529**	0.007*	−0.474*	2.310	3.24	29.89	1.122	1.969**	0.131	10.202*	5.48
2-L 20	0.816	1.075	−0.001	0.075	0.513	0.24	26.21	0.786	0.672	−0.230	4.027*	9.15
3-L 19	0.690	1.274**	0.000	0.276*	0.770	1.91	31.38	0.891	1.438*	−0.117	7.685*	2.24
4-L 235	0.786	1.054	−0.003	0.054	0.256*	0.39	32.03	1.635**	1.324*	0.682	6.886*	1.55
5-L 350	0.301	0.762**	−0.002	−0.239*	0.513	1.17	34.83	0.797	0.013	−0.218	0.907	1.84
6-L 11	0.239	0.542**	−0.004	−0.460*	0.000*	2.49	30.96	1.257*	1.856*	0.276	9.629**	4.82
7-L 990	0.789	1.833**	0.003	0.837*	1.283	5.07	24.73	0.397*	0.834	−0.647	4.592*	6.12
8-L 770	0.521	0.567**	0.001	−0.435*	1.026	4.58	27.17	0.761*	0.120	−0.257	1.404	2.08
9-L 8	0.835	1.288**	−0.003	0.289*	0.256*	1.78	25.18	0.654**	0.484	−0.372	3.090*	6.04
10-L 460	0.815	1.087	−0.003	0.088	0.257*	0.30	30.65	1.284*	4.264**	0.305	21.039**	5.18
11-Giza102	0.857	1.244**	−0.001	0.245*	0.770	1.70	32.86	1.370**	0.216	0.397	1.808	0.79
12-Sakha53	0.636	0.745**	0.002	−0.256*	1.284	1.93	28.30	1.047	0.361	0.051	2.583	3.32
L.S.D $_{0.05}$	0.181						0.189					

$\bar{X}_i$ = Mean yield; bi = regression of coefficient and S2di = Mean square deviations from linear regression. (Eberhart and Russel 1966); α = linear response to environmental effects and λi = the deviation from linear response (Tai 1971); ASV = AMMI stability value. S*, ** ignificant at 0.05 and 0.01 levels of probability, respectively

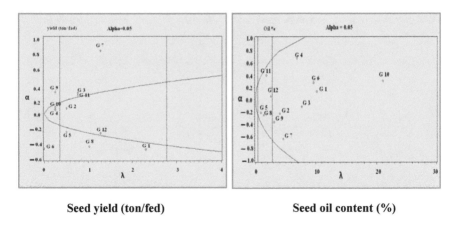

Seed yield (ton/fed)   **Seed oil content (%)**

Fig. 8.4 Illustration of stability parameters for seed yield (ton/fed) and seed oil content (%) of sunflower genotypes using Tai 1972 model (Salem et al. 2012)

8.5   Additive Main Effects and Multiplicative Interaction

Additive main effects and multiplicative interaction AMMI model (Crossa 1990) provide a visual inspection and interpretation of genotype × environment interaction and stability. It separates the additive variance from the multiplicative variance (interaction) and then applies PCA (principal components analysis) to the interaction (residual) portion from the analysis of variance to extract a new set of co-ordinate axes which account more effectively for the interaction patterns (Cravero et al. 2010). The additive main effects and multiplicative interaction model combines the analysis of variance for the genotype and environment main effects with the principal components analysis of the G × E. A genotype with the smaller AMMI Stability Value (ASV) is expressed as more stable.

In this concern, Khan et al. (2017) evaluated 63 single crosses under heat stress situation during two years with the two commercial hybrids. GGE biplot indicated that numerous single hybrids gave greater seed yield rather than standard check. GGE biplot for seed yield/plant and its components indicated that single crosses were categorized into two major groups. Group I was further categorized into two sub group. Group Ia include crosses with great 100-seed weight, whereas group Ib contain crosses with greater number of seeds/head and head diameter. Group II had the crosses with high seed weight and kernel to seed ratio.

Moreover, twelve sunflower genotypes were tested for seed yield and oil content under six different environments by Salem et al. (2012). The analysis of variance (Table 8.4) showed that environment (E), genotype (G) and G × E mean squares were significant for seed yield (ton/fed) and seed oil content (%). Interaction principle component analysis (IPCA) scores of a genotype were significant for IPCA1 for seed yield (ton/fed) and IPCA 1, IPCA2, IPCA3 and IPCA4 for seed oil content (%). Variance component % of sunflower genotypes were 38.15% for seed yield

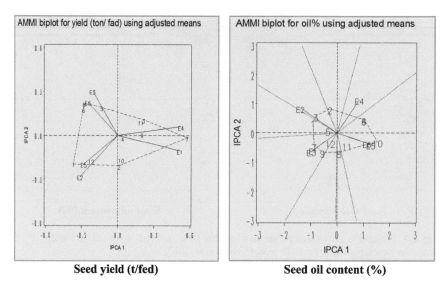

Fig. 8.5 Illustration of stability parameters for seed yield (ton/fed) and oil content (%)of sunflower genotypes using AMMI (C) model (Salem et al. 2012)

(ton/fed) and 75.28% for seed oil content (%). Whereas, variance component % due to environments were 44.32 and 10.02%, in the same respective order. The IPCA 1 exhibited the highest component of variance, since it represents 86.5% for seed yield/fed and 75.29% for seed oil content %, followed by the other scores (Fig. 8.5). According to the ASV ranking (Table 8.6 and Fig. 8.5), the most stable sunflower genotypes were L20, L460 and L235 for seed yield (ton/fed) as well as Giza102, L235 and L350 for seed oil content %. They mentioned that there were agreement between Eberhart and Russell statistics with AMMI for measuring stability in genotypes L20, L235 and L460 for seed yield (ton/fed) and L350 and Sakha 53 for seed oil %. The two stability measures were equivalent for measuring stability in the previous cases. Meantime, there was harmony between Tai (1971) method and AMMI for assessing stability in sunflower genotypes L20 for seed yield, and L350 as well as Giza 102 for seed oil content (%).

Under different mega-environments in Uruguay, González-Barrio et al. (2017) regarding GGE biplot analysis indicated that the first two main components accounted 61.3% of the total phenotypic variability. Specially, three mega-environments were defined which grouped two environments in one of them, three in the second, and the rest environments on the other. The environments 2005YG1, 2006LE1 and 2006YG1, and the cultivar 'NK 55 RM' being the best cultivar, defined the first mega-environment (ME1). The best cultivar in the ME2 was 'Pannar Pan 7355', whereas in ME3 was 'Dekasol 3810'. The mega-environments have been defined through agricultural procedures of sowing date and differences between years, using meteorological data, precipitation and temperature at flowering time.

Table 8.6 Mean squares (M.S.) from AMMI analysis for yield (ton/ fed) and seed oil content (%)of 12 sunflower genotypes across 6 environments (Salem et al. 2012)

Source of variation	d.f	Seed yield (t/fed)			Seed oil content (%)		
		S.S	M.S	%	S.S	M.S	%
Environment (E)	5	17.009	3.402**	44.32	358.887	71.777**	10.02
Reps / Env	18	0.279	0.016	0.73	88.698	4.928**	2.48
Genotype (G)	11	14.641	1.331**	38.15	2696.940	245.176**	75.28
G × E	55	3.102	0.056*	8.08	292.321	5.315**	8.16
IPCA1	15	2.686	0.179**	86.50	220.094	14.673**	75.29
IPCA2	13	0.306	0.024	9.85	37.052	2.850**	12.68
IPCA3	11	0.075	0.007	2.41	20.643	1.877*	7.06
IPCA4	9	0.026	0.003	0.82	14.177	1.575*	4.85
IPCA5	7	0.013	0.002	0.41	0.357	0.051	0.12
Pooled error	198	3.342	0.017		145.769	0.736	
Total	287	38.37			3582.62		

*, ** Significant at 0.05 and 0.01 levels of probability, respectively

8.6 Gene Action, Genetic Behavior and Heritability for Sunflower

Genetic components of variance have been estimated by several investigators for morpho physiological and yield characters related to environmental stresses. In this regard, additive gene action was more pronounced in controlling leaf chlorophyll concentration, net photosynthesis, internal CO_2 concentration and transpiration rate rather than the environmental variance. Narrow sense heritability was low for relative water content (0.22), moderate for transpiration (0.40), and high for chlorophyll concentration (0.57) as mentioned by Hervé et al. (2001). The importance of both additive and non-additive gene action for seed yield and other related characters have been mentioned by Goksoy and Turan (2004) showed that non-additive gene action was accumulated the most part of the genetic variation for seed yield and plant height. Thus, hybrid breeding method could be used aiming to improve these characters. On the other hand, neither additive nor non-additive variances were found to be significant for head diameter and 1000-seed weight. However, Ortis et al. (2005) indicated the major role of additive component in controlling the genetics of plant height, 1000-seed weight and seed oil content. While, both additive and dominance effects were governed 1000-seed weight and oil yield. Moreover, plant height and oil content were controlled by additive effects. However, over dominant effects were detected for seed yield (Ghaffari et al. 2011). Furthermore, Bakheit et al. (2010) indicated that the dominance gene action was more imperative in controlling plant height, head diameter, 100-achene weight, achene yield/plant and achene oil %. Encheva et al. (2015) recorded positive and significant heterosis in the direction of both relative to parental average and better parent for plant height, diameter of head and seed yield per plant. Moreover, under three levels of water regimes i.e. control supplemented with 3000 m^3, moderate drought 2000 m^2 and severe drought 1000 m^3 of drip irrigation system, Salem and Ali (2012) found that variance of general combining ability GCA was greater in magnitude than that of SCA for the evaluated characters, showing the superiority of additive gene action. Inbred lines L350, L460, L990 and L770 proved to be good combiners for seed yield, whereas the parents L38, L990 and L235 were promising for oil % content. Inbred lines L38, L11 and L235 were good candidates for drought tolerance. L350 was a good combiner for seed yield, but L38 proved to be good combiner for oil content and drought tolerance. The cross L38 × L350 was identified as promising for seed yield and oil content based on SCA effects. Furthermore, Awaad et al. (2016) found that cross combination L38 × L350 scored desired and significant heterosis for leaf chlorophyll content at moderate stress; transpiration rate at severe stress; achene yield/plant and achene oil content at adequate water supply and moderate stress. Genetic parameters were fluctuated from adequate water supply, moderate to severe drought stress (Tables 8.7 and 8.8). Since the dominance genetic variance was the main type controlling the inheritance of leaf water content under adequate water supply, resulting in an average degree of dominance (H1/D) 0.5 was more than unity. Hereby, pedigree method might be exploited and superior genotypes could be identified from its phenotypic expression and selected in F2. Both additive (D) and dominance (H1 and H2) genetic components were significant and involved in the inheritance of achene yield/plant and achene oil content under the three levels of water regimes with the predominant of additive gene action in controlling achene oil content. Whereas, dominance gene action was the prevailed type governing achene yield/plant under various water regimes. Otherwise, the additive genetic component was the prevailed type controlling leaf water content under moderate; leaf chlorophyll content under moderate and severe stress and transpiration rate under severe stress condition, resulting in an average degree of dominance (H1/D)$^{0.5}$ was less than unity. Both additive (D) and dominance (H1 and H2) genetic components were involved in the inheritance of transpiration rate under moderate stress, with the predominant of additive gene action. Increasing alleles were more frequent than the decreasing ones (based on positive F value) in the parental populations for transpiration rate under adequate water supply and moderate stress, and for achene yield/plant under the three levels of water regimes. The environmental variance had significant effect on gene expression of physiological characters in most cases, explaining the changes in the genetic components and their derived parameters from condition to another. Narrow sense heritability (T(n)) which reflect the fixable type of gene action transmissible from the parents to the progeny was high (>50%) for transpiration rate and achene oil content under the three levels of water regimes; leaf water content and leaf chlorophyll content under severe water stress. However, moderate narrow sense heritability have been registered for leaf water content under moderate water stress and leaf chlorophyll content under adequate water supply. Whereas, it was low for leaf water content under adequate water supply and leaf chlorophyll content under moderate water stress as well as achene yield/plant under the three levels of water

Table 8.7 Genetic components for physiological characters of sunflower genotypes of half-diallel analysis under three environments (Awaad et al. 2016)

Character	Leaf water content (%)			Leaf chlorophyll content (SPAD)			Transpiration rate (mg $H_2O/cm^2/h$)		
Water supply	Adequate water supply	Moderate stress	Severe stress	Adequate water supply	Moderate stress	Severe stress	Adequate water supply	Moderate stress	Severe stress
Genetic parameters									
D	0.06	11.36**	3.71*	2.02**	2.81*	1.71**	0.024**	0.032**	0.012**
H_1	6.09*	4.80	7.62	5.16*	4.70	0.84	0.004	0.013**	0.0002
H_2	4.99*	4.36	6.66	4.73*	2.78	0.11	0.003	0.008**	0.0002
F	−0.41	1.03	−6.79	−0.04	2.81	−2.21	0.010**	0.016**	−0.006
h^2	3.17*	9.30*	7.76**	4.92**	2.43	−0.28	0.001	0.010**	0.001**
E	1.69**	5.01**	1.69**	0.56	1.57**	1.55**	0.001**	0.001	0.0004**
Derived parameters									
$[H_1/D]^{0.5}$	10.18	0.65	1.43	1.60	1.29	0.70	0.41	0.64	0.14
$[H_2/4H_1]$	0.20	0.23	0.22	0.23	0.15	0.03	0.16	0.15	0.22
$[h^2/H_2]$	0.64	2.13	1.16	1.04	0.87	−2.66	0.45	1.26	3.00
$[KD/KR]$	0.49	1.15	0.22	0.99	2.26	0.04	3.32	2.33	0.67
$T_{(n)}$	21.13	46.88	63.08	41.70	29.70	59.67	83.68	79.96	92.99

*, **, Significant at P = 0.05 and P = 0.01, respectively

Table 8.8 Genetic components for achene yield/head and achene oil content (%) characters of sunflower genotypes of half-diallel analysis under three environments (Awaad et al. 2016)

Character Water supply	Achene yield /plant (g)			Achene oil content (%)		
	Adequate water supply	Moderate stress	Severe stress	Adequate water supply	Moderate stress	Severe stress
Genetic parameters						
D	109.53^{**}	99.46^{**}	28.78^{**}	16.60^{**}	15.90^{**}	14.98^{**}
H_1	155.19^{**}	167.82^{*}	44.30	8.33^{**}	4.75^{**}	5.76^{*}
H_2	133.59^{**}	119.31	31.57	7.74^{**}	4.64^{**}	4.82^{**}
F	99.02^{*}	130.96	40.10	0.02	−0.11	1.63
h^2	97.47^{**}	29.51	29.18^{*}	10.70^{**}	4.29^{**}	1.97
E	25.32^{**}	15.16	14.32^{**}	0.36	0.38^{*}	0.35
Derived parameters						
$[H_1 / D]^{0.5}$	1.19	1.30	1.24	0.71	0.55	0.62
$[H_2 / 4H_1]$	0.22	0.18	0.18	0.23	0.24	0.21
$[h^2 / H_2]$	0.73	0.25	0.92	1.38	0.92	0.41
$[KD / KR]$	2.22	3.06	3.56	1.00	0.99	1.19
$T_{(n)}$	21.48	15.90	3.07	78.90	84.01	82.14

*, **, Significant at P = 0.05 and P = 0.01, respectively

regimes. Therefore, selection for the aforementioned characters must extend for late segregating generations.

Razaq et al. (2017) showed that pollen fertility index under heat stress exhibited moderate to high heritability. Pollen fertility was governed by dominant type of gene action, revealing that selection for the pollen fertility might be enhanced through recurrent selection. General combining analysis indicated that gametophytic type of heat tolerance was imperative in the genetics of pollen viability.

Parental genotypes and their hybrids were subjected to two water regimes, normal irrigation and drought stress for number of leaves, leaf area, specific leaf weight, leaf area index, leaf water potential, relative leaf water content, photosynthetic efficiency, proline content, biological yield, harvest index, seed yield/plant and oil content by Tyagi et al. (2018). Analysis of variance for combining abilities of the twelve traits in a line × tester design was highly significant for all the traits except for the proline content. Mean squares due to lines (female), testers (males), and female × male interactions were highly significant for all the traits under both the environments and over the years. Parents and their hybrids exhibited highly significant general combining ability (GCA) and specific combining ability (SCA) effects for all the traits at both growth conditions. The ratios of GCA/SCA effect were > 0.5 for biological yield and oil content under normal water environment, while under water stress environment leaf water potential, photosynthetic efficiency and harvest index, suggesting the predominance of additive over non-additive gene effects. The ratio was < 0.5 for most of the traits at both growth conditions, suggesting an important role of non-additive gene action in controlling these traits.

By analyzing DEGs, Liang et al. (2017) registered 17 genes that play roles in sunflower response to abiotic stresses. Among them, nine genes might be related with responses to water stress. Under various types of abiotic stresses, Najafi et al. (2018) found that the maximum expression was correlated with HaAP2/ERF-047gene and the lowest expression was found in gene HaAP2/ERF-114. In all the applied abiotic stresses, HaAP2/ERF-047 and HaAP2/ERF-120 genes were up-regulating the effect on sunflower tolerance. In cold stress, HaAP2/ERF-047 and HaAP2/ERF-039 were up-regulated, respectively, and played an operative role in tolerance to cold stress. Furthermore, HaAP2/ERF-047 and HaAP2/ERF-120 genes can be considered as effective factors in salinity tolerance as a result of its high expression under salinity pressure. The gene expression of HaAP2/ERF-067 was more pronounced under heat stress situations.

8.7   Role of Recent Approaches?

8.7.1   Biotechnology

The use of biotechnological techniques for improving sunflower characters is limited principally by the difficulty of regenerating plants in a reproducible and efficient way (Moghaddasi 2011).

Biochemical markers like proteins are potent tools for classification of the genetic diversity and differentiation in crop species. Protein banding patterns reveal information about similarities and dis-similarities among varieties and reflecting genome relations in the breeding germplasm. These tools also offer a basis for recognizing the different cultivars of a given species genome identification and selection have progressed quickly with help of PCR technology. An enormous number of molecular markers that are rapid and require small amounts of DNA have been established. The four widely-used PCR-based markers are RAPDs, ISSR, SSRs or microsatellites and AFLP. In this concern, Hervé et al. (2001) used the AFLP linkage map for the identification of QTL related to net photosynthesis, stomatal movements and water status i.e. transpiration and leaf water potential. They detected 19 QTL explaining 8.8–62.9% of the phenotypic variance for each trait. Among these, two major QTL for net photosynthesis were identified on linkage group IX. One QTL co-location was initiated on linkage group VIII for stomatal movements and water status. Coincidental locations for QTL regulating photosynthesis, transpiration and leaf water potential were defined on linkage group XIV. On the other hand, Kiani et al. (2007) discovered 24 QTLs under well-watered environments, of which 5 (about 2 (were identified at water-stressed treatment. Phenotypic variance % explained by the QTLs varied from 6 to 29%. Among the eight QTLs detected for osmotic adjustment OA, four of them (50%) were co-located with the QTLs for turgor potential (Psi(t)) on linkage group IV (OA.4.1), with the QTL for osmotic potential at full turgor (Psi(sFT)) in well watered RILs on linkage group VII (OA. 12.2) and with QTLs of several traits on linkage group V (OA.5.1 and OA.5.2). The four other QTLs for OA (50%) were very specific. Enas (2012) utilized molecular (RAPD and ISSR) and the biochemical (protein) markers to study the genetic relationships among six sunflower genotypes based on genetic similarity using the Dice coefficient. The results indicated that the genetic similarity varied from 9 to 77% and 24.1 to 96.6% for the RAPD and ISSR, respectively. Cluster analysis of ISSR and combined data isolated Giza 102 from all the remaining genotypes, while the other genotypes located in the second cluster. The RAPD analysis gave 29 unique markers for five of the six genotypes under evaluation, while ISSR technology has revealed the capability to differentiate 22 unique markers in the four genotypes of the six genotypes. Biochemical analysis exhibited slight differences among the six evaluated genotypes. Furthermore, SNP approach is useful to recognize genomic regions governed in genotype × environment interaction of complex characters of multiple stresses in various circumstances, in this respect, among the 65,534 verified SNP, Mangin et al. (2017) recognized nine QTL governing oil yield tolerance to cold stress. Associated SNP are localized in genes previously discovered to be responsible in cold stress responses; oligopeptide transporters, LTP, cystatin, alternative oxidase, or root development.

Several tissue culture variables determining the tissue culture success in sunflower i.e. genotype, explant, development stage of the explant, the time of culture initiation, culture media composition, growth conditions, acclimatization, plant establishment and fertility analysis have been reported by Dhaka and Kothari (2002) and Yordanov et al. (2002) and Moghaddasi (2011). In continuous, Enas (2012) used tissue culture technique to determine the best genotypes and the best explant suit for different genotypes. Diverse regeneration media with different concentrations of BA and NAA were tested. The results revealed that the shoot tips of developing the highest rate of response and El Wadi El Gidid showed the best genotype for shoot formation in various concentrations of hormones when compared with other genotypes.

The response of sunflower cultivars to drought stress under both in vitro and in vivo conditions was investigated by Turhan and Başer (2014). They used Murashige and Skoog basal medium supplemented with a range of polyethylene glycol (PEG-1000) concentrations for in vitro drought evaluation. Both in vitro and in vitro results indicated that plant growth decreased with increasing PEG concentrations. Significant differences were observed between cultivars in relations of their reaction to drought. They recorded significant relationships between in vitro and in vivo characters, so in vitro methodology might be valuable in evaluating and selecting for drought response prior to field trial. All in vitro characters excluding number of roots can provide clues for performance of sunflower genotypes against drought in vivo.

Breton et al. (2010) showed that wild Helianthus species display morphological variation for resistance to stresses due to which they thrive in diverse environments. They recorded several attempts to introgression agronomic traits from these species to sunflower, and detected an introgressed progenies from *H. mollis*, a diploid species with sessile small leaves. They constructed a primary genetic map with AFLP markers in 21 BC1 plants, some progenies display 6 to 44% of introgression from *H. mollis*. Moreover, yeast metallothionein gene (CUP1) derived from yeast was transferred into sunflower to enhance tolerance of transgenic plants to heavy metals at the callus stage and selected heavy metal-tolerant lines of the transgenic sunflower calli. The results showed use of transgenics to obtain abiotic stress tolerance in sunflowers (Watanabe et al. 2005).

LBA4404 line harboring T-DNA containing dsRNA-suppressor of proline dehydrogenase gene, produced based on the ProDH1 gene of Arabidopsis, was integrated into the genome of sunflower plants transformed in vitro to raise sunflower tolerance level to water stress and salt tolerance (Tishchenko et al. 2014). Transformed procedures have been reported, more efficient transformation involving transgenic sunflower plants to increase yield, oil content, insect/fungal resistance, stress tolerance and production of biopharmaceutical proteins (Çaliskan and Dangol 2016).

8.7.2 Nano Technology

Sustainable agricultural systems depend on environmentally friendly tools based on physical and biological managements in crop production (Vashisth and Nagarajan 2010). Nanotechnology plays a vital role in promoting a sustainable green revolution by enhancing agricultural productivity with fewer inputs. Nano technology can be used in crop production to increase yield and improve tolerance to environmental stress factors. Nanotechnology plays an important role in improving field crop management (Nair et al. 2010). Using Nano fertilizer is an approach to release nutrients into the soil gradually and in a controlled system, thus avoiding pollution of water resources (Naderi and Abedi 2012). In this respect, Yaseen and Wasan (2015) showed that spray sunflower shoots with 50 ml/l of Nano silver, 1.5 g/l of organic fertilizer (Algastar) and 120 mg/l of salicylic acid provided the maximum positive effects of unsaturated fatty acids. Shukla et al. (2017) found that sunflower plants treated with ZnONPs exhibited early flowering, improved pollen viability and high starch content in pollens with respect to control. Therefore, ZnONPs has displayed noteworthy result on reproductive and biochemical parameters of sunflower. Seghatoleslami and Forutani (2015) tried two irrigation treatments (full irrigation and 50% water requirement) and 7 ZnO fertilizer levels (control, three bulk ZnO treatments and three Nano ZnO treatments). They found that application of ZnO treatments improved significantly seed yield by improving No. of seeds/head. Water stress significantly increased seed and biomass WUE by 43.6 and 40.2 percent, correspondingly. Under full irrigation, the maximum seed and biomass WUE were associated with bulk ZnO treatments, however under water stress situation the maximum biomass WUE was associated with NZnO treatment. So, application of nano ZnO increased seed yield and water use efficiency.

8.8 How Can Measure Sensitivity of Sunflower Genotypes to Environmental Stress

8.8.1 Stress Sensitivity Measurements

Drought sensitivity indices were estimated by many investigators for determining tolerance of sunflower genotypes to environmental stress. Response of six sunflower genotypes viz., G-101, SF-187, Hysun-33, Hysun-38, 64-A-93 and S-278 to water stress imposed at germination and seedling growth stages was investigated by Ahmad et al. (2009). They established five water stress levels of zero (control), -0.35, -0.6, -1.33, and -1.62 MPa using polyethyleneglycol-6000 (PEG-6000). Germination stress tolerance index, plant height stress index, root length stress index and dry matter stress index were measured to evaluate the genotypic response to PEG-induced water stress. Plant height and dry matter stress tolerance indices in all sunflower hybrids declined with increasing water stress. Otherwise, an increase in root length stress index was detected in all sunflower hybrids. Sunflower hybrids G-101 and 64-A-93 performed well and were categorized as drought tolerant. The differences between sunflower hybrids for dry matter stress index was found to be a reliable indicator of drought tolerance.

Drought sensitivity index (DSI) as regards of achene yield/plant was assessed for identifying tolerance of sunflower genotypes to water stress. Awaad et al. (2016) indicated that P1 (L38) and P5 (L990) and the F1 crosses P1 × P5, P1 × P6, P1 × P7, P2 × P7, P3 × P4, P3 × P5, P5 × P7 and P6 × P7 displayed DSI less than unity, hence these genotypes were considered as more tolerant to severe stress. Whereas, sunflower cross P2 × P5 exhibited DSI value more than 1.0 (1.32), and classified as sensitive to water stress.

Sixty sunflower genotypes were tested to drought stress at normal and drought stress treatments by Razzaq et al. (2017). Results showed that highest dry matter stress index % was detected in the genotype A-79 followed by genotype 017,583 at both treatments, respectively rather than control. Minimum DSI was found in genotype 017,571 (51.49%) followed by CM-612 (50.86%) at -1.33 MPa, while the genotype PEMS-R had lowest DSI (0.69%) followed by 017,570 (14.51%) at -1.62 MPa. Genotypes 017,583, A-75, A-79, 017,592, G-33, A-48, A-23, G-61, HBRS-1 and 017,566 were classified as drought tolerant whereas, CM-621, 017,577, HA-124, HA-133, HA-342 and HA-341 were as drought sensitive.

8.9 Agricultural Practices to Mitigate Environmental Stress on Sunflower

On the light of climate change, adaptation to environmental changes requires new crop genotypes modified to new opportunities. In order to achieve sustainable sunflower production and partially mitigate the harmful effects of environmental stress on crop growth, the following crop management systems should be followed:

– Altering planting dates and developing new genotypes with appropriate phenology and proper agricultural practices (Acosta-Gallegos and White 1995; Rauf 2008). In this respect, Hamza and Safina (2015) showed that Egyptian sunflower cultivars Sakha-53 and Giza 102 could be planting from February up to August in new reclaimed sandy soil. The highest seed and oil yields/fed were obtained from 1st March planting followed by April planting with superiority of Sakha-53 compared to Giza-102.

– Nutrients also play a vital role in reducing the environmental impact of the sunflower production. Application of inorganic nitrogen and biofertilizers as a source of N_2 fixing bacteria for sunflower increased plant height, head diameter, 100-seed weight, seed yield/fed and seed oil content (Keshta and El-Kholy 1999). Under New Valley conditions, Egypt, Abd El-Gwad and Salem (2013) indicated that spraying silicon and biofertilization treatments gave significant effect on plant height, number of leaves, leaves surface area, fresh and dry weight of leaves/plant, stem diameter, head diameter, seeds number/head, 100-seed weight as well as seed and straw yields. Also, seed oil percentage and oil yield were increased. Abdelaal (2020) found that application of 100% mineral NPK + bio fertilizers inoculation + humic acid on sunflower cultivar Sakha 53 increased growth, yield components and quality compared to the control. Foliar application of humic acid at rate of 1 or 2 g/l significantly increased most characters. Interaction between 100% mineral NPK + bio fertilizers inoculation + humic acid x humic acid treatment had the highest values of plant height, specific leaf area, seed number and weight/head, seed index, oil %, seed and oil yields/fed as well as total and net return/fed. Attia et al. (2020) found that sunflower cultivar Sakha 53 with 25 kg N/fed besides foliar spraying twice with the mixture of amino acids, yeast extract and commercial NPK + micro elements (Fe, Mn and Cu) gave high growth and seed quality and reduce the environmental pollution under the environmental conditions of El-Muhandis village, Sherbin Centre, Dakahlia Governorate, Egypt.

– Under sandy soil stress conditions, it is preferable to add farmyard manure at 20 m^3/fed to be mixed with soil with super phosphate at 150 kg 15.5% P_2O_5/fed during seed bed preparation and add 2 bags of phosphorin to the seeds to be mixed well before planting directly. Addition 45 kg N/fed in five doses starting from planting until the formation of flowering buds. Also, 50 kg/fed of potassium sulphate fertilizer is added after thinning with the second dose of nitrogen fertilization. Spraying with a mixture of micro elements 45 g Fe + 25 g Zn + 25 g Manganese + 20 g copper on two stages gives good results (Anonymous 2020).

– Because of sunflower is sensitive to irrigation, irrigation should be regularly taken care from the bud formation stage and during the flowering period as it is considered the critical period in plant life. In the case of sprinkler irrigation, regular irrigation shall be observed especially during the flowering period (Anonymous 2019). Awaad et al. (2016) found that the best results for various physiological characters, achene yield/plant and achene oil % have been obtained under adequate water supply with 7140 m³/ha, compared to moderate (4760 m²/ha) and severe stress (2380 m³/ha.) of drip irrigation.

– It is possible to enrich soil by using humic and folic acids and affordable organic fertilizer to increase the efficiency, strengthen the plant, and to preserve its nutrients. Also reduce the negative effects of water stress on the yield and quality (Poudineh et al. 2015).

8.10   Conclusions

Based on the previous review obtained in this chapter, the following conclusions could be drawn:

– Evaluation of genotypes in different environments is important in light of the genetic × environment interaction and determines the most adapted and stable sunflower genotypes.
– The use of biotechnology techniques to improve sunflowers from auxiliary tools for the genetic improvement of adaptability and stability.
– The adverse effects of environmental stress on sunflower growth can be mitigated through altering planting dates or by developing genotypes with appropriate phenology and appropriate agricultural practices.

8.11   Recommendations

Due to the sunflower genotypes fluctuated in their performance of seed yield and oil content under environmental stresses, hence, the G × E interaction and appropriate agronomic practices strategies have a significant impact on sunflower yield and selection of superior cultivars. For this reason, previous studies showed that adaptation to high temperature mitigates the impact of water deficit during combined heat and drought stress in sunflower. Authors emphasized that to evaluate the impression on plant gas exchange, physiology and morphology, biogeographic analysis, adaptation and stability must be performed by its degree of interaction with diverse growing environments. The development of genotypes with appropriate agricultural procedures leads to mitigate the impact of environmental stress on sunflower.

References

Abd El-Gwad AM, Salem EMM (2013) Effect of biofertilization and silicon foliar application on productivity of sunflower (*Helianthus annuus* L) under New Valley Conditions. Egypt J Soil Sci 53 (4):509–536

Abdelaal MSM (2020) Effect of mineral and bio fertilization and humic acid on growth productivity and quality of sunflower. In: 16th International conference for crop science. Argon Dept Fac Agric Al-Azhar Univ Egypt, Oct 2020 Abstract, p 79

Abd El-Mohsen AA (2013) Analyzing and modeling the relationship between yield and yield components in sunflower under different planting dates. World J Agric Res Food Safety 1(2):46–55

Acosta-Gallegos J, White JW (1995) Phenological plasticity as an adaptation by common bean torainfed environments. Crop Sci 35:199–204

Ahmad S, Ahmad R, Ashraf MY, Ashraf M, Waraich EA (2009) Sunflower (Helianthus annuus L) response to drought stress at germination and seedling growth stages. Pak J Bot 41(2):647–654

Ahmed SBM, Abdella AH (2017) Genetic yield stability in some sunflower (Helianthus annuus L) hybrids under different environmental conditions of Sudan. Int J Plant Breed Genet 4(3):259–264

Ali SS, Manzoor Z, Awan TH, Mehdi SS (2006) Evaluation of performance and stability of sunflower genotypes against salinity stress. J Anim Pl Sci 16(1–2):47–51

Anonymous (2004) What is agricultural biotechnology? Copyright © 2004, PBS and ABSP II, BRIEF #6: developing a biosafety system

Anonymous (2019) Recommendations techniques in sunflower cultivation. Agricultural Research Center, Giza, Egypt

Anonymous (2020) Recommendations techniques in sunflower cultivation. Agricultural Research Center, Giza, Egypt

Ardiarini RN, Kufswanto K (2013) The path analysis on yield due to the sunflower's (Helianthus annuus L) oil under drought stress. J Basic Appl Sci Res 3(4):1–7

Attia ANS, Badawi MA, Seadh SE, Amal EE Awad (2020) Effect of N-levels and foliar application treatments on growth traits and seed quality of sunflower. 16th International conference for crop science, Argon Dept Fac of Agric Al-Azhar Univ, Egypt, Oct 2020 Abstract, p 107

Awaad HA, Salem AH, Ali MMA, Kamal KY (2016) Expression of heterosis, gene action and relationship among morpho-physiological and yield characters in sunflower under different levels of water supply. J Plant Prod Mansoura Univ 7(12):1523–1534

Baghdadi A, Halim RA, Nasiri A, Ahmad I, Aslani F (2014) Influence of plant spacing and sowing time on yield of sunflower (Helianthus annuus L) J Food Agric Environ 12(2):688–691

Bakheit BR, Mahmoud AM, El-Shimy AA, Attia MA (2010) Combining ability for yield and yield components in sunflower. Egypt J Plant Breed 14(1):173–186

Berrios EF, Gentzbittel L, Alibert G, Grievau Y, Sarrafi AB (1999) Genetic control of in vitro organogenesis in recombinant inbred lines of sunflower (Helianthus annuus L). Pl Breed 118:359–361

Bosnjak DJ (2004) Drought and its relation to field crops production in Vojvodina province (Serbia, Serbia & Montenegro). Zbornik-radova-Naucni-institut-za-ratarstvo-i-povrtarstvo. (Serbia and Montenegro) 40:45–55

Breton C, Serieys H, Berville A (2010) Gene transfer from wild Helianthus to sunflower: topicalities and limits. OCL 17(2):104–114

Burli AV, Pawar BB, Jadhav MG (2001) Combining ability studies of some mal sterile lines and restore in sunflower. J Maharashtra Agric Univ 26(2):190–191

Çaliskan ME, Dangol SD (2016) Genetic engineering studied on sunflower. In: 19th Internationals sunflower conference, Edirne, Turkey 651–658

Chachar MH, Chachar NA, Chachar Q, Mujtaba SM, Chachar S, Chachar Z (2016) Physiological characterization of six wheat genotypes for drought tolerance. Int J Res-Granthaalayah 4:184–196

Cravero V, Martin E, Anido FL, Cointry E (2010) Stability through years in a non-balanced of globe artichoke varietal types. Sci Hortic 126(2):73–79

Crossa J, Gauch HG, Zobel RW (1990) Additive main effect and multiplicative interaction analysis of two international maize cultivar trials. Corp Sci 30:493–500

Debaeke P, Casadebaig P, Flenet F, Langlade N (2017) Sunflower crop and climate change: vulnerability, adaptation, and mitigation potential from case-studies in Europe. OCL 24(1):1–15

Dhaka N, Kothari S (2002) Phenyl acetic acid (PM) improves bud elongation and in vitro plant regeneration efficiency in *Helianthus annuus* L. Plant Cell Rep 21:29–34

Eberhart SA, Russel WW (1966) Stability parameters for comparing varieties. Crop Sci 6:36–40

Enas AGM (2012) Genetic studies on sunflower using biotechnology. PhD thesis Agri Sci (Genetics) Fac Agric Zagazig Univ Egypt.

Encheva J, Georgiev G, Penchev E (2015) Heterosis effects for agronomically important traits in sunflower (*Helianthus annuus* L.). Bulg J Agric Sci 21:336–341

FAOSTAT (2019) Food and agricultural organization statistical database. http://www.fa.org/faostat/en/#data/QC

Ghaffari M, Farokhi I, Mirzapour M (2011) Combining ability and gene action for agronomic traits and oil content in sunflower (*Helianthus annuuns* L) using F1 hybrids. Crop Breed J 1 (1):75–87

Ghafoor A, Arshad IA, Muhammad F (2005) Stability and adaptability analysis in sunflower from eight Locations in Pakistan. J Appl Sci 5:118–121

Goksoy AT, Turan ZM (2004) Combining abilityties of certain characters and estimation of hybrid vigour in sunflower (*Helianthus annuus* L). Acta Agronomica Hungarica, 52 (4):361–368

González-Barrios P, Castro M, Pérez O, Vilaró D, Gutiérrez L (2017) Genotype by environment interaction in sunflower (*Helianthus annus* L) to optimize trial network efficiency. Span J Agric Res 15(4):1–13

Hamza M, Safina SA (2015) Performance of sunflower cultivated in sandy soils at a wide range of planting dates in Egypt . J Plant Prod Mansoura Univ 6(6):821–835

Hassan MB, Sahfiqu FA (2010) Current situation of edible vegetable oils and some propositions to curb the oil gap in Egypt. Nat Sci 8:1–12

Hernández FA, Poverene MM, Presotto A (2018) Heat stress effects on reproductive traits in cultivated and wild sunflower (*Helianthus annuus* L): evidence for local adaptation within the wild germplasm. Euphytica Aug, 214, p 146

Hervé D, Fabre F, Berrios EF, Leroux N, Chaarani GA, Planchon C, Sarrafi, Gentzbittel L (2001) QTL analysis of photosynthesis and water status traits in sunflower (*Helianthus annuus* L) under greenhouse conditions. J Experimental Bot 52(362):1857–1864

Hussain S, Saleem MF, Iqbal J, Ibrahim M, Atta S, hmed T, Rehmani MIA (2014) Exogenous application of abscisic acid may improve the growth and yield of sunflower hybrids under drought. Pak J Agric Sci 51:49–58

Hussain S, Saleem MF, Iqbal J, Ibrahim M, Ahmad M, Nadeem SM, Ali A, Atta S (2015) Abscisic acid mediated biochemical changes in sunflower (*Helianthus annuus* L.) grown under drought and wellwatered field conditions. J Anim Plant Sci 25:406–416

Ibrahim ME, El-Absawy EA, Selim AH, Gaafar NA (2003) Effect of nitrogen and phosphorous pigments, yield and yield attributes of some sunflower varieties (*Helianthus annuus* L). Zagazig J Agric Res 30 (4):1223–1271

Iocca AFS, Dalchiavon FC, Malacarne BJ, Carvalho CGPC (2016) Content and oil productivity in sunflower genotypes produced in campo novo do parecis—Mt Brazil. In: 19th International sunflower conference, Edirne, Turkey, pp 1136–1141

Iqbal M, Ijaz U, Smiullah M, Iqbal K, Mahmood, Najeebullah M, Abdullah S, Niazand, H. A. Sadaqat (2013). Genetic divergence and path coefficient analysis for yield related attributes in sunflower (*Helianthus annuus* L) under less water conditions at productive phase. Plant Knowl J 2(1):20–23

Keshta MM, El-Kholy MH (1999) Effect of inoculation with N2-fixing bacteria, nitrogen fertilizer and organic manure on sunflower. In: Proceedings of the international symposium of biological nitrogen fixation and crop production, Cairo, Egypt, May 11–13, 181–187

Khan M, Rauf S, Munir H, Kausar M, Hussain MM, Ashraf E (2017) Evaluation of sunflower (*Helianthus annuus* L) single cross hybrids under heat stress condition. J Arch Agron Soil Sci 63(4):525–535

Kiani SP, Grieu P, Maury P, Hewezi T, Gentzbittel L, Sarrafi A (2007).Genetic variability for physiological traits under drought conditionsand differential expression of water stress-associated genes in sunflower (*Helianthus annuus* L). Theor Appl Genet 114(2):193–207

Killi D, Bussotti F, Raschi A, Haworth M (2017) Adaptation to high temperature mitigates the impact of water deficit during combined heat and drought stress in C3 sunflower and C4 maize varieties with contrasting drought tolerance. Physiol Plant 159(2):130–147

Liang C, Wang W, Wang J, Ma J, Li C, Zhou F, Zhang S, Yu Y, Li Zhang W, Li XH (2017) Identification of differentially expressed genes in sunflower (*Helianthus annuus*) leaves and roots under drought stress by RNA sequencing. Bot Stud Int J 58(42):1–13

Lobell DB, Burke MB, Tebaldi C, Mastrandrea MD, Falcon WP, Naylor RL (2008) Prioritizing climate change adaptation needs for food security in 2030. Science 319:607–610

Mangin B, Casadebaig P, Cadic E, Blanchet N, Boniface M, Carrère S, Gouzy J, Legrand L, Mayjonade B, Pouilly N, André T, Coque M, Piquemal J, Laporte M, Vincourt P, Muños S, Langlade NB (2017) Genetic control of plasticity of oil yield for combined abioticstresses using a joint approach of crop modeling and genome-wide association. Plant Cell Environ 40(10):1–32

Maury P, Berger M, Mojayad F, Planchon C (2006) Leaf water characteristics and drought acclimation in sunflower genotypes. Plant Soil 223(1–2):155–162

Moghaddasi MS (2011) Sunflower tissue culture. Adv Environ Biol 5(4):746–755

Monotti M (2004) Growing non-food sunflower in dry land conditions. Ital J Agron 8:3–8

Naderi MR, Abedi A (2012) Application of nanotechnology in agriculture and refinement of environmental pollutants. J Nanotech 11(1):18–26

Nair R, Varghese S, Nair B, Maekawa T, Yoshida Y, SakthiKumar D (2010) Nanoparticulate material delivery to plants. Plant Sci 179(3):154–163

Najafi S, Sorkheh K, Nasernakhaei F (2018) Characterization of the APETALA2/Ethylene-responsive factor (AP2/ERF) transcription factor family in sunflower. Scientific Reports. Vol. 8 Article Number 11576:1–16

of promising drought tolerant hybrids in breeding program. The present study aimed to select sunflower parents with

Ortis L, Nestares G, Frutos E, Machado N (2005) Combining ability analysis for agronomic traits in sunflower (*Helianthus annuus* L). Helia 28(43):125–134

Pasda G, Diepenbrock W (1990) The physiological yield analysis of sunflower. Part II Clim Factors Wissenschfat Technol 93:155–168

Poudineh Z, Moghadam ZG, Mirshekari S (2015). Effects of humic acid and folic acid on sunflower underdrought. Biol Forum An Int J 7(1):451–454

Rauf S (2008) Breeding sunflower (*Helianthus annuus* L) for drought tolerance. Commun Biometry Crop Sci 3(1):29–44

Rauf S, Sadaqat HA (2007) Screening sunflower (*Helianthus annuus* L) germplasm for drought tolerance. Commun Biometry Crop Sci 2(1):8–16

Ravi R, Sheoran RK, Rakesh K, Gill HS (2004) Combining ability analysis in sunflower (*Helianthus annuus* L). Natl J Plant Improv 6(2):89–93

Razaq K, Rauf S, Shahzad M, Ashraf E, Shah F (2017) Genetic analysis of pollen viability: an indicator of heat stress in sunflower (*Helianthus annuus* L). Int J Innov Approaches Agric Res 1(1):40–50

Razzaq H, Tahir MHN, Sadaqat HA, Sadia B (2017) Screening of sunflower (*Helianthus annuus* L) accessions under drought stress conditions, an experimental assay. J Soil Sci Plant Nutr 17(3):662–671

Reddy AR, Chaitanya KV, Vivekanandan M (2004) Drought-induced responses of photosynthesis and antioxidant metabolism in higher plants. J Plant Physiol 161:1189–1202

Salem AH, Awaad HA, Ali MMA, Omar AEA, Kamal KY (2012) Some stability parameters in sunflower (*Helianthus annuus* L) Genotypes at various applications. Egypt J Agron 34(2):141–153

Salem AH, Ali MA (2012) Combining ability for sunflower yield contributing characters and oil content over different water supply environments. J Am Sci 8(9):227–233

Santhosh B (2014) Studies on drought tolerance in sunflower genotypes. M.Sc in Agricultural Department of crop physiology college of Agriculture. Acharya NG Ranga Agricultural University, Rajendranagar, Hyderabad-500030, Crop Physiology

Santhosh B, Reddy SN, Prayaga L (2017) Physiological attributes of sunflower (*Helianthus annuus* L.) as influenced by moisture regimes. Green Farming 3:680–683

Sarvari M, Darvishzadeh R, Najafzadeh R, Maleki HH (2017) Physio-biochemical and enzymatic responses of sunflower to drought stress. J Plant Physiol Breed 7(1):105–119

Schultea LR, Ballardb T, Samarakoonc T, Yaoc L, Vadlanid P, Staggenborge S, Rezac M (2013) Increased growing temperature reduces content of poly unsaturated fatty acids in four oilseed crops. Ind Crops Prod 51:212–219

Seghatoleslami M, Forutani R (2015) Yield and water use efficiency of sunflower as affected by nano ZnO and water stress. J Adv Agric Technol 2(1):1–4

Shukla S, Kumar Sh P, Himansh P, Ramteke PW, Pragati M (2017) Effect of different modes and concentrations of ZnO nano particles on floral properties of sunflower variety SSH6163. Vegetos 30(Special):307–314

Sial MM, Arian M, Ahmad (2000) Genotype x environment interaction on bread wheat grown over multiple sites and years in Pakistan. Pak J Bot 32(1):85–92

Tabrizi HZ (2012) Genotype by environment interaction and oil yield stability analysis of six sunflower cultivars in Khoy. Iran Adv Environ Biol 6(1):227–231

Taha MA, Ali AAG, Zeiton OA, Geweifel HGM (2010) Some agronomic factors affecting prodc-tivity and quality of sunflower (*Helianthus annuus* L) in newly cultivated sandy soil. Zagazig J Agric Res 37(3):505–532

Tai GCC (1971) Genotypic stability analysis and it's application to potato regional trials. Crop Sci 11:184–190

Tishchenko OM, Komisarenko AG, Mykhalska SI, Sergeeva LE, Adamenko NI, Morgun BV, Kochetov AV (2014) Agrobacterium-mediated transformation of sunflower (*Helianthus annuus* L.) in vitro and in planta using LBA4404 strain harboring binary vector pBi2E with dsRNA-suppressor of proline dehydrogenase gene. Cytol Genet 48:218–226

Turhan H, Başer I (2014) In vitro and In vivo water stress in sunflower (*Helianthus annuus* L). HELIA 27(40):227–236, ISSN (Online) 2197–0483, ISSN (Print) 1018–1806. https://doi.org/10.2298/hel0440227t

Turhan H, Citak N, Pehlivanoglu H, Mengul Z (2010) Effects of ecological and topographic conditions on oil content and fatty acid composition in sunflower. Bulg J Agric Sci 16(5):553–558

Tyagi V, Dhillon SK, Kaushik P, Kaur G (2018) Characterization for drought tolerance and physio-logical efficiency in novel cytoplasmic male sterile sources of sunflower (*Helianthus annuus* L). Agron 8(232):1–20

USDA (2019) World agricultural production. United States department of agriculture, foreign agri-cultural service, circular series WAP 6–19 June 2019, Office of Global Analysis, FAS, USDA, Foreign Agricultural Service/USDA

Vashisth A, Nagarajan S (2010) Effect on germination and early growth characteristics in sunflower (*Helianthus annuus*) seeds exposed to static magnetic field. J Plant Physiol 167:149–156

Virgona JM, Hubick KT, Rawson HM, Farquhar GD, Downes RW (1990) Genotypic variation in transpiration efficiency, carbon isotope discrimination and carbon allocation during early growth in sunflower. Aust J Plant Physiol 17:207–214

Watanabe M, Shinmachi F, Noguchi A, Hasegawa I (2005) Introduction of yeast Metallothionein gene (CUP1) into plant and evaluation of heavy metal tolerance of transgenic plant at the callus stage. Soil Sci 51:129–133

Yaseen AAM, Wasan MH (2015) Response of sunflower (*Helianthus annuus* L) to spraying of nano silver, organic fertilizer (Algastar (and salicylic acid and their impact on seeds content of fatty acids and vicine. AJEA 9(1):1–12

Yordanov Y, Yordanova E, Atanassov A (2002) Plant reqeneration From interspecific hybrid and backcross progeny of *Helianthus eggertii* X *H. annuus*. Plant Cell Tissue Org Cult 71:7–14

Part IV
Conclusions and Recommendations

Chapter 9
Approaches in Cotton to Mitigate Impact of Climate Change

Abstract A combination of two or more abiotic stresses, such as drought, heat and salinity results in more yield loss than normal condition. Therefore, genotype × environment interaction is considered a major challenge in crop genotype recommendations under Egyptian circumstances and similar conditions in other countries. Then identify the magnitude and nature of genotype × environment interaction and determine stability of yield potentiality are of great important. Results of literature showed high degree of genetic differentiation in response to climate change based on mean performance and yield stability parameters in cotton varieties for economic traits. Hereby, this chapter focused on assess behavior of cotton genotypes in multi-environment experiments to evaluate the adaptability and stability and to recommend the genotypes to be grown diverse environments. Also role of molecular markers, gene transfer, Nanotechnology in the field of cotton improvement under various environments, and limp on some of the agricultural procedures used to reduce environmental stress and sustain cotton production.

Keywords Cotton · Genotype × environment · Genetic diversity · Adaptability · Heritability · Gene transfer · Stress sensitivity · Alleviate stress

9.1 Introduction

Cotton is considered as commercial crop of world importance. It is the world's leading textile fiber crop and it is also a source of secondary products like oil, livestock feed (cotton seed cake) and cellulose (Anderson 1999; Frelichowski et al. 2006). Cotton plays a major role in economy and national income of the produced countries. Globally, the highest produced of cotton are China, India, United States, Brazil, Pakistan, Turkey and Australia (FAOSTAT 2018). The development of novel, high-yielding, stress-tolerant varieties of cotton with adequate fiber quality characteristics are among the major goals of cotton breeders in cotton breeding programs. The first step to achieve that is choosing suitable parents for generating promising genotypes with higher heterosis (Hussain et al. 2019).

As cotton is grown in warm climates, production occurs in more than 80 countries, and cotton is an economic crop in more than 30 of these countries. With rapidly increasing world population and climate change, abiotic and biotic pressures represent the major challenges in crop production worldwide. So it is necessary to increase crop yields by at least 40% in arid and semi-arid regions (Nakashima et al. 2014; Shaar-Moshe et al. 2017). Therefore, crop varieties should be produced more suitable for climate change.

Drought stress is one of the most challenging problems in cotton production in the Egypt and worldwide. Drought only affects 45% of the world's agricultural land, additional, 19.5% of irrigated lands are saline. Drought with salinization is ordinary to cause up to 50% of arable land loss worldwide in the next 40 years (Wang et al. 2003). For example, drought in Texas in 1998 and 2009 caused more than 500 million dollars in cotton losses (Phillips 1998 and Fannin 2006). Improvement of drought heat and/or salt stress tolerant genotypes represents one of the most useful solutions. Genetic variation in abiotic stress tolerance exists within cotton genotypes. In 2015, USDA predicted that a future decline in cotton production might occur in the presence of drought stress. Indeed, cotton industry has been affected by drought and heat stress, leading to a loss of fiber yield by 34% (Ullah et al. 2017). The total area of cotton seed in 2018 reached about 32.95 million hectares worldwide with average production 1.37 metric tons/ha. gave total production 45.15 million metric tons. Whereas, the total area of cotton reached about 33.58 million hectares worldwide with average yield 803 kg/ha. gave total production 123.78 (million 480 Ib. bales). In the meantime, in Egypt, the area of cotton seed cultivation during 2018/2019 is 0.14 million hectares with average production 1.08 metric tons/ha. gave total production 0.15 million metric tons (USDA 2019). Figure 9.1 show production/yield quantities of cotton seed worldwide (FAOSTAT 2019).

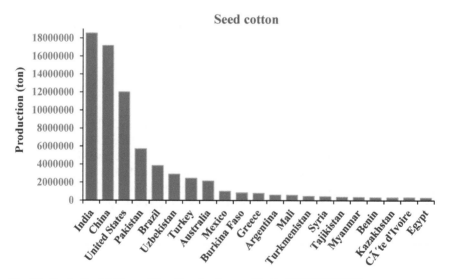

Fig. 9.1 Production/yield quantities of seed cotton worldwide (FAOSTAT 2019)

9.2 Genotype × Environment Interaction and Its Relation to Climatic Change on Cotton Traits

Genotype × environment interaction is considered a major challenge in crop genotype recommendations.

A major challenge for plant breeders is selection of high yielding genotype with wide adaptation. Stability is a key word for plant breeders who analyze G × E interaction data because it enhances the progression of selection in any single environment. So, evaluating genotypes in diverse environments is essential to obtain information regarding phenotypic stability, which is beneficial for the selection of promising genotypes and improvement programs. Most often, plant breeders face an interaction between G × E when assessing breeding program outcomes across environments. To identify stable genotypes, breeders should divide the G × E interaction into the stability statistics assigned to each genotype evaluated across a range of environments. Under Egyptian conditions, for evaluation heat stress and other environments, eleven cotton varieties (*Gossypium barbadense* L.) were evaluated under eighteen environments. The eighteen environments were the combinations of three planting dates, i.e., 21th March (recommended planting date), 11th April (relatively late planting date) and 2nd May (late planting date "heat stress"), and two locations under Ghazaleh and El-Khattara regions, Sharkia governorate by Awaad and Nassar (2001). Pooled analyses of variance indicated highly significant differences among cotton genotypes (G), years (Y), locations (L) and planting dates (D), as well as their first-order interactions between genotypes and the environmental factors for most studied characters. Second-order (G × Y × L) and (G × L × D) and third order interaction (G × Y × L × D) for most yield and fiber characters were significant. This, providing evidence for necessarily of testing cotton genotypes in multiple environments. Location effect accounted for most part of the total variation on the studied characters, followed by planting data, year and then genotype effects. Since the contribution of these items were 62.96% for locations, 15.99% for planting dates, 12.60% for years and 8.45% for genotypes from the total variance of seed cotton yield (Table 9.1). Stability analysis of variance provided evidence for significant (G × E) interactions (linear) for the studied yield and quality characters, indicating that genotypes differed in their response to changes in environments. The (G × E) interactions (linear) was also significant when tested against pooled deviation in all studied characters, except seed oil%, revealing that the linear regression and the deviation from linearity were important for describing stability for most characters. Whereas, Thirty six cotton genotypes (six parents, 15 F1 hybrids and 15 F1 reciprocals) were tested by Abd El-Bary (1999) under two locations. Genotypes, locations and genotype × location interaction were highly significant for number of days to first flower and number of days to first open boll. Hassan et al. (2006) detected significant or highly significant effect of genotype, year, location, location × year, genotype × year, genotype × location and genotype × location × year on fiber length (2.5%) span length, micronaire reading, fineness, fiber strength and fiber elongation, but genotype × year interaction was not significant in micronaire reading.

Table 9.1 Mean squares for stability analysis of seed cotton yield, some economic characters and fiber properties (Awaad and Nassar 2001)

Source	d.f	Mean squares										
		Seed cotton yield	Lint percentage	Boll weight	Seed oil content	Earliness percentage	2.5% span length	50% span length	Length uniformity (%)	Fiber fineness	Fiber strength (g/tex)	Fiber elongation (%)
Genotypes	10	37.763**	80.456**	2.342**	23.812**	607.957**	151.957**	50.653**	174.545**	4.545**	23.219**	114.165**
Environment + (Genotype × environment)	187	2.281**	9.957**	0.171**	0.320	68.355**	1.596**	0.733**	4.947**	0.053**	0.947**	0.236**
Environment (linear)	1	273.561**	453.802**	20.080**	5.768**	908.123**	33.454**	9.853**	13.965**	2.599**	16.078**	3.832**
Genotype × environment (linear)	10	0.453**	12.972**	0.099**	2.149**	139.625**	6.942**	2.431**	10.924**	0.169**	4.137**	0.571**
Pooled deviation	176	0.843**	7.263**	0.062**	0.185	18.687**	1.111**	0.584**	4.379**	0.031**	0.679**	0.196**
Pooled error	384	0.183	0.724	0.022	0.264	2.508	0.180	0.121	0.558	0.0096	0.161	0.063

** Highly significant at 1% level of probability

Kafr El-sheikh location provided the highest values for utmost characters in cotton cultivars Giza 45, Giza 70, Giza 87 and Giza 88. Also, highly significant effect of the environments, genotypes and genotype × environment interactions were found on seed cotton yield and lint yield by El-Shaarawy et al. (2007). El-Adly and Eissa (2008) investigated genetic stability for several long staple cotton strains and showed that genotypes, environments and genotype × environment interactions were highly significant for boll weight, seed index, lint percentage and lint index.

Nine Egyptian cotton genotypes were evaluated by Shaker (2009) for long and extra-long staples. He recorded highly significant mean squares due to genotypes, years, locations, genotypes × locations and the second order interaction genotypes × locations × years in boll weight, lint percentage, seed index and lint index. Whereas, the first order interaction genotypes × years did not reach the level of significance in respect to boll weight. Under different environments, Hassan et al. (2012) showed that, the effects of each of genotypes, locations and years were differed significantly in seed cotton yield/fed, lint cotton yield/fed, lint percentage, boll weight, seed index, lint index, fiber length, fiber strength and micronaire reading. Locations × years interaction was highly significant for the aforementioned traits. The first order interaction of genotypes × locations was highly significant regarding seed cotton yield/fed, lint cotton yield/fed, lint index, fiber length, fiber strength, and micronaire reading except boll weight, lint percentage and seed index. Whereas the interaction between genotypes × years was significant for the studied characters excluding lint % and seed index. The second order interaction of genotypes × years × locations had highly significant for yield and fiber traits excluding seed index.

To study the magnitude and nature of genotype × environment interaction and determine of stability of yield potentiality for five Egyptian cotton varieties, Dewdar (2013) showed highly significant mean squares due to environment × genotype interactions for seed cotton yield/plant, lint yield/plant, boll weight, number of open bolls, lint % and lint index. Mean squares of environment (linear) and linear interaction of genotypes × environment were highly significantly for the studied characters. The first effect means that differences on environments i.e. sowing dates and years produce disparities on cultivar responses, while the late effect shows the exists of genetic variation between cultivars and their differential responses to environmental conditions. The unpredictable constituents of the interaction were more important compared to the predictable one.

Evaluation of four cotton genotypes i.e., (Giza 80, Giza 90, the promising strain Giza 90 × Australy and [Giza 83 (Giza 75 × 5844)] × Giza 80 under six locations (Giza, El-Fayoum, Beni-Suef, El-Minya, Assiut and Sohag) during two seasons of 2011 and 2012 was performed by Sahat (2015). Results showed that the differences were significant among genotypes, locations, seasons, genotypes × locations, genotypes × seasons, locations × seasons and genotypes × locations × seasons interaction on earliness, seed cotton yield/fed, lint cotton yield/fed, yield components, fiber quality properties (upper half mean, length uniformity index, fiber strength, micronaire reading and yarn strength) and fiber cross section characters (area $\mu^{0.2}$, perimeter μ, secondary wall area μ^2 and degree of thickening %). Under four locations of the middle and north Delta during the years of 2011 and 2012, Abd El-Rhman (2016) revealed that analysis of variance showed highly significant differences for each of year (Y), location (L) and genotype (G) for seed and lint cotton yields, relevant components and fiber traits. Suggesting the presence of wide range of differences among genotypes and locations. The first order as well as the second order (Y × L × G) interactions were signification for all studied traits except (Y × G) with seed cotton and lint yields. Furthermore, combined analyses of variance of the six cotton genotypes, based on Wald Chi-Square tests, across 3 environments are computed by Djaboutou et al (2017). The main effect differences between genotypes, environments and the interaction effects were verified. The results showed that density and cotton seed yield were significantly influenced by environment. Significant effect of genotypes was detected with average bolls weight. However, insignificant differences were recorded among genotypes for variables density and cotton seed yield. Also, genotypes × environment interaction effect on density was highly significant. Recently, under two sowing dates i.e. normal (April 15) and late (May 15), Shaker et al. (2020) found that cotton yield characters and fiber properties showed significant mean squares for varieties in normal, late and combined analyses, except fiber strength and micronaire reading. Mean squares due to sowing dates were significant for all characters, except fiber length and length uniformity ratio. The varieties × dates was significant for seed index, lint index and fiber strength, also the interaction of variety × year × date was significantly only for fiber strength. Furthermore, years mean squares under normal and late sowing dates were not significant for fiber quality, fiber length, length uniformity ratio and micronaire reading. The interaction between year × date, variety × date and variety × year × date were not significant for fiber length and length uniformity ratio, also variety × date and variety × date × year for the both characters and micronaire reading, indicating that the different characters were stable from year to year either for normal or for late sowing dates.

9.3   Performance of Cotton Genotypes in Response to Environmental Changes

Under Egyptian conditions, assessment the contribution % of maximum air temperature to lint yield variation of cotton cultivars was done by Hassan (2000). Results showed that Giza 85 cultivar revealed the maximum response during the end of August. However, adapted cotton cultivar Giza 83 to high temperature gave little reaction to the maximum air temperature through the beginning of April, and showed the highest response to the minimum air temperature through the start of May to the end of July. Furthermore, Giza 89 cultivar revealed the highest reaction to the minimum relative humidity through beginning of April, middle of June, start of July, middle of August and middle of September. Whereas, under Ras suder, South Sinai, Egypt during three seasons, Moustafa (2006) revealed highly significant differences among ten cotton genotypes for earliness, yield contributing characters and fiber quality traits under four salinity levels of irrigation water i.e. 2000, 4000, 6000 and

8000 ppm. The genotypes Giza 70 and Ashmouni were the most superior in most traits. Increasing salinity levels resulted in a reduction of all yield traits, except fiber fineness and concentration of magnesium, sodium and proline which were increased due to salinity elevation.

To demonstrate the genetic variances between cotton genotypes in their tolerance to heat stress, Arafa et al. (2008) observed that, the highest accumulation of heat units DD15's have been recorded for boll weight, lint %, seed index, seed cotton yield and lint cotton yield in all cultivars except, Giza 87 which was adapted to a wide range of accumulated heat units. The higher estimates of earliness percentage 88.55, 95.15, 96.17, 94.96, 90.88 and 98.16 for Giza 45, Giza 87, Giza 70, Giza 85, Giza 86 and Giza 89, respectively were recorded at the maximum DD15's. However, the highest value of earliness percentage (96.84%) was verified at the lowest DD15's in Giza 88 cultivar only. Thus Giza 88 need heat units less than the other cultivars.

Mean performance and yield stability parameters in cotton varieties for the yield traits are computed by Dewdar (2013) through three field experiments at El-Fayoum, Egypt on five Egyptian cotton varieties. Results indicated that Giza 90 followed by Giza 80 exhibited high mean performances for all the studied yield traits. Also, Giza 86 displayed moderate mean performance of seed cotton yield, lint yield, boll weight, No. of open bolls/plant, lint % and lint index, whereas Giza 88 and Giza 70 provided the lowest values. Hereby, both genotypes Giza 90 and Giza 80 might be exploited in cotton breeding programs. Sekmen et al. (2014) measured the physiological and biochemical responses between two cotton cultivars differ in water and heat stress tolerance. The sensitive cotton cultivar 84-Shad decreased activities of catalase (CAT), and peroxidase (POX) under stress condition rather than control condition. Conversely, the drought-tolerant cultivar M-503 preserved levels of enzyme activities of superoxide dismutase (SOD) and ascorbate peroxidase (APX) and increased the levels of CAT, POX, and proline content under drought and heat stresses.

Four cotton genotypes namely (Giza80, Giza 90, the promising strain Giza 90 × Australy and [Giza 83 (Giza 75 × 5844)] x Giza 80 were cultivated in six locations (Giza, El-Fayoum, Beni-Suef, El-Minya, Assiut and Sohag) in two seasons (2011 and 2012) by Sahat (2015). The promising strain [G.83 (G.75 × 5844)] × G.80 and Giza 90 × Australy surpassed the commercial varieties Giza 80 and Giza 90 in earliness, yield, yield components and fiber quality which recorded lower number of days to first flower and first open bolls, higher number of open bolls/plant, heavier boll weight as well as higher averages of lint percentage, upper half mean length, fiber uniformity index, fiber strength, with lower micronaire reading and better values for cross section characters Under Sohag region conditions compared to the other locations. Cotton plant was the highest efficient in the use of heat units under the circumstances of Sohag region compared to the other locations. In continuous, the best values were attained by genotypes grown in Sohag region for most characters excepting reflectance degree and yellowness. Abdel-Kader et al (2015) tested 21 cotton genotypes (6% and their 15 F_1 crosses) under two irrigation treatments i.e., 100% ETc, 1269 mm/season (normal) and 60% ETc, 761 mm/season (drought). They showed highly significant differences between genotypes for all physiological and morphological traits under normal and water stress treatments. While the studied traits in all genotypes were significantly influenced by water stress, however some genotypes like Tamcot C. E. × Deltapine, Giza 90 × (Giza 90 × Australian) and Giza 80 × Deltapine were drought tolerance by maintaining the highest values of morphological and physiological traits.

The overall mean performance for six cotton varieties and lines across the eight environments (4 locations × 2 years) in Egypt were investigated by Abd El-Rhman (2016). He demonstrated that Gharbia location was superior to other locations in seed cotton and lint yields and Kafr El-Sheikh came in second rank followed by Damietta, while Dakahlia produced the lowest value. Damietta location was superior to other locations in fiber strength, fiber maturity, yarn strength and yarn elongation of both spinning system (ring and compact), while Dakahlia location surpassed the other locations in fiber length. Kafr El-Sheikh location ranked second in the superiority of fiber and yarn traits in all governorates. The long stable promising strain 10,229 × Giza 86 surpassed variety Giza 86 in seed cotton and lint yields, fiber strength, degree of maturity and yarn strength of both ring and compact spinning. The extra-long stable G84 (G70 × 51b) × P 62 and Giza 88 showed superiority in fiber length followed by strain Giza77 × Pima S6. Mudada et al (2017) found that performance of the individual genotypes and locations was highly significant. LS9219 showed the earliest maturity index of 78.41%, whilst CRIMS2 had the lowest maturity index of 66.41%. Variety 917/5/7 exhibited the least score and was found to be generally more stable across the two locations. Additional cotton yield was picked at Chitekete than Kadoma in the first picking of mature seed cotton. Variety LS9219 revealed the earliest maturity index of 78.41%, however CRIMS2 was the lowest maturity index of 66.41%. This indicated that there was more heat units for the crop per day in Chitekete than in Kadoma. Evaluated cultivars 644/98/01 and 648/01/04 ranked 1 and 3 at Chitekete yet 10 and 6 at Kadoma.

Assessment the participating five new selected genotypes in the three cotton growing region of Benin and compare their performance with the commercial variety was done by Djaboutou et al. (2017). The experiment was carried out in three cotton growing regions in farmers' fields during two years 2011 and 2012. The results showed that environments Savalou and Kandi have more favorable conditions for cotton seed yield in respect to all genotypes. Genotypes Djougou 8/5, H-279–1, Kandi ¾ and Okpara 3/5 were classified as adapted to the environments of Kandi and Savalou and found to be more stable in the production of average bolls weight and cotton seeds yield. Hence, the previous genotypes can be exploited as breeding stock to be incorporated in crosses for improving the aforementioned traits.

Finally, performance of three Egyptian cotton varieties as affected by drought stress at three irrigation regimes; no stress 14 (S-0), 21(S-1) and 28 (S-2) days that were started after the first irrigation at El-Fayoum, Egypt. Dewdar (2019) showed that mean squares of combined data due to irrigation regimes were highly significant for earliness traits, yield and its components. Most of fiber properties did not affect by water stress conditions. Significant differences were found between no stress (S-0) and the stress treatments (S-1 and S-2) on the earliness traits. Treatment S-2 caused significant reduction in yield and yield components rather than S-0. Giza 85 variety gave the highest fiber length, fiber strength and were finer cultivars giving the lower micronaire values. The interaction between genotypes and stress treatments was significant for most tried traits. Giza 90, Giza 83 and Giza 85 cotton varieties displayed highest seed cotton yield under non- stress environment (S-0). Shaker et al. (2020) showed that Egyptian cotton varieties Giza 97, Giza 95 and Giza 94 are response to late sowing date also, Giza 96 is somewhat response tolerant but, it had few seed cotton yield to both sowing dates. Fiber quality characters were less affected by late sowing date as stress environment.

9.4 Adaptability and Yield Stability

A combination of two or more abiotic stresses, such as drought heat and salinity results in more yield loss than a solitary stress. Assessment of stability and variability is very important for breeders to choice of genotypes that have the high level of stability, yield and most of the economic traits. Also, the select of parents that have a high level of stability in the start of the breeding program is a very important step to the success in cotton improvement. So understanding the nature of genotype × environment interaction gave breeders the possibility to select the more efficient genotypes. Evaluation of the stability and adaptability of genotypes crosswise diverse environment circumstances is important for release and recommendation of new cultivars with high adaptability. Plant breeders assess genotypes in multi-environment experiments to evaluate the stability and adaptability of genotypes and to recommend the genotypes to diverse environments (Maleia et al. 2017). Stability parameters helps to understand the response of the genotypes to different environmental conditions and to identify the more closely genotypes adaptable with environmental changes. Mudada et al. (2017) showed that cotton yield and fiber quality parameters are reliant on the environment in which the crop is grown. The identification of cultivars with high degree of adaptability and stability is one of the best methods to cope this challenge.

Under Egyptian conditions, for evaluation heat stress and other stress environments, eleven cotton varieties (*Gossypium barbadense* L.) were evaluated under eighteen environments. Awaad and Nassar (2001) showed that stability parameters indicated that the new local cultivar Giza 89 proved to be more stable for seed cotton yield, boll weight and length uniformity %, while the Russian cotton genotype Karshenscki-2 was the best for earliness percentage and fiber fineness. Also, high degree of adaptability (b < 1) for improved environments were detected for Giza 76 in seed cotton yield, lint percentage, boll weight, earliness %, 2.5% and 50% span length, length uniformity% and fiber elongation as well as Giza 85 in seed cotton yield, lint %, earliness % and fiber elongation. Whereas, cotton genotypes Ashmouni and Dandra recorded (b > 1) for seed cotton yield, seed oil content, boll weight, earliness%; Giza 75 for seed cotton yield, lint percentage, earliness%, 50% span length, fiber fineness, fiber strength and fiber elongation as well as Giza 77 for seed cotton yield, boll weight, earliness%, 2.5% span length, length uniformity% and fiber strength. Therefore, they could be grown under late planting date stress and

Khattara region as less favorable environments. Thus the above mentioned genotypes may be used as a breeding stock for incorporating in any cross for producing new recombinants more stable with high yield and fiber properties (Tables 9.2 and 9.3 and Figs. 9.2, 9.3, 9.4 and 9.5). Rank correlation results showed that selection for stability in lint percentage is possible through selection for stability in earliness percentage. Also, selection for stability in fiber elongation percentage could be achieved through selection for stability in lint and earliness percentages. In continuous to improve stable cotton genotypes characterized by length uniformity percentage and fiber fineness, selection could be practiced for stability in 2.5% span length (Tables 9.4 and 9.5).

Under Ras Suder as salinity condition, Moustafa (2006) showed that phenotypic stability parameters indicated that Egyptian cotton genotypes Giza 45 and Giza 80 were classified as highly adapted to salinity environment for seed cotton yield/fed. Whereas, cotton genotype Ashmouni attained regression coefficient value deviated significantly from unity (b < 1) indicating higher production potential in favorable environments. The most desired and stable genotypes were Giza 83 followed by Dandara. Based on genotypic stability, the most stable genotypes were Giza 83, Giza 85, Giza 86 and Dandara however, Giza 45, Giza 70, Giza 80 and Giza 89 were unstable (Table 9.6).

Results of lint yield/fed. based on phenotypic stability parameters showed that cotton genotypes Giza 45, Giza 70, Giza 80 and Giza 89 were categorized as highly adapted to salinity condition. Whereas cotton genotype Ashmouni could be grown under improved environments. The most desired and stable genotypes were Dandara and karashensekki-2. Based on genotypic stability, Giza 86 showed above average stability, while Giza 85 and Giza 86 was nearly perfect tability (Table 9.7).

Cotton genotypes responded differently to changes in environments. So, breeding genotypes with wide adaptability has long been a worldwide aim to the plant breeders. Dewdar (2013) showed that Giza 86 was stable for seed cotton yield, No. of open bolls, and lint index. But, Giza 88 was stable only for boll weight. Both genotypes Giza 90 and Giza 80 are expressed as ideal stable ones for improving yield traits. Whereas, Abd El-Rhman (2016) sowed that the promising extra-long stable strain Giza 84 (Giza 70 × 51B) P62 had the highest seed and lint cotton yields, regression coefficient equals to one and deviation from the regression line did not significant deviate from zero, so it is characterized by good stability and convenience for all environment. The strain 10,229 × Giza 86 (long stable category) had the highest seed and lint yield and adaptability to different environments. Therefore, these two promising stains are recommended to be developed as new elite cultivars.

Furthermore, among 40 genotypes of cotton evaluated under various environments in Golestan province of Iran, Sadabadi et al. (2018) recorded high degree of adaptability for improved environments for G 17, G 18, G 29, G 34 and G 37 as they exhibited (b < 1) in seed cotton yield. Whereas, cotton genotypes G1, G 4 and G10 recorded (b > 1) for seed cotton yield. Moreover, G 8 was classified as ideal stable genotype. Meanwhile, Shaker et al. (2020) revealed that Egyptian cotton varieties Giza 97, Giza 95 and Giza 94 are average stable and favorable to late sowing date for seed cotton yield according to Eberhart and Russiell and GGE- Biplot analyses. However, under semiarid region of the Northeast of Brazil, Silva et al. (2020) showed

Table 9.2 Stability parameters for seed cotton yield and some economic characters of eleven cotton cultivars grown under 18 environments (Awaad and Nassar 2001)

Genotype	Seed cotton yield (Kentars/fad)			Lint percentage			Boll weight (gm)			Seed oil content%			Earliness percentage		
	$\overline{X}$	bi	S^2di	$\overline{X}$	bi	S^2di	$\overline{X}$	bi	S^2di	$\overline{X}$	bi	S^2di	$\overline{X}$	bi	S^2di
1-Ashmouni	6.123	0.805	0.643**	36.362	0.751	2.695	2.367	0.855	0.40	21.150	0.605	0.208	61.871	0.032	2.372
2-Dandara	6.745	0.811*	0.424	37.621	1.002	11.049**	2.346	0.849	0.009	23.360	0.953	0.129	67.872	0.871	4.313
3-Giza 45	5.607	1.060	1.006**	34.204	0.959	8.68**	2.164	1.375*	0.075**	25.231	3.54**	0.157	65.370	1.037	21.148**
4-Giza 70	6.329	1.086	0.548**	36.197	0.823	3.131**	2.316	1.302*	0.086	20.329	3.955**	0.113	62.989	1.193	14.485**
5-Giza 75	7.414	0.912	0.378	36.791	0.695	8.843**	2.584	0.509	0.025	21.950	2.141*	0.107	68.931	1.212	35.316**
6-Giza 76	6.200	1.278*	0.236	38.113	1.695*	2.584	2.476	1.291	0.039	25.018	0.460	0.115	68.685	1.332*	29.522**
7-Giza 77	7.018	0.873*	0.365	37.464	1.896*	2.795	2.274	0.862	0.004	22.850	1.957*	0.215	67.106	1.303	7.568**
8-Giza 80	6.887	0.991*	0.346	38.218	0.589	2.899	2.136	1.207	0.034	23.625	0.815	0.133	68.741	0.826	3.792
9-Giza 85	7.525	1.203	0.424	37.419	1.830*	2.868	2.389	0.626	0.068**	19.889	0.627	0.186	68.944	1.335*	2.290
10-Giza 89	7.090	1.062	0.346	36.354	0.516	6.143**	2.304	0.961	0.042	21.567	0.579	0.107	65.935	0.289	4.709
11-Karshenseki-2	6.662	1.043	1.049**	36.362	0.239	4.272**	2.69	0.851	0.037	19.157	2.224*	0.141	70.159	0.980	4.234
General mean	6.669			36.855			2.329			22.193			66.964		
L.S.D$_{0.05}$	0.606			1.278			0.114			0.283			2.853		

*, ** significant at 5 and 1% levels of probability, respectively

Table 9.3 Stability parameters for fiber properties of eleven cotton cultivars grown under 18 environments (Awaad and Nassar 2001)

Genotype	2.5% span length (mm)			50% span length (mm)			Length uniformity (U%)			Fiber fineness Mic. Reading			Fiber strength gm/tex			Fiber elongation		
	$\bar{X}$	Bi	S²di	$\bar{X}$	bi	S²di	$\bar{X}$	bi	S²di	$\bar{X}$	bi	S²di	$\bar{X}$	bi	S²di	$\bar{X}$	bi	S²di
1-Ashmouni	32.328	0.115	1.68**	16.19	0.608	0.399**	49.860	1.840	1.624*	3.726	0.057	0.008	31.302	0.294	0.095	7.441	0.207	0.15*
2-Dandara	29.344	1.039	0.301	15.212	1.081	0.035	51.840	0.621	1.002	3.835	2.041**	0.046**	30.851	1.326	0.212	7.079	1.626	0.489**
3-Giza 45	34.554	2.956**	1.144**	17.228	1.398*	0.497**	47.858	2.774*	1.775**	3.307	0.414	0.021	36.719	4.481**	1.858**	5.681	0.143	0.028
4-Giza 70	34.500	0.534	0.923**	17.185	2.577*	0.224	49.812	0.741	7.167**	3.206	1.512	0.019*	38.394	3.540**	1.035	2.586	0.401	0.113
5-Giza 75	31.120	2.926**	0.456*	15.106	0.472	0.315	48.541	1.422	3.330**	3.883	0.713	0.002	33.908	0.892	1.428**	6.505	0.405	0.026
6-Giza 76	33.943	2.725**	0.358	16.963	1.495*	0.054	49.975	1.930	2.526**	3.165	0.636	0.013	37.398	0.073	0.187	5.13	1.471	0.038
7-Giza 77	34.294	0.688	0.332	17.629	4.004**	0.194	51.415	0.857	1.353	3.339	1.700*	0.074**	37.140	0.164	0.317	5.049	4.016**	0.184**
8-Giza 80	29.451	0.121	0.256	15.239	3.170**	0.226	51.743	0.620	1.481	3.887	2.520**	0.032**	31.778	0.172	0.181	7.244	1.372	0.097
9-Giza 85	30.743	0.531	0.542**	15.132	0.005	0.083	49.220	0.429	0.104	3.700	0.078	0.004	31.628	0.639	0.052	8.000	1.411	0.069
10-Giza 89	31.659	1.657	1.142**	16.328	1.219	0.012**	51.575	1.085	1.058	3.900	0.558	0.008	31.989	1.205	0.019	7.169	0.195	0.177**
11-Karshenseki-2	32.417	2.296*	2.202	15.982	1.207	0.055**	49.301	3.355*	4.809**	3.857	1.025	0.002	33.507	0.0208	0.368*	7.302	0.385	0.084
General mean	32.214			16.193			50.285			3.617			34.056			6.738		
L.S.D₀.₀₅	0.695			0.504			1.381			0.1147			0.544			0.292		

*, ** significant at 5 and 1% levels of probability, respectively

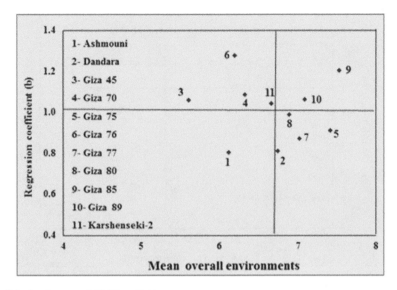

Fig. 9.2 Seed cotton yield (Kent./fad.)

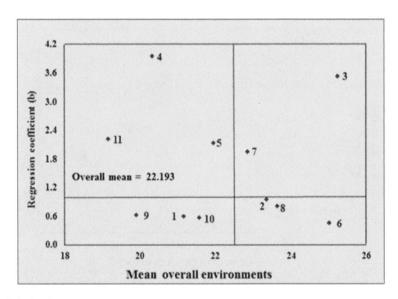

Fig. 9.3 Seed oil content (%)

that according to the stability analysis, Brazilian cotton genotypes CNPA BA 2011–4436, CNPA BA 2011–1197, and CNPA BA 2010–1174 were stable and presented great adaptability to the environmental changes.

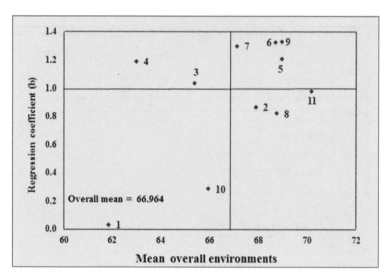

Fig. 9.4 Earliness percentage (%)

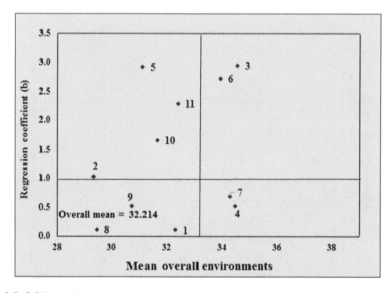

Fig. 9.5 **2.5**% span length

9.5   Additive Main Effects and Multiplicative Interaction

The additive main effects and multiplicative interaction (AMMI) model has appeared as a powerful analysis tool for genotype × environment studies. Thus, AMMI genotypic scores could provide an adequate way of resolving the magnitude and nature

Table 9.4 Rank correlations between regression coefficients stability parameters for yield characteristics in eleven cotton cultivars (Awaad and Nassar 2001)

Character	Lint percentage	Boll weight	Seed oil percentage	Earliness percentage
Seed cotton yield	0.118	0.414	−0.118	0.527
Lint percentage		0.050	−0.100	0.673*
Boll weight			0.332	0.105
Seed oil percentage				0.118

*, ** Significant at 0.05 and 0.01 levels of probability, respectively

Table 9.5 Rank correlations between regression coefficients stability parameter for some yield characters and fiber properties in eleven cotton cultivars (Awaad and Nassar 2001)

Character	2.5% span length	50% span length	Length uniformity (%)	Fiber fineness	Fiber strength	Fiber elongation (%)
Seed cotton yield	0.255	0.109	0.098	−0.245	0.045	−0.091
Lint %	0.073	0.109	−0.239	0.055	−0.227	0.782**
Boll weight	0.405	0.677	0.439	0.159	0.386	−0.114
Earliness %	0.282	0.009	−0.273	−0.036	−0.464	0.490
2.5% span length		0.036	0.634*	0.809**	0.36	−0.345
50% span length			0.052	0.564	−0.064	0.236
Length uniformity%				−0.275	−0.102	−0.343
Fiber fineness					0.100	0.418
Fiber strength						−0.400

*, ** Significant at 0.05 and 0.01 levels of probability, respectively

of genotype × environment interactions. Seed cotton yield is a feature controlled by polygenes that cause changes in the performance of genotypes based on the cultivation environment. Breeding programs test the genotype × environment interaction (GE) using accurate statistical procedures, like AMMI (additive main effects and multiplicative interaction) and GGE biplot (genotype main effects + genotype × environment interaction). The AMMI method combines the analysis of variance and principal components, to adjust the main effects (genotypes and environments) and the effects of GE interaction, respectively. The GGE biplot groups the genotype additive effect together with the multiplicative effect of the GE interaction, and submits both of these to the principal components analysis (Farias et al. 2016). They assessed the association between the AMMI and GGE biplot methods and select cotton genotypes that simultaneously showed high productivity of seed cotton yield

Table 9.6 Means, phenotypic and genotypic stability parameters of seed cotton yield/fed of ten cotton varieties under eight environments (Moustafa 2006)

Variety	Seed cotton yield/fed				
	Phenotypic stability			Genotypic stability	
	$\overline{X}$	b	S^2d	α	λ
1-Giza 45	219.8	0.894*	306.08	−1.762	4.15
2-Giza70	200.9	0.939	991.16*	−1.005	6.28
3-Giza80	201.7	0.874*	1349.65**	−2.046	7.80
4-Giza83	294.5	1.099	614.41	1.651	7.64
5-Giza85	206.2	0.970	718.69*	−0.551	1.55
6-Giza86	215.5	0.990	658.98	−0.171	0.26
7-Giza89	237.4	0.954	254.75	−0.753	3.36
8-Dandara	265.2	1.039	335.65	0.649	2.59
9-Ashmouni	382.4	1.197*	2255.64**	3.266	7.60
10-karashensekki-2	235.6	1.043	785.88*	0.720	3.03
General means	235.99				
LSD 5%	33.40				

Table 9.7 Means, phenotypic and genotypic stability parameters of lint yield/fed of ten cotton varieties under eight environments (Moustafa 2006)

Variety	Lint yield/fed				
	Phenotypic stability			Genotypic stability	
	$\overline{X}$	B	S^2d	α	λ
1-Giza 45	131.195	0.086**	7689.65**	− 6.519	7.520
2-Giza70	120.804	0.890*	323.03	− 4.979	5.330
3-Giza80	117.965	0.884*	341.16	− 5.236	6.772
4-Giza83	196.080	1.111	1069.14**	4.981	5.508
5-Giza85	125.620	0.962	416.70	− 1.721	0.892
6-Giza86	123.250	0.989	81.24	− 0.496	0.614
7-Giza89	137.840	0.878*	261.04	− 5.518	6.971
8-Dandara	175.110	1.077	117.33	3.446	5.500
9-Ashmouni	185.260	1.302**	934.47*	3.617	5.614
10-karashensekki-2	149.390	1.054	74.07	2.425	2.203
General mean	146.252				
LSD 5%	25.27				

*, ** Significant at $P < 0.05$ and $P < 0.01$ respectively

and stability in Mato Grosso environments. Trials were conducted with cotton cultivars in eight environments across Mato Grosso State, Brazil in the 2008/2009 crop
season. Results for seeds cotton productivity according to AMMI and GGE biplot
showed that both methods were compatible in distinguishing between environments
and genotypes for phenotypic stability. The genotypes BRS ARAÇÁ and LD 05 CV
exhibited high seed cotton productivity and phenotypic stability, and could be grown
under all environments across Mato Grosso State.

Additive main effects and multiplicative interaction of six cotton genotypes based
on three cotton growing regions in farmers' fields during two years 2011 and 2012 are
computed by Djaboutou et al. (2017). Effects of AMMI analysis of variance reveal
the main effect means on the abscissa and principal component PC scores of six
genotypes as well as three environments based on the boll average weight and cotton
seed yield indicated significant effects for genotypes, but the effects of environments
and G × E interaction did not reach the level of significance. For cotton seed yield,
the AMMI analysis of variance revealed high PC1 and PC2 scores for the genotypes
and environments. The results from analysis of multiplicative effects indicated that
PC1 scores varied from −16.893 to + 10.591 among genotypes and from −11.766 to
+ 19.393 for the environments. PC2 scores varied from −13.01 to + 13.128 among
the six genotypes and from −14.447 to +16.66 for the three environments. Three
genotypes i.e. G2, G3 and G4 as well as single environment E1 displayed positive
PC1 scores, whereas the genotypes G5, G6 and environment E2 exhibited negative
PC1 scores. Regarding PC2, the genotypes G1, G2 and environment E1 revealed
high positive scores. The highest mean cotton seed yield response was shown in
environments E2 and E3 than grand mean yield. Whereas, environment E1 display
cotton seed yield response smaller than the grand mean yield, but the genotype G1
and genotype G2 appear to be adapted to the environment E1.

Maleia et al. (2017) evaluated 11 genotypes of cotton during 3 seasons in 3 locations in Mozambique. The AMMI analysis model showed significant influence of
genotype, environment and G × E interaction. The first two principal components
described about 80% of the interaction. Genotypes FK 37, BA 919 and Flash were the
most adaptable, while BA 2018 and BA 320 were the most stable across the variation
of environments. Genotypes G3 (BA 919) and G4 (Flash) were grouped together on
the biplot of PC1 against PC2 (Fig. 9.6). The biplot graphic also revealed that there
are 4 mega-environments: two main ones represented by 2 environments (E3; E6,
in the 2nd quadrant and E4; E5, in the 3rd quadrant), while the 2 minor ones were
represented by 1 environment (E1, in the 1st quadrant and E2 in the 2nd quadrant).
The genotypes displayed a different genetic performance once they were positioned
in opposing quadrants. Varieties BA919, Flash, FK37 performed well and differently
rather than the most of the check varieties (Churedza, CA324 and ISA 205). The pair
of environments comprising Namialo 2014 (E3)/Nhamatanda 2014 (E6) and Balama
2014 (E5)/ Balama 2012 (E4) was similar and appropriate for BA320 (G2), QM 301
(G7), Albar SZ9314 (G9) and BA2018 (G1), Churedza (G8), CA324 (G10) and ISA
205 (G11), respectively they gathered into the same quadrant.

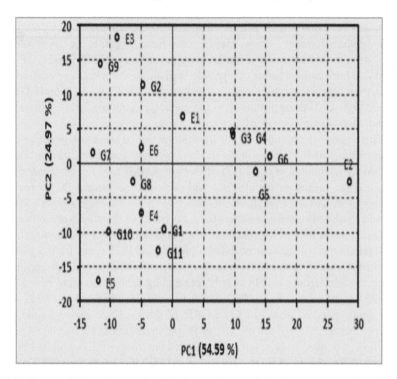

Fig. 9.6 Graphics biplot of PC1 against PC2 for seed cotton yield of e genotypes (G1 to G11) in 6 environments (E1 to E6). G1 genotype BA2018; G2: genotype BA320; G3: genotype BA919; G4: genotype Flash; G5: genotype BA525; G6: genotype FK37; G7 genotype QM 301; G8: genotype Chureza; G9: genotype Albar SZ9314; G10: genotype CA324; G11: genotype ISA205 (Maleia et al. 2017)

Mudada et al. (2017) investigated genotype × Environmental interaction on *Gosypium hirsutum* of ten genotypes. Additive Main Effect and Multiplicative Interaction model showed significant genotype × environment interactions for lint yield, boll weights, staple length and fuzzy seed grade. However, cultivar × location interactions were insignificant on total seed cotton yield. Cotton varieties differ in their response to different growth conditions, which means that the right choice of varieties by farmers for specific production conditions, is important to avoid losses caused by the interaction between the genetic makeup and the environment. They showed that both cultivars 644/98/01 and 648/01/04 were less stability across two locations Chitekete and Kadoma in Zimbabwe. Adaptation of the test varieties was more inclined to Chitekete than Kadoma location. CRIMS2 was more stable for earliest maturity index over the two locations. CRIMS2 ranked 9 for both locations and its maturity values were rated second on the mean separation. Added that the varieties were more adapted to Kadoma than Chitekete for seed cotton yield. Cotton varieties CRIMS2 and 644/98/01 were classified as the most stable ones. However, genotype 644/98/01 was more adapted to Kadoma environment, also characterized

Table 9.8 AMMI-1 Analysis of variance and decomposition of degrees of freedom for total seed cotton yields of cotton genotypes at two sites in 2011–12 (Mudada et al. 2017)

Source	Degree of freedom	Sums of squares	Means squares
Total	59	38,582,932	
Treatments	19	28,590,093	1,504,742**
Genotypes	9	9,856,870	1,095,208**
Environments	1	16,468,452	16,468,452**
Block	4	212,766	53,191[ns]
Interaction (G × E)	9	2,264,771	251,641[ns]
IPCA (1)	9	2,264,771	251,641[ns]
Residual	0	0	*
Error	36	9,780,074	

*, ** Significant at $P < 0.05$ and $P < 0.01$ respectively; ns: not significant

by excellent in stability and yield performance, otherwise, BC853 was less adapted to environments. IPCA scores of genotype BC853 were very far away from the origin displaying its less stability across both locations. The AMMI-1 analysis model showed that differences between the environments represents (57.6%) of total sum of squares (SS) followed by genotypes (34.5%) and then GE interaction (7.9%). The first interaction PCA was insignificant. The AMMI method recognized 648/01/4 and CRIMS2 as good performing cotton genotypes with positive interaction on total seed cotton yield (Table 9.8). Moreover, Shaker et al. (2019) showed that the biplot analysis cleared that two environments, Beni-Souif and EL-Sharkeia with Egyptian cotton lines No.1 and 2 as the winning genotypes and the lines No. 4 and 5 as the winning in the two environments EL- Sharkeia and Kafr El-Sheikh. Lines No. 1 and 2 were the most stable, the highest yielding and closest to the ideal genotypes. Line No. 4 had average stable with favorable yield, however line No. 5 had high average mean, but it more variable in environments. Lines No. 1, 2 and 4 were average stable according to AMMI method generally, the two lines No. 1 and 2 should be selected as the best and ideal genotypes. Also, the best environments were EL- Sharkeia and Beni-Souif.

9.6 Gene Action, Genetic Behavior and Heritability for Cotton Traits Related to Environmental Stress Tolerance

Across eighteen diverse environments, Awaad and Nassar (2001) recorded low heritability estimates in broad sense (>50%) for seed cotton yield, lint percentage, earliness percentage; moderate (< 50% and >75%) for boll weight, seed oil content, 50%

span length, length uniformity, fiber fineness as well as high (equal or < 75%) for 2.5% span length, fiber strength and fiber elongation (Table 9.9). At irrigation water salinity 2000 ppm under Ras suder, South Sinai, Egypt Moustafe (2006) evaluated six crosses and found that the additive gene action played an important role in the inheritance number of fruiting branches/plant, concentration of magnesium, sodium and proline in most studied crosses. Otherwise, the dominance gene effects and its digenic type play the major role in the genetics of seed cotton yield, number of open bolls/plants and boll weight in almost crosses. Narrow sense heritability ranged from low to moderate for the studied traits.

Whereas, Abdelraheem et al. (2015a) showed that heritability estimates under stress treatment were consistently lower than those under the control one. The broad-sense heritabilities for plant height, fresh shoot weight and root weight were 0.795 and 0.947, 0.708 and 0.913, and 0.643 and 0.913 in Test 1 and Test 2 under the control conditions, respectively, and 0.741 and 0.751, 0.559 and 0.719, and 0.575 and 0.791 in Test 1 and Test 2 under the PEG conditions, respectively. Abid et al (2016) evaluated eight parents and their F_1 crosses under control "100% field capacity" and drought "50% FC" conditions. The biochemical traits i.e. total soluble proteins, chlorophyll a, chlorophyll b, total chlorophyll, carotenoids, total phenolic contents and enzymatic antioxidants (superoxide dismutase, peroxidase and catalase) seemed to be governed by non-additive gene action with low narrow sense heritability assessments. Both general combining ability and specific combining ability were significant for all biochemical traits under control and drought conditions.

Most studies indicate that drought tolerance in *G. hirsutum* is governed by additive, dominant and epistatic gene action in early generations. Abdelraheem (2006) performed a diallel analysis among nine Egyptian cotton cultivars under well-watered conditions and drought stress conditions. Under well-watered conditions, a simple additive model was appropriate to describe the inheritance of seed cotton yield, lint yield, boll weight, number of bolls/plant, lint percent and lint and seed indices. However dominance gene action was greater than additive one under drought stress conditions in clay and sandy soils. Similar results were noticed for salt tolerance in Upland cotton by Gholamhossein and Thengane (2007) showed the involvement of both additive and non-additive gene effects in the genetics of salinity related characters and high heritability in narrow-sense estimates were detected for root length, plant height root length/shoot height and tolerance index. In the same time, Ahmed (2007) utilized crosses from four Egyptian cultivars and one Upland cotton to estimate gene action under drought situations. The non-additive effect was greater than the additive ones for all traits excluding for days to first flower. Nabi et al. (2010) advising that additive and dominant gene effects were involved in the genetics of seed cotton yield, boll weight and number of bolls/plant with great role of additive gene action under low and high salinity levels. Estimates of narrow sense heritabilities for previous characters were high. Whereas, Soomro et al. (2012) stated that mean squares of general combing ability (GCA) attributed to additive effects were higher than specific combing ability (SCA) due to non-additive effects for plant height, leaf area, leaf dry weight, number of bolls per plant and seed cotton yield under drought and non-drought regimes.

Table 9.9 Heritability in broad sense (T_b%) for seed cotton yield, some economic characters and fiber properties (Awaad and Nassar (2001)

Parameter	Seed cotton yield	Lint percentage	Boll weight	Seed oil content	Earliness percentage	2.5% span length	50% span length	Length uniformity (%)	Fiber fineness	Fiber strength (g/tex)	Fiber elongation (%)
Hb%	26.37	23.53	55.72	69.74	46.74	82.8	66.31	62.19	72.57	92.31	96.45

Table 9.10 Salinity sensitivity index of seed cotton yield/fed and lint yield/fed based on control (2000 ppm) and salinity stress (80000 ppm) treatments of water irrigation for ten cotton varieties under Rus Sudr condition (Moustafa 2006)

Variety	Seed cotton yield/fed. (Kg)		Lint yield/fed. (Kg)	
	2001	2002	2001	2002
1-Giza 45	1.05	1.02	1.04	1.02
2-Giza70	0.95	1.01	0.95	1.02
3-Giza 80	0.97	0.99	0.98	1.03
4-Giza 83	0.95	082	0.97	0.79
5-Giza 85	0.92	1.00	0.92	1.02
6-Giza 86	1.08	1.08	1.07	1.08
7-Giza 89	0.98	0.99	0.97	0.98
8-Dandara	0.97	0.95	0.98	0.92
9-Ashmouni	1.05	1.04	1.05	1.04
10-karashensekki-2	1.07	1.06	1.06	1.08

Moreover, Nasimi et al. (2016) exploited four drought-tolerant and four sensitive lines in a diallel analysis, differed in their root length. Results revealed the presence of additive and non-additive effects, with the predominant of additive in controlling fiber quality traits i.e., fiber length, strength and fineness with moderate to high narrow-sense heritability estimates for the characters under drought conditions. Furthermore, thirteen cotton varieties were evaluated under both moisture stress and non-stress field environments. Koleva and Dimitrova (2018) showed that genetic gain from indirect selection under moisture stress environment would improve yield better than selection from non-moisture stress environment. The varieties Viki and Avangard-264 had the highest yields under non-stress conditions. Vega and Chirpan-539 varieties exhibited a low yield potential and revealed a high stress tolerance to drought.

9.7   Role of Recent Approaches?

9.7.1   Biotechnology

In the latest 10–15 years, significant progress has been done in understanding the genetic basis of drought, heat and salt tolerance through molecular marker technologies.

Evaluation of genetic markers and genetic diversity is an important part of any successful breeding program. Several molecular marker technologies have emerged in recent years and have been successfully used in the field of cotton improvement. DNA based molecular markers such as RFLP, AFLP, SSR, ESTs, SNP and RAPD have been extensively utilized in genetic and breeding analyses and examinations of genetic diversity and the relationship between crop genotypes. The RAPD markers were applied in cotton for the assessment of genetic diversity and fingerprinting cotton genotypes (Hussein et al. 2006 and 2007; Muhammad et al. 2009; Zahid et al. 2009) as well as for the detection of variation between closely related cultivars (Rahman et al. 2002; El-Defrawy et al. 2004; Masoud et al. 2007).

The RAPD method displayed substantial possible for recognizing and differentiating cotton genotypes. Abdel-Fattah (2010) utilized RAPD analysis of nine Egyptian cotton varieties (*Gossypium barbadense* L.), namely, Dandara, Giza 75, Giza 83, Giza 85, Giza 86, Giza 88, Giza 89, Giza 90 and Giza 91. Results showed that 8 out of the 10 tried primers exhibited polymorphism among the evaluated varieties, while 2 primers were monomorphic. 36.4% out of 151 bands amplified were polymorphic. They identified unique DNA fragments with diverse sizes in Giza 75, Giza 85, Giza 86, Giza 88, Giza 90 and Giza 91 varieties, while the unique markers were not detected for Dandara, Giza 83 and Giza 89. The genetic similarity among the nine cotton varieties was high, and ranged from 87.3 to 96.1%. The highest similarity was evidenced between Giza-89 and Giza-75. Whereas, the lowest genetic similarity was registered between Giza 85 and Giza 86.

Abdelraheem (2017a) assessed a total of 376 Upland cotton genotypes to abiotic and biotic stress tolerances using 26, 301 polymorphic SNPs from the CottonSNP63 K array to recognize abiotic stress tolerance QTL. Major QTL were discovered for abiotic and biotic stress tolerance. Main-effect QTL were discovered on the basis of weights of dry shoot and root for drought on chromosomes c8, c15, c21, c24, c25, and c26, and for salt tolerance on c1, c9, c11, c12, c13, c14, c18, c21, and c24. Abdelraheem et al. (2017b) completed analysis to identify QTL clusters and hotspots among 661 abiotic and biotic stress tolerances QTL for physiological responses, yield and its components, also fiber quality characters at abioitc and biotic stress circumstances. The distribution of tolerance QTL are not uniform across the cotton genome, where 23 QTL clusters were recognized on 15 diverse chromosomes (c3, c4, c5, c6, c7, c11, c14, c15, c16, c19, c20, c23, c24, c25, and c26) for tolerance to stress. Moreover, 28 QTL hotspots were known for various stress tolerance characters. Also, two QTL were recognized on chromosome c24 for chlorophyll content at drought and salt tolerance.

Several markers have been developed that include candidate genes responsible for abiotic stress (Tiwari et al. 2013; Abdelraheem et al. 2015b; Rodriguez-Uribe et al. 2014). Transgenic crop development with desired traits like biotic and abiotic stress resistance has become a reality due to the development of genetically modified (GM) crop technology (Gasser and Fraley 1992). Gene transfer in crop plants was succeeded by a several methods such as *Agrobacterium* mediated transformation, direct gene transfer by imbibition and biolistic transformation, chemical method, microinjection and pollen tube pathway, liposome method, shoot apex method of transformation, infiltration, and silicon carbide mediated transformation (Rao et al 2009).

In Egyptian cotton, Momtaz et al. (2010) aimed to increasing levels of polyamine accumulation via expressing S-adenosyl methionine decarboxylase (SAMDC) gene. SAMDC cDNA isolated from *Saccharomyces cerevisiae* isolate was isolated and genetically engineered into Egyptian cotton cultivars Giza 88 as an extra-long staple and Giza 90 as a long staple through particle bombardment by meristem transformation, and screened T(0) transgenic plants. Control plants from Giza 88 and Giza 90 were exposed to drought regime by various concentrations of PEG6000 (2.5, 5, 7.5, 10, 12.5, 15, 17, and 20%) for 9 h to recognize drought stress symptoms and identify the potential concentration level for inducing polyamine accumulation. T(1) transgenic plants grown under induced drought stress regime were tested positive for gene integration and expression and tested with HPLC study to identify levels of spermine as polyamine accumulated in response to water stress regime. They compared elevated spermine accumulation in Egyptian cotton varieties Giza 88 and Giza 90 as non-transgenic plants at the same induced drought conditions with T(1) transgenic plants using reverse-phase HPLC analysis. They found that elevated spermine accumulation expressing SAMDC gene was main cause for enhancing water stress tolerance in both transgenic cultivars.

Hamoud et al. (2016) identified the genes expression related to drought stress tolerance in cotton. The gene flowering locus T-like protein 1 (FTL1) and the Gossypium heat shock protein 1 (GHSP1) were expressed in most genotypes with diverse degrees of intensity except in the pattern of genotype "Giza 85"under control and stress which propose that these genes clearly discriminate the drought sensitive cotton genotypes. No discriminate trend was achieved for the genes Lea4 D9a and Lea4 D9b, while, diverse band patterns were detected. The most affected genes with drought stress are FTL1and HSP1 and could be utilized to determine the drought sensitive cotton genotypes. For upland cotton, a new method of transgenic improvement called "In-planta" transformation method, were investigated by Kalbande and Patil (2016) using *Agrobacterium tumefaciens* EHA-105 harbouring recombinant binary vector plasmid pBinAR with Arabidopsis At-NPR1 gene. Transgenic events with At-NPR1 gene and promoter nptII gene were obtained. They recorded a transformation frequency of 6.89% for the cultivar LRK-516.).

Transcription factors play a significant role in plant reaction to abioitc stress. For example, when GhABF2, a bZIP transcription factor gene, was overexpressed in cotton, the transgenic genotypes showed drought and salt tolerance through regulation of ABA-related genes. Also GhABF2-overexpressing cotton genotypes demonstrated an enhanced yield rather than control plants (Liang et al. 2016). Moreover, Shen et al. (2015) showed that transgenic cotton genotypes with co-overexpressing AVP1 and AtNHX1gave 24 and 35% more fiber than non-transgenic genotypes under low irrigation and dryland circumstances, respectively. Recently, Mishra et al. (2017) confirmed that overexpression in cotton of the SUMO E3 ligase gene from rice, OsSIZ1, enhanced tolerance to heat and drought stresses, and increased significantly fiber yield in a dryland condition.

9.7.2 Nano-Technology

Nano-fertilizers play an significant role as favorable approaches for increasing water use efficiency and land resources and decreasing environmental pollutions. The biosynthesis of Nanomaterials using bacteria, algae, yeast, fungi, actinomycetes and plants has led to a novel area of research for eco-friendly fertilizers. Currently, use of Nano fertilizers in plant nutrition is one of the main roles of Nanotechnology in crop production. Henceforth, this leads to sustainable agriculture by placing fewer inputs, producing less waste, reducing nutrient loss and release nutrients at an appropriate rate to the plant compared to conventional agriculture (El-Ghamry et al. 2018).

In diverse fields, comprising agriculture, nanotechnology has great potential to make it at ease for the subsequent phase of accuracy farming approaches. The agricultural sector will be consuming more of nanotechnology in the future to attain greater yields in eco-friendly method even in challenging environment (Ali et al. 2014 and Prasad 2014).

The use of Nano fertilizers aims to improve the efficiency of nutrient use by exploiting the unique properties of molecules of Nano-particles with range of Nano-dimension from1 to 100 nm (Suppan 2017). Nanotechnology is minimizing the risk of environmental pollution (Manjunatha et al. 2016). In Nano fertilizers, particles are absorbed by crop plants quickly and completely, saving fertilizer consumption and reducing environmental pollution (DeRosa et al. 2010). Soaking of cotton seeds in Nano fertilizers gave favorable effects and reduced the quantity of fertilizers applied by half (Vakhrouchev and Golubchikov 2007).

Moreover, Sohair et al. (2019) evaluated NPK Nano-fertilizers application, times, methods and rates on yield and fiber properties of Egyptian cotton cultivar Giza 90. Treatments comprised two application times were used in main plots, two application methods are foliar and soil in sub-plots and four rates applications of control (100% soil application normal recommended NPK fertilizer dose (RFD) and Nano NPK fertilizers 12.5, 25 and 50% RFD) were applied in sub-sub-plots. They found significant increases in open bolls/plant, boll weight and seed cotton yield due to the application of three times NPK Nano-fertilizers compared to two times. Application of foliar Nano-fertilizers on the previous traits gave greater values than soil application. However, the maximum values of former traits were attained with 50% RFD Nano NPK with split 3 times and applied by foliar application in most cases. Application rates of NPK Nano fertilizers increased significantly fiber properties.

From another point of view, Nanoparticles pesticide is successful in reducing pest resistance to pesticides and reducing environmental pollution. In this respect, Ahmed et al. (2019) developed a new synthetic structure to produce pesticide Nano composite of very high effectiveness rather than the original ones. They used silver nanoparticles (AgNPS) as a pesticide carrier by loading the pyrethroid pesticide Lambda-cyhalothrin (L-CYN) on the surface of prepared AgNPS. They showed that silver lambda-cyhalothrin Nanocomposite approved to be more effective on cotton leaf worm larvae compared to the formulated lambda-cyhalothrin alone. The required concentration for control of cotton leaf worm decreased more than 37 times.

9.8   How Can Measure Sensitivity of Cotton Genotypes to Environmental Stress

9.8.1   Stress Sensitivity Measurements

The association between variation with differing genotypes or genotype × environment interactions showed that genotypes with low values of stress sensitivity index are recognized to be tolerant, because they exhibited lesser reductions in yield under stress environment. According to salinity sensitivity index of seed cotton yield and lint yield, Moustafa (2006) showed that the most tolerant cotton genotypes to salinity (SSI > 1) were Giza 85 and Giza 83, whereas the most sensitive ones (SSI < 1) were Giza 86 and karashensekki-2 while, the remaining cotton genotypes were moderate tolerant to salinity. However, based on drought sensitivity index, Hamoud et al. (2016) revealed that the Suvin, 10,229 and Giza 75 were the most drought tolerant. Meanwhile, the genotypes Giza 92, [Giza 84 × (Giza 70 × Giza 51B) × S 6] and Giza 93 were moderate tolerant. Whereas, Giza 85 was the most sensitive one.

On the basis of yield under both moisture stress and non-stress field environments, Koleva and Dimitrova (2018) evaluated thirteen cotton varieties using six drought tolerance indices i.e. stress sensitivity index, stress tolerance index, tolerance index, mean productivity, geometric mean productivity and mean harmonic productivity were used. Based on drought sensitivity index, the most tolerant cotton genotypes (DSI > 1) were Vega followed by Chirpan-539, Natalia, Nelina, Rumi. While, the most sensitive ones (DSI < 1) were Dorina, Viki and Colorit. Whereas, the remaining cotton genotypes were moderate ones. Furthermore, Dewdar (2019) showed that Giza 90 was relatively stress sensitive and took similar trend of those obtained using data of relative productivity (%). Whereas, genotypes Giza 83 and Giza 85 are more drought tolerance and could be used as sources of drought stress tolerance in cotton breeding programs.

9.9   Agricultural Practices to Mitigate Environmental Stress on Cotton

- The farmers can partially cope with the adverse effects of high temperature and drought on photosynthesis efficiency and avoid yield reduction through the following practices:
- Selecting cotton varieties more tolerant to environmental changes.
- Planting cotton at the appropriate time to avoid the effect of heat stress on physiological processes during the critical stages of plant life.
- Applying recommended agricultural management i.e. adequate fertilizer, planting population and irrigation are necessary to increase photosynthesis and mitigate

environmental stress effects and sustain crop production in cotton (Sawan 2014; Anonymous 2021).

– Application of putrescine and humic acid enhanced cotton plants under salt stress and increased number of open bolls, seed cotton yield/plant, lint% and seed index. Also, putrescine and humic acid increased chemical constitutes correlated to salt tolerance such as N, P and K, proline, total free amino acids, total sugars, total soluble phenols, chlorophyll a, b, total chlorophyll and total carotenoids (Ahmed et al. 2013).

Azza, El-Hendawy (2020) indicated that drought tolerance inducers (Humic acid, Potassium silicate and Potassium application) to Giza 94 cotton cultivar significantly increased plant height, number of fruiting branches/plant, open bolls/plant, boll weight, seed index, seed cotton yield/fed, fiber length and fiber strength as compared with untreated plants. The drought tolerance inducers could improve plant growth and yield particularly under water stress.

– Double-cropped cotton and wheat, Arkansas wheat producers have recently increased area due to higher market prices and value of wheat production. Similarly, increases in market prices for cotton have spurred producers' interest in considering double-cropping cotton after wheat harvest. Planting an early-maturing variety of wheat with plans to harvest by the end of May is crucial to ensure that cotton can be planted on time. The optimum cotton planting dates in Arkansas are from April 20 to May 20 most years. Only varieties that mature early and are determinant in growth characteristics should be planted following wheat due to the compressed growing season (Anonymous 2007). On the other hand, the planted area of Egyptian cotton and yield per unit area is decreasing due farmers delay the planting date for a month after March 30 to get complete winter crop especially wheat before cotton, and the Egyptian cotton cultivars were bred as a full season crop (180 days) grown from mid-March to mid-September. Consequently, breeding Egyptian cotton cultivars able to tolerate the environmental stress of late sowing is considered important in this regard. Cotton breeders advocated the need for considering the concept of double cropping of summer Egyptian cotton with winter crops like wheat to increase the amount of production, the economic return and achieve food security (Baker and Eldessouky 2019).

9.10 Conclusions

The adaptability and stability of cotton genotypes crosswise diverse environment circumstances are important for release and recommendation of new cultivars with high adaptability. Genetic markers and gene manipulation appeared to be more important of any successful breeding program in the field of cotton improvement. Nano-fertilizers play a significant role as favorable approaches for increasing water use efficiency and land resources and decreasing environmental pollutions. The farmers can partially cope with the adverse effects of high temperature, drought and salinity through applying recommended agricultural management i.e. tolerant cultivar, adequate fertilizer, irrigation and other ideal transactions.

9.11   Recommendations

The scientists interested in environmental studies recommend conducting adaptation and stability analyzes to distinguish between environments and genotypes. Also, it is important to assess the stability of cotton genotypes and their ability to adapt to the diverse environmental conditions intersecting their release and to recommend new varieties with high adaptability. Also, it is important to evaluation stability and adaptability of genotypes under diverse environmental conditions in order to release new cultivars with high adaptability. Genetic engineering studies recommend the development of several markers including candidate genes responsible for abiotic stress. This is in addition to the development of genetically modified cotton with desirable traits such as tolerance to biotic and abiotic stress, which enhances tolerance to stress conditions, and significantly improves fiber productivity in dry lands. This besides applying optimum agronomic management. Evaluation of the stability and adaptability of genotypes crosswise diverse environment circumstances is important for release and recommendation of new cultivars with high adaptability.

References

Abd El-Bary AM (1999) Inheritance of quantitative traits of Egyptian cotton (*G. barbadense* L) MSc thesis Fac of Agric, Mansoura Univ, Egypt

Abd El-Rhman ERE (2016) Effect of locations and growing seasons on performance and stability of some Egyptian cotton genotypes for agronomic, fiber and spinning quality traits. PhD thesis Fac of Agric, Ain Shams Univ, Egypt

Abdel-Fattah BE (2010) Genetic characterization and relationships among Egyptian cotton varieties as revealed by biochemical and molecular markers. Egyptian J Genetic Cytol 39(1):157–178

Abdel-Kader MA, Esmail AM, El Shouny KA, Ahmed MF (2015) Evaluation of the drought stress effects on cotton genotypes by using physiological and morphological traits. Int J Sci Res 4(11):1358–1362

Abdelraheem A (2006) Genetic analysis of drought tolerance in Egyptian Cotton (*Gossypium barbadense* L). MSc thesis. Agr Dept, Assiut University, Egypt

Abdelraheem A (2017a). Joint genetic linkage mapping and genome-wide association study of drought and salinity tolerance and verticillium wilt and thrips resistance in cotton. PhD dissertation. New Mexico State Univ, Las Cruces, NM, USA

Abdelraheem A, Liu F, Song M, Zhang JF (2017b) A meta-analysis of quantitative trait loci for abiotic and biotic stress resistance in tetraploid cotton. Mol Genet Genom 292(6):1221–1235

Abdelraheem A, Hughs SE, Jones DC, Zhang JF (2015a) Genetic analysis and quantitative trait locus mapping of PEG-induced osmotic stress in cotton. Plant Breed 134:110–120

Abdelraheem A, Mahdy Z, Zhang JF (2015b) The first linkage map for a recombinant inbred line population in cotton (*Gossypium barbadense*) and its use in studies of PEG induced dehydration tolerance. Euphytica 205:941–958

Abid MA, Malik W, Yasmeen A, Qayyum A, Zhang R, Liang C, Guo S, Ashraf J (2016) Mode of inheritance for biochemical traits in genetically engineered cotton under water stress AoB Plants 8:plw008

Ahmed AHH, Darwish E, Hamoda SAF, Alobaidy MG (2013) Effect of putrescine and humic acid on growth, yield and chemical composition of cotton plants grown under saline soil conditions. Am-Eurasian J Agric Environ Sci 13(4):479–497

Ahmed Kh, Mikhail WZA, Sobhy HM, Radwan EMM, El Din TS, Youssef A (2019) Effect of Lambda-Cyahalothrin as nanopesticide on cotton leafworm. Spodoptera littoralis (Boisd). Article 4, 62(7):1663–1675

Ahmed MF (2007) Cotton diallel cross analysis for some agronomic traits under normal and drought conditions and biochemical genetic markers for heterosis and combining ability. Egypt J Plant Breed 11(1):57–73

Ali MA, Rehman I, Iqbal A, Din S, Rao AQ et al (2014) Nanotechnology, a new frontier in agriculture. Adv Life Sci 1:129–138

Anderson CG (1999) Cotton marketing. In: Smith CW, Cothren JT (eds) Cotton: origin, history, technology and production. Wiley, New York, pp 659–679

Anonymous (2021). Recommendations Techniques in Cotton Cultivation. Agricultural Research Center, Giza, Egypt.

Anonymous (2007) Double-Cropped cotton and wheat. Agriculture and Natural Resources, FSA2163

Arafa AS, Nour ODM, Hassan ISM (2008) Impact of accumulated heat units (DD15'S) on the performance behavior of some Egyptian cotton cultivars. Egypt J Agric Res 86(5):1945–1956

Awaad HA, Nassar MAA (2001) Genotype × environment interactions for yield and fiber quality in cotton (*Gossypium barbadense,* L). J Adv Agric Res 6(2):337–361

Azza AME-H (2020) The effect of spraying with drought tolerance inducers on growth and productivity of cotton plants growing under water stress conditions in Delta Egypt. In: 16th international conference for crop science, Agron. Dept, Fac of Agric, Al-Azhar Univ, Egypt, Oct 2020 Abstract, p 61

Baker KMA, Eldessouky SEI (2019) Blend response of four Egyptian cotton population types for late planting stress tolerance. Bull Nat Res Centre 43, Article number:12, 1–9

DeRosa MC, Monreal C, Schnitzer M, Walsh R, Sultan Y (2010) Nanotechnology in fertilizers . Nat Nanotechnol 5:91

Dewdar MDH (2013) Stability analysis and genotype × environment interactions of some Egyptian cotton cultivars cultivated. Afr J Agric Res 8(41):5156–5160

Dewdar MDH (2019) Productivity; cotton; stress susceptibility index; relative productivity. Int J Plant Soil Sci 27(5):1–7

Djaboutou MC, Serge SH, Angelo CD, Marius GS, Florent JBQ, Gilles HC, Ahanhanzo C (2017). Adaptability and Stability of six cotton genotypes (*Gossypium hirsutum* L) in three cotton growing regions of Benin. Int J Curr Res Biosci Plant Biol 4(2):26–33

El-Adly HH, Eissa AEM, Nagib MAA (2008) Genetic analysis of some quantitative traits in cotton. Egypt J Agric Res 86(1):167–177

El-Defrawy MM, Mervat MH, Elsayed EN (2004) Molecular polymorphism in egyptian cotton (*Gossypium barbadense* L). Assiut J Agric Sci 35:83–96

El-Ghamry et al. (2018).El-Ghamry AM, Mosa1 AA, Alshaal TA, El-Ramady HR (2018) Nanofertilizers versus Biofertilizers: new insights. Environ Biodiv Soil Secur 2:51–72

El-Shaarawy SA, Abd El-Bary AMR, Hamoud HM, Yehia WMB (2007) Use of high efficient AMMI method to evaluate new Egyptian cotton genotypes for performance stability. The world cotton Res Conf 4, 10–14 Sept

Fannin B (2006) Texas Drought Losses Estimated at $4.1 Billion (Aug 15). AgriLife News. http://agnews.tamu.edu

FAOSTAT (2018) Food and Agricultural Organization Statistical Database. http://www.fa.org/fao stat/en/#data/QC

FAOSTAT (2019) Food and Agricultural Organization Statistical Database. http://www.fa.org/fao stat/en/#data/QC

Farias FJC, Carvalho LP, Silva Filho JL, Teodoro PE (2016). Biplot analysis of phenotypic stability in upland cotton genotypes in Mato Grosso. Genet Mol Res 15(2):gmr15028009

Frelichowski Jr MB, Palmer DM, Tomkins JP, Cantrell RG, Stelly DM, Yu J, Kohel RJ, Uloa M (2006) Cotton genome mapping with new microsatellites from Acala "Maxxa" BAC-ends. Mol Gen Genomics 275:479–491

Gasser CS, Fraley RT (1992) Transgenic Crops. Sci Am 266:62–69

Gholamhossein H, Thengane RJ (2007) Estimation of genetic parameters for salinity tolerance in early growth stages of cotton (*Gossypium hirsutum* L) genotypes. Int J Botany 3(1):103–108

Hamoud HM, Soliman YAM, Samah MME, Abdellatif KF (2016). Field performance and gene expression of drought stress tolerance in cotton (*Gossypium barbadense* L). BBJ 14(2):1–9

Hassan ISM (2000) Evaluation of two new Extra long staple cotton genotypes and three commercial cultivars grown at North Delta. Ann Agric Sc Mashtohor, Zazazig Univ 38(4):1839–1846

Hassan ISM, Abou Tour HB, Badr SSM (2006) Evaluation of two new extra long staple cotton varieties with commercial cultivars grown in North Delta. Egypt J Agric Res 84(5):1561–1576

Hassan ISM, Badr SSM, Hassan ISM (2012) Study of phenotypic stability of some Egptian cotton gynotypes under different environment. Egypt J Appl Sci 27(6):298–315

Hussain A, Zafar ZU, Athar HR, Farooq J, Ahmad S, Nazeer W (2019) Assessing gene action for hypoxia tolerance in cotton (*Gossypium hirsutum* L). Agron Mesoam 30(1):51–62

Hussein EHA, Amina Mohamed A, Attia S, Adawy SS (2006) Molecular characterization and genetic relationships among cotton genotypes 1-RAPD, ISSR and SSR analysis. Arab J Biotech 9:313–328

Hussein EHA, Osman MHA, Hussein MH, Adawy SS (2007) Molecular characterization of cotton genotypes using PCR-based markers. J Appl Sci Res 10:1156–1169

Kalbande BB, Patil AS (2016) Plant tissue culture independent *Agrobacterium tumefaciens* mediated *In-planta* transformation strategy for upland cotton (*Gossypium hirsutum*). J Genetic Eng Biotech 14(1):9–18

Koleva M, Dimitrova V (2018) Evaluation of drought tolerance in new cotton cultivars using stress tolerance indices. AGROFOR Int J 3(1):11–17

Liang C, Meng Z, Meng Z, Malik W, Yan R, Lwin KM, Lin F, Wang Y, Sun G, Zhou T, Zhu T, Li J, Jin S, Guo S, Zhang R (2016) GhABF2, a bZIP transcription factor, confers drought and salinity tolerance in cotton (*Gossypium hirsutum* L). Sci Rep 6:1–13

Maleia MP, Raimundo A, Moiana LD, Teca JOC, Jamal E, Dentor JN, Adamugy BA (2017) Stability and adaptability of cotton (*Gossypium hirsutum* L) genotypes based on AMMI analysis. AJCS 11(04):367–372

Manjunatha SB, Biradar DP, Aladakatti YR (2016) Nanotechnology and its applications in agriculture. J Farm Sci 29:1–13

Masoud S, Zahra H, Shahriari HR, Noormohammadi Z (2007) RAPD and cytogenetic study of some tetraploid Cotton (*Gossypium hirsutum* L) cultivars and their hybrids. Cytologia 72:77–82

Mishra N, Sun Li, Zhu X, Smith J, Srivastava AP, Yang X, Pehlivan N, Esmaeili N, Luo H, Shen G, Jones D, Auld D, Burke J, Paytonand P, Zhang H (2017) Overexpression of the Rice SUMO E3 Ligase Gene OsSIZ1 in Cotton Enhances Drought and Heat Tolerance, and Substantially Improves Fiber Yields in the Field under Reduced Irrigation and Rainfed Conditions. Plant Cell Phys 58(4):735–746

Momtaz OA, Hussein EM, Fahmy EM, Ahmed SE (2010) Expression of S-adenosyl methionine decarboxylase gene for polyamine accumulation in Egyptian cotton Giza 88 and Giza 90. GM Crops 1(4):257–266

Moustafa EA (2006) Gene action and stability of Egyptian cotton yield (*Gossypium barbadense* L) under Wady Seder conditions. PhD thesis, Dept of Agron Fac of Agric Zagazig Univ, Egypt

Mudada N, Chitamba J, Macheke TO, Manjeru P (2017) Genotype × Environmental interaction on seed cotton yield and yield components. Open Access Lib J 4(11):1–22

Muhammad A, Ur M, Rahman JI, Mirza YZ (2009) Parentage confirmation of cotton hybrids using molecular markers. Pak J Bot 41:695–701

Nabi G, Azhar FM, Khan AA (2010) Genetic mechanisms controlling variation for salinity tolerance in upland cotton at plant maturity. Int J Agric Biol 12:521–526

Nakashima K, Yamaguchi-Shinozaki K, Shinozaki K (2014) The transcriptional regulatory network in the drought response and its crosstalk in abiotic stress responses including drought, cold and heat. Front Plant Sci 5:170

Nasimi RA, Khan IA, Iqbal MA, Khan AA (2016) Genetic analysis of drought tolerance with respect to fiber traits in upland cotton. Genet Mol Res 15(4):1–16

Phillips K (1998) Cotton drought losses pegged at $1.8Billion for Texas. Texas A&M Agriculture News

Prasad R (2014). Synthesis of silver nanoparticles in photosynthetic plants. J Nanoparticle 2014: Article ID 963961, 8 pages

Rahman M, Hussain D, Zafar Y (2002) Estimation of divergence among elite cotton cultivarsgenotypes by DNA fingerprinting technology. Crop Sci 42:2137–2144

Rao AQ, Bakhsh A, Kiani S, Shahzad K, Shahid A, Husnain T, Riazuddin S (2009) The myth of plant transformation. Biotechnol Adv 27:753–763

Rodriguez-Uribe L, Abdelraheem A, Tiwari R, Goplan CS, Hughs SE, Zhang JF (2014) Identification of drought-responsive genes in drought-tolerant cotton (*Gossypium hirsutum* L) cultivar under reduced irrigation field conditions and development of candidate gene markers for drought tolerance. Mol Breed 34:1777–1796

Sadabadi MF, Ranjbar GA, Zangi MR, Kazemi Tabar SK, Najafi Zarini H (2018) Analysis for stability and adaption of cotton genotypesusing GGE biplot method. Trakia J Sci 1:51–61

Sahat SA (2015) Effect of climatic changes on yield and quality properties of some Egyptian cotton genotypes. PhD thesis, Dept of Agron Fac of Agric Cairo Univ, Egypt

Sawan ZM (2014) Nature relation between climatic variables and cotton production. J Stress Physiol Biochem 9:251–278

Sekmen AH, Ozgur R, Uzilday B, Turkan I (2014) Reactive oxygen species scavenging capacities of cotton (*Gossypium hirsutum*) cultivars under combined drought and heat induced oxidative stress. Environ Exp Bot 99:141–149

Shaar-Moshe L, Blumwald E, Peleg Z (2017) Unique physiological and transcriptional shifts under combinations of salinity, drought, and heat. Plant Physiol 174:421–434

Shaker SA (2009) Genotypic stability and evaluation of some Egyptian cotton genotypes. PhD thesis, fac of Agric Kafr El-Sheikh Univ, Egypt

Shaker SA, Habouh MAF, El-Fesheikwy ABA (2019) Analysis of stability using AMMI and GGE-Biplot methods in some Egyptian cotton genotypes. Menoufia J Plant Product 4:153–163

Shaker SA, El-Mansy YM, Darwesh AEI, Badr SSM (2020) Evaluation and stability of some Egyptian cotton varieties under normal and late sowing conditions. Menoufia J Plant Product 5(4):91–105

Shen G, Wei J, Qiu X, Hu R, Kuppu S, Auld D, Blumwald E, Gaxiola R, Payton P, Zhang H (2015) Co-overexpression of AVP1 and AtNHX1 in cotton further improves drought and salt tolerance in transgenic cotton plants. Plant Mol Biol Rep 23:167–177

Silva RS, Farias FJC, Teodoro PE, Cavalcanti JJV, de Carvalho LP, Queiroz DR (2020) Phenotypic adaptability and stability of herbaceous cotton genotypes in the Semiarid region of the Northeast of Brazil 24(12):800–805

Sohair EED, Abdall AA, Amany AM, Hossain MF, Houda RA (2019). Evaluation of nitrogen, phosphorus and potassium Nano-fertilizers on yield, yield components and fiber properties of Egyptian cotton (*Gossypium barbadense* L). J Plant Sci Crop Protect 1(2):1–10

Soomro MH, Markhand GS, Mirbahar AA (2012) Estimation of combining ability in F2 population of upland cotton under drought and non-drought regimes. Pak J Bot 44:1951–1958

Suppan S (2017) Applying nanotechnology to fertilizer: rationales, research, risks and regulatory challenges. The Institute for Agriculture and Trade Policy works locally and globally, Brazil

Tiwari RS, Picchioni G, Steiner RL, Hughs SE, Jones DC, Zhang JF (2013) Genetic variation in salt tolerance at the seedling stage in an interspecific backcross inbred line population of cultivated tetraploid cotton. Euphytica 194:1–11

Ullah A, Sun H, Yang X, Zhang X (2017) Drought coping strategies in fcotton: increased crop per drop. Plant Biotech J 15:271–284

USDA (2019) World agricultural production. United States Department of Agriculture, Foreign Agricultural Service, Circular Series WAP 6–19 June 2019, Office of Global Analysis, FAS, USDA, Foreign Agricultural Service/USDA

Vakhrouchev AV, Golubchikov VB (2007) Numerical investigation of the dynamics of nanoparticle systems in biological processes of plant nutrition. J Phys: Conf Ser 61(1):31–35

Wang WX, Vinocur B, Altman A (2003) Plant responses to drought, salinity and extreme temperatures towards genetic engineering for stress tolerance. Planta 128:1–14

Zahid M, Raheel F, Dasti AS, Shahzadi S, Athar M, Qayyum M (2009) Genetic diversity analysis of the species of Gossypium by using RAPD markers. Afr J Biotech 8:3691–3697

Chapter 10
Update, Conclusions, and Recommendations of "Sustainable Agriculture in Egypt: Climate Change Mitigation"

Abstract This chapter highlights the foremost conclusions and recommendations of the chapters presented in the current book. Moreover, selected results from a few newly published research works interrelated with the crop productivity, biotechnology and climate change, vulnerability, adaptation, stability and mitigation potentials from case-studies in Egypt and the world. Consequently, the present volume contains information in Part I: Introduction. Part II about Climate change and its impact on crop production. Part III is about foundations of crop tolerance to environmental stress and plant traits relevant to stress tolerance and contains mechanisms of adaptation to environmental stress conditions. Whereas, Part IV explained improve crop adaptability and stability to climate change and modern technology, and contain approaches to mitigate impact of climate change in strategic field crops i.e. wheat, rice, faba bean, sesame, sunflower and cotton. Finally, a set of recommendations for future research work is pointed out to direct the future research towards sustainability which is a main strategic theme of the Egyptian Government.

Keywords Environmental stresses · Critical stages · Impact of climate change · Mechanisms of adaptation · Genotype × environment · Performance · Additive main effects · Mitigation · DNA-Markers · Genetically modified plants · Nano technology · Genetic behavior · Stress sensitivity measurements · Agricultural practices

10.1 Introduction

Climate change in the Mediterranean region is hampering crop production and these changes are expected to increase in the future (Mansour et al. 2018; Awaad et al. 2021). So, understanding crop responses to climate change is essential to cope with anticipated changes in environmental conditions. Crop plants suffer from stress due to higher temperatures, water stress, salinity, pollution and others particularly during critical periods plant life cycle.

The World Bank has identified Egypt as one of the countries at risk of environmental stress in the Mediterranean region (Hereher 2016). In the light of these considerations, the study of the interaction between the genotype × environment has a significant impact on the expression of the phenotype and the response to environmental variables. This determines the decision of plant breeders in breeding programs to develop more adaptable and stable varieties. Therefore, in-depth study of the climate change and its impact on crop production, foundations of crop tolerance to environmental stresses, and identifies the characteristics of plants associated with mechanisms to tolerate stress environmental conditions is extremely important. The next section will present a transitory of the important results of some of the recently published research on improve crop adaptability and stability to climate change and modern technology on wheat, rice, faba bean, sesame, sunflower and cotton. Then the main conclusions and recommendations of the book chapters in addition to recommendations for researchers and decision makers.

10.2 Update

Adaptability, genetic stability and diversity are three of the key factors for the improvement of many crop plants to environmental changes. A major challenge for plant breeders is selection of high yielding genotype with wide adaptation. Stability is a key word for plant breeders who analyze G × E interaction data because it enhances the progression of selection in any single environment. So, evaluating genotypes in diverse environments is essential to obtain information regarding phenotypic stability, which is beneficial for the selection of promising genotypes and improvement programs. Most often, plant breeders face an interaction between G × E when assessing breeding program outcomes across environments. To identify stable genotypes, breeders should divide the G × E interaction into the stability statistics assigned to each genotype evaluated across a range of environments.

Stability is defined as the ability of genotype to avoid substantial fluctuation in yield over a range of environments. Yield stability of promising genotypes in diverse environments are very important in breeding approach for new cultivars that are to be released regionally or nationally. Therefore, it is of important to ascertain stable genotypes of wider adaptability or high yielding genotypes for a specific environment. In plant breeding programs and because a significant interaction with G × E can seriously obstruct the efforts of the plant breeder to select superior genotypes of new crop introductions and the breeding development program, the stability assessment is important (Shafii and Price 1998).

The global mean surface temperature change for the period 2016–2035 is will likely be in the range 0.3–0.7 °C by mid-twenty-first century, the magnitude of the expected climate change is significantly affected by the emissions scenario (IPCC 2014). The new statistically-based projections, published July 31 in Nature Climate Change, show a 90 percent chance that temperatures will increase this century by 2.0–4.9 C according to data of FAO (2015). Several studies have estimated the likely impacts of future climate on crop production in the developing countries, the impact of climate change is often locally specific and hard to predict (Morton 2007). The mitigation of greenhouse gases in the atmosphere of carbon dioxide, methane and nitrous oxide that contribute significantly to global warming is a major task in recent agriculture. Up to 90% of the methane gas generated in the cultivation of submerged rice is released into the atmosphere. The remaining 10% of CH4 in soil is often re-oxidized to CO_2 and differs broadly between rice genotypes. Therefore, the selection of the varieties and the treatments that will reduce the anthropogenic GHG emissions is considered an important goal in this regard (Wu et al. 2018). Recently, Liu et al. (2016) showed that, with a 1 °C global temperature increase, global wheat production is projected to decrease between 4 and 6.4%. This underscores the effects of climate on global food security.

In Egypt, the results of several investigators demonstrate significant impact of temperature and drought stress during growing season on crop physiology biochem-istry and production (Awaad 2009). We observe the trends of warming in the average temperature and water stress during the growing season and then its effects over the last few decades with clear evidence of negative impacts on phenology, physiology, biochemistry and yield of various strategic crops (Lobell et al. 2011; Hereher 2016; Ali et al. 2018; Li et al. 2018; Fatma Farag 2019; Awaad et al. 2021).

The results of the Agricultural Meteorology and Climate Change Research Unit of the Soil, Water and Environment Research Institute of the Agricultural Research Center–Egypt of long-range prediction using simulations and different climate change scenarios, revealed that climate change and the resulting rise in tempera-ture of the surface of the earth will damagingly affect the productivity of Egyptian agricultural crops. This causing a severe shortage in the productivity of the main food crops in Egypt in addition to increasing water consumption. The most impor-tant results of studies conducted in this regard showed that the yield of the wheat crop will be decreased by about 9% if the temperature rises 2 °C and the rate will decrease to about 18% if the temperature is about 3.5 °C. The water consumption of this crop will increase by about 2.5% rather than those under current weather conditions. The productivity of the rice crop could be declined by 11% and its water consumption will increase by about 16%. Sunflower crop will decline about 27% and will increase water consumption by about 8%. Whereas, climate change will have a positive effect on cotton yield productivity, and its productivity will increase by about 17% when the temperature of the air is rises about 2 °C. The rate of increase in this crop will rise to about 31% at 4 °C. On the other hand, its water consumption will increase by about 10% compared to its water consumption under current weather conditions (El-Marsafawy Samia 2007). Also, environmental stress adversely affected growth and yield of faba bean (Abdelmula and Abuanja 2007; Tayel and Sabreen 2011; Hegab et al. 2014) and productivity and physiological parameters of sesame, endogenous phytohormones including Salicylic Acid and kinetin (Hussein et al. 2015) and yield (Awaad and Ali 2002; Mostafa et al. 2016; Ayoubizadeh et al. 2019).

A fluctuation in productivity of field crops in response to the environmental changes contributes to the increase of potential risk in the agricultural sector. Possible climate change impact could have crucial influence on agricultural production and consequent socio-economic impacts. Potential impacts of climate change on crop production have received immense attention over the last decades (Tao et al. 2008). Therefore, understanding the relationship between climate and crop production is essential to identify possible impacts of future climate and to develop adaptation of crop plants. Additionally, assessing the impressions of historic climate tendencies on crop production helps to assess the possible impacts of prospect climate, review the ongoing efforts of adaptation and evaluate the resulting change in crop production (Lobell et al. 2011).

Based on all of the above, adaptability, stability and biotechnology are important indicators in the light of climate change in the world, especially in the Mediterranean region and in Egypt. Therefore, modification of plant metabolism and biological processes associated with the formation of protoplasm by gene manipulation, both in traditional breeding methods and in biotechnology, is an important procedure for improving the performance of the crop varieties under environmental stress conditions. Where advanced studies indicate the possibility of genetic improvement of the crop product under stress conditions. In practice, the selection of the crop cultivar, the appropriate planting date for the crop growth stages in the region and the combination of fertilizers with available water are considered appropriate agricultural practices to improve plant growth and minimize the adverse effect of high temperature and water stress on crop productivity.

Studies have shown that stability analyses according to various measures can result in better identification of stable genotypes (Akter and Islam 2017; Iqbal et al. 2017 and Fatma Farag et al. 2019). Furthermore, biotechnology as an important tool in the scientific research system provides solutions to overcome environmental stresses. In this respect, by using SNPs markers, ElBasyoni et al. (2017) registered highly significant differences among 2111 spring wheat accessions and association of functional genes related to abiotic stress tolerance in wheat conditions. In the field, Stokstad (2020) showed that the transgenic rice yielded up to 20% more grain and resist to heat waves. In sesame plants, some changes in amount and number of protein bands and also new protein bands were detected in both stressed and normal condition by Hussein et al. (2015), 287 MYB genes are highly active in growth and adaptation to the major abiotic stresses specifically drought and waterlogging in sesame were identified by Ali Mmadi et al. (2017) and detected 132 and 120 significant Single Nucleotide Polymorphisms (SNPs) resolved to 9 and 15 Quantitative trait loci (QTLs) for drought and salt stresses, respectively by Li et al. (2018). Establishment of regeneration and transformation system in Egyptian sesame cv Sohag 1 has been performed using Agrobacterium tumefaciens by Al-Shafeay et al. (2011). In Egyptian cotton, Momtaz et al. (2010) isolated SAMDC cDNA from Saccharomyces cerevisiae isolate and genetically engineered into Egyptian cotton varieties Giza 88, as an extra-long staple, and Giza 90, as a long staple by means of particle bombardment through meristem transformation, and screened T0 transgenic plants using basta herbicide (200 mg/l). T1 transgenic plants grown under induced drought stress regime and showed accumulation expressing of SAMDC gene reflect main cause for increasing drought tolerance. In sunflower, Ali et al. (2018) differentiated 100 inbred sunflower lines under natural and water-limited states by using 30 simple sequence repeat (SSR) primer pairs and 14 inter-retrotransposon amplified polymorphism (IRAP) also 14 retrotransposon-microsatellite amplified polymorphism (REMAP) primer combinations. They identified 22 and 21 markers for the studied characters under natural and water-limited states, respectively. Moreover, in faba been, 28 clones were identified as drought stress induced by Abid et al. (2015). Ten unique expressed sequence tags showed homology to known drought responsive genes including heat shock protein (HSP). Eight genes were consistently up-regulated in Hara compared to Giza 3 cultivar, known as drought-tolerant and sensitive respectively under water deficit treatment. Alghamdi et al. (2018) showed that water stress differentially expressed genes through the root transcriptome analyses of the drought-tolerant genotype Hassawi 2 at vegetative and flowering stages. A total of 36,834 and 35,510 unigenes were differentially expressed at the vegetative and flowering stages of Hassawi 2 under drought stress, respectively. Therefore, the study of the impact of climate change is necessary so that appropriate measures can be taken to adapt with expected changes in the environmental conditions.

The projected worldwide warming, with an anticipated increase of 1–3 °C through the twenty first century, might consequently have a great impact on agriculture and decreases crop production. Multiple investigations affirm that the frequent oscillations of the high and low temperatures and in combination with drought, salinity in soils problems and pollution have negative effects on the plants through the following main changes:

(a) Germination, via reduced seed vigor and subsequent germination and plant development.
(b) Morphological, such as inhibition of plant growth i.e. roots, shoot system and reproductive organs.
(c) Phenological, for example modifications in flowering, and both grain filling rate and period.
(d) Physiological, such as alteration in cell turgor pressure, stomatal conductance, transpiration rate, Photo System II, hinders the process of phosphorylation and inhibition of photosynthesis.
(e) Biochemical, for instance chlorophyll and cell membrane damage, alteration in levels of nucleic acids RNA and DNA, osmotic substances i.e. proline, glycine betaine, benitol and sugars, heat shock proteins and a decrease in antioxidant enzyme activity.
(f) Productive parameters, such as alterations in yield, its relevant traits and quality of the crop plants.

Therefore, in this book the climate–crop yield relationship and the impact of environmental change will be focused on wheat, faba bean, sesame, sunflower and cotton crops in Egypt.

10.3 Conclusions

The following conclusions are mainly extracted from the presented chapters:

1. In the introduction, the main parts and important practical of the book are presented. The chapters deal with areas of research and results that represent various

challenges facing the production of a number of important crops, including strategic crops wheat and rice; a basic food crop faba bean; oil crops sesame and sunflower and finally the fiber crop cotton under environmental changes. The book also discusses how to cope with changes in environmental pressures, through the study of the adaptation and stability of the abovementioned crops using appropriate statistical models and the role of biotechnology and Nano-technology techniques in addition to agricultural procedures to mitigate the impact of climate change on agricultural crops in Egypt and the world. The main themes included in the present book are (a) Climate change and its impact on crop production, (b) Foundations of crop tolerance to environmental stress and plant traits relevant to stress tolerance and (c) Improve crop adaptability and stability to climate change and modern technology. The extracted conclusions from the chapters under each theme will be presented in the next subsections.

10.3.1   Climate Change and Its Impact on Crop Production

2. Strategies for improving economic crops to tolerate environmental stress are important goal in light of the problems facing agricultural production, represented by climate change, which negatively affect crop productivity. Therefore, identifying the the most sensitive periods of plant life to extreme environment conditions is of great importance. It is also considered important to analyze the crop productivity under different environmental conditions. In this chapter, several reference studies showed variation in the amount of reduction in the crop production and quality under different stress environments, which reflects the variance of associations between plant traits.

10.3.2   Foundations of Crop Tolerance to Environmental Stress and Plant Traits Relevant to Stress Tolerance

3. Improving field crops to environmental stress tolerance is the most important problem facing agricultural production. So, it is very important to determine the plant properties associated with environmental stress conditions. This will help the breeder a lot in understanding the relevant physiological, biochemical and molecular basics and to tolerate climate changes and produce more adaptive varieties. Therefore, this chapter focused on identifying the most important characteristics of crop plants related to crop tolerance to environmental stresses.

10.3.3 Improve Crop Adaptability and Stability to Climate Change and Modern Technology

4. From the reality of the strategic importance of the wheat crop and in the light of climate change and its negative impacts on the agricultural sector, came this chapter. It is of important to develop and release new wheat genotypes and cultivars and determine their adaptability and stability to cope with environmental changes. Also, understanding the relationship between climate and wheat crop yield is fundamental to identify possible impacts of future climate and to develop adaptation measures. Therefore, multiple environmental trials are necessary in order to determine the promising and adapted cultivars.

5. Development of adaptability and stability in rice is a thought-provoking task that requires a comprehensive thoughtful of the various morphological, biochemical, physiological and molecular characters. Although remarkable progresses have been achieved through rice breeding, we still have several critical problems to overcome the effects of climate change on rice. Moreover, the complex nature and multigenic control of abiotic stress would be a major bottleneck for the current and coming future research in this respect. Maintenance of yield in rice under stress conditions is a multifaceted phenomenon governed by the cumulative effects of several traits. Transgenic approaches play a pivotal role in improving agronomic traits and yield characteristics of rice and it would be an efficient way to boost the rice-breeding program for stress tolerance and yield stability. In addition, appropriate agricultural techniques play a significant role in mitigating the impact of climate change on rice production of recent cultivars.

6. Previous studies show the importance of studying the interaction between genotype x environmental on faba bean crop production. This is useful in assessing the adaptation and stability of the genotypes and the outcome of breeding programs. In this chapter, the role of genetic differences in responding to different geographical and environmental conditions, the importance of planting date, irrigation, fertilization and biotechnology in mitigating the harmful impact of environmental stress on the faba bean yield was clearly demonstrated.

7. Measuring the adaptation and stabilization of genotypes is an important strategy in sesame crop production. In this chapter, the role of varieties in responding to different environmental changes, and the genetic behavior of different traits associated with stress tolerance, have been illustrated. Besides the importance of agricultural treatments and biotechnology in improving the tolerance of sesame varieties to environmental stress.

8. Based on the previous review obtained in this chapter, it could be concluded that evaluation of sunflower genotypes under different environments is important to reduce the genetic x environment interaction. So, determine adapted and stable sunflower genotypes for yield. Biotechnology techniques played the importance role in improving sunflowers as auxiliary tools for the genetic improvement of adaptability and stability. The adverse effects of environmental stress on sunflower growth can be mitigated through altering planting dates or by developing hybrids with appropriate phenology or appropriate agricultural practices.

9. The adaptability and stability of cotton genotypes crosswise diverse environment circumstances are essential for release and recommendation of new cultivars with high adaptability. Genetic markers and gene manipulation appeared to be more important of any successful breeding program in the field of cotton improvement. Nano-fertilizers play a significant role as favorable approaches for increasing water use efficiency and land resources and decreasing environmental pollutions. The farmers can partially cope with the adverse effects of high temperature, drought and salinity through selecting the tolerant cultivar, applying recommended agricultural management i.e. adequate fertilizer, irrigation and other ideal practices.

10.4 Recommendations

The following recommendations are mainly extracted from the chapters:

1. Generally, based on the results of research studies in this book, it is recommended to study the nature of the interaction between the genotype and environmental factors and identify the most environmental elements affecting gene expression of crop varieties using appropriate statistical models, which gives a strong indication of the adaptation and stability of crop strains and then recommend cultivation the most stable genotypes in the face of changing environments. It also recommends the application of appropriate agricultural procedures and the use of environmentally friendly technologies such as nanotechnologies. This besides the application of DNA markers and gene transfer techniques in improving plant productivity. The author recommends the importance of working further books in these areas of research to provide more ideas to deal with climate change in the light of fast scientific progresses in the world.

10.4.1 Climate Change and Its Impact on Crop Production

2. Climate change represents the most important challenges facing the world today that have effect on agriculture and human health. The Intergovernmental Panel on Climate Change (IPCC 2018) Special Report on Global Warming of 1.5 °C showed how keeping temperature increases below 2 °C would decrease the hazards to human well-being, ecosystems and sustainable progress. The negative effects of climate change on agricultural crops can be mitigated by implementing integrated adaptation strategies and understanding the nature of the relationships between different crop traits, and cultivating more adapted genotypes to environmental changes.

10.4.2 Foundations of Crop Tolerance to Environmental Stress and Plant Traits Relevant to Stress Tolerance

3. Based on the studies reviewed in this regard, it is recommended to deepen the research work on the nature of gene action controlling the inheritance of traits associated with tolerance to environmental stress factors. Also, it is also necessary to use modern trends in the process of genetic improvement. This facilitates the possibility of transferring tolerant genes from genetic resources to the commercial varieties to cope with climate change and sustain the productivity in the face of expected changes in environmental conditions.

10.4.3 Improve Crop Adaptability and Stability to Climate Change and Modern Technology

4. Utilized Joint regression analysis and additive main effects and multiplicative interaction (AMMI) and GEE-biplot analyses are considered more important in determining the nature and significant of genotype x environment effects. Due to the differences existed in wheat genotypes in adaptation and stability and tolerance to various environmental stresses of high temperature, drought, salinity and others, it is necessity testing wheat varieties at multiple locations. Therefore, the possibility to recommend the cultivation of promising, adaptable and stable genotypes is worthy of attention to overcome climate variability. Also, integration of recent improvement procedures as DNA-marker assisted selection and gene manipulation with traditional crop breeding aids in improving adaption in wheat genotypes.

5. There are various strategies and options for adaptation to climate change in the rice sector. The measures to take advantage of the positive impacts and mitigate the negative ones, including: (1) adjusting cropping patterns to expand rice cultivation; (2) breeding new rice varieties with short growing periods that are tolerant to environmental stress; (3) improving irrigation systems, and (4) improving pest and disease control, fertilizer application and mechanization.

6. Because of the sensitivity of faba bean to climate change, it seems necessary to identify the size and nature of the interaction between the genotype x environment, and determine the extent of adaptability and stability of faba bean genotypes under different conditions by applying appropriate statistical models. Also study the genetic behavior of various characteristics that could be used as selection criteria for improving yield under different environments. Focus on the role of DNA -markers and biotechnology in improving crop stability beside some agricultural practices to reduce the impact of environmental stress on faba bean production.

7. Extensive variations in seed yield due to genotype, location, and genotype × location interaction were recorded, so demonstrating that genotypes performed differently across locations and the essential of analyzing adaptability and stability of genotypes. Hereby it is of important to pyramiding the favorable alleles for enhancing

stress tolerance in sesame. The results of previous studies have recommended that negative effects of severe climate change on quantitative and qualitative characters could be compensated by foliar application of iron Nano-chelates and organic compounds such as fulvic acid.

8. Due to the sunflower genotypes fluctuated in their performance of seed yield and oil content under environmental stresses, accordingly, the genotype-environment interaction and appropriate agronomic practices strategies have a significant impact on sunflower yield and selection of superior cultivars. For this reason, previous studies showed that adaptation to high temperature mitigates the impact of water deficit during combined heat and drought stress in sunflower. Therefore, the development of genotypes with appropriate agricultural techniques reduces the effect of environmental stress on sunflower.

9. The scientists interested in environmental studies recommend conducting adaptation and stability analyzes to distinguish between environments and genotypes. Genetic engineering studies recommend the development of several markers including candidate genes responsible for abiotic stress. This is in addition to the development of genetically modified cotton with desirable traits such as tolerance to biotic and abiotic stress, beside improves seed cotton yield and fiber quality in drylands. This besides applying optimum agronomic managements.

References

Abdelmula1 AA, Abuanja IK (2007) Genotypic responses, yield stability, and association between traits among some of Sudanese Faba bean (*Vicia faba* L) genotypes under heat stress. Tropentag 2007 university of Kassel-Witzenhausen and university of Göttingen, Oct 9–11 conference on international agricultural research for development, pp 1–7

Abid Gh, Yordan M, Dominique M, Bernard W, André´ T, Guy M, Mahmoud M, Khaled S, Moez J (2015) Identification and characterization of drought stress responsive genes in faba bean (*Vicia faba* L) by suppression subtractive hybridization. Plant Cell Tiss Organ Cult121 (2):1–13

Akter N, Islam RM (2017) Heat stress effects and management in wheat: a review. Agron Sustain Dev 37(37):1–17

Alghamdi SS, Khan MA, Ammar MH, Sun Q, Huang L, Migdadi HM, El-Harty EH, Al-Faifi SA (2018) Characterization of drought stress-responsive root transcriptome of faba bean (*Vicia faba* L) using RNA sequencing. Abstract 3 Biotech 8(12):502

Ali Mmadi M, Dossa K, Wang L, Zhou R, Y Wang, Cisse N, Oureye Sy M, Zhang X (2017) Functional characterization of the versatile MYB gene family uncovered their important roles in plant development and responses to drought and waterlogging in sesame. Genes 8(12):362

Ali SG, Darvishzadeh R, Ebrahimi A, Bihamta MR (2018) Identification of SSR and retrotransposon-based molecular markers linked to morphological characters in oily sunflower (*Helianthus annuus* L) under natural and water-limited states. J Genet 97(1):189–203

Al-Shafeay AF, Ibrahim AS, Nesiem M, Tawfik MS (2011) Establishment of regeneration and transformation system in Egyptian sesame (*Sesamum indicum* L) cv Sohag 1. GM crops 2(3):182–92

Awaad HA (2009) Genetics and breeding crops for environmental stress tolerance I: drought, heat stress and environmental pollutants. Egyptian library, Egypt

Awaad HA, Aly AA (2002) Genotype x environment interaction and interrelationship among some stability statistics in sesame (*Sesamum indicum* L). Zagazig J Agric Res 29(2):385–403

Awaad HA, Abu-hashim M, Negm A (2021) Mitigating environmental stresses for agricultural sustainability in Egypt. Springer Nature Switzerland AG. ISBN 978–3–030–64322–5

Ayoubizadeh N, Laei G, Dehaghi MA, Sinaki JM, Rezvan S (2019) Seed yield and fatty acids composition of sesame genotypes as affecet by foliar application of iron Nanochelate and fulvic acid under drought stress. Appl Ecol Environ Res 16(6):7585–7604

ElBasyoni I, Saadalla M, Baenziger S, Bockelman H, Morsy S (2017) Cell membrane stability and association mapping for drought and heat tolerance in a worldwide wheat collection. Stainability 9:1–16

El-Marsafawy S (2007) Climate change and its impact on the agriculture sector in Egypt. http://www.radcon.sci.eg/environment2/ArticlsIdeasDetails.aspx?ArticlId=35

FAO (2015) Climate change and food security, UN Food & Agricultural Organization FAO. http://www.climatechange-foodsecurity.org/fao.html

Fatma Farag M (2019) Breeding parameters for grain yield and some morpho-pysiological characters related to water stress tolerance in bread wheat. MSc thesis, Agron Dept Fac of Agric Zagazog Univ Egypt

Hegab ASA, Fayed Maha MTB, Hamada MA, Abdrabbo MAA (2014) Productivity and irrigation requirements of faba-bean in North Delta of Egypt in relation to planting dates. Ann Agric Sci 59(2):185–193

Hereher ME (2016) Time series trends of land surface temperatures in Egypt: a signal for global warming. Environ Earth Sci 75:1218

Hussein Y, Gehan Amin, Adel Azab, Hanan Gahin (2015) Induction of drought stress resistance in sesame (Sesamum indicum L) plant by salicylic acid and kinetin. J Plant Sci 10(4):128–141

IPCC (2014) Climate Change 2014 Synthesis report contribution of working groups I, II and III to the fifth assessment report of the intergovernmental panel on climate change. Core writing team Pachauri RK, Meyer LA(eds) IPCC, Geneva, Switzerland, p 151

IPCC (2018). Climate change report is a "wake-up" call on 1.5 °C global warming. https://public.wmo.int/en/media/press-release/climate-change-report-%E2%80%9Cwake-%E2%80%9D-call-15%C2%B0c-global-warming

Iqbal M, Raja NI, Yasmeen F, Hussain M, Ejaz M, Shah MA (2017) Impacts of heat stress on wheat: a critical review. Adv Crop Sci Technol 5:251

Li D, Dossa K, Zhang Y, Wei X, Wang L, Zhang Y, Liu A, Zhou R, Zhang X (2018) Differential genetic bases for drought and salt tolerances in sesame at the germination stage. Genes 9:87

Liu B, Asseng S, Müller C, Ewert F et al (2016) Similar estimates of temperature impacts on global wheat yield by three independent methods. Nat Clim Chang 6(12):1130–1137

Lobell DB, Schlenker W, Costa-Roberts J (2011) Climate trends and global crop production since 1980. Science 333:616–620

Mansour E, Moustafa ED, Qabil N, Abdelsalam A, Wafa HA, El Kenawy A, Casas AM, Igartua E (2018) Assessing different barley growth habits under Egyptian conditions for enhancing resilience to climate change. Field Crop Res 224: 67-75

Momtaz OA, Hussein EM, Fahmy EM, Ahmed SE (2010) Expression of S-adenosyl methionine decarboxylase gene for polyamine accumulation in Egyptian cotton Giza 88 and Giza 90. GM Crops 1(4):257–266

Morton JF (2007) The impact of climate change on smallholder and subsistence agriculture. Proc Natl Acad Sci USA 104:19680–19685

Mostafa H, Maryam G, Hadi G, Mahdi BF (2016) Effect of drought stress and foliar application of iron oxide nanoparticles on grain yield, ion content and photosynthetic pigments in sesame (Sesamum indicum L). Iranian J Field Crop Sci Iran J Agric Sci Winter 46 (4): 619–628

Shafii B, Price WJ (1998) Analysis of genotype-by-environment interaction using the additive main effects and multiplicative interaction model and stability estimates. J Agric Biol Environ Stat 3:335–345

Stokstad E (2020) Rice genetically engineered to resist heat waves can also produce up to 20% more grain. Science https://www.sciencemag.org/news/2020/04/rice-genetically-engineered-resist-heat-waves-can-also-produce-20-more-grain

Tao F, Yokozawa M, Liu J, Zhang Z (2008) Climate-crop yield relationship at provincial scales in China and the impact of recent climate trends. Clim Res 38:83–94

Tayel MY, Sabreen KhP (2011) Effect of irrigation regimes and phosphorus level on two Vica faba varieties: 1-growth traits. J Appl Sci Res 7(6):1007–1015

Wu S, Hu Z, Hu T, Chen J, Yu K, Zou J, Liu S (2018) Annual methane and nitrous oxide emissions from rice paddies and inland fish aquaculture wetlands in southeast China Environ https://doi.org/10.1016/j.atmosenv.2017.12.008